Michalis Konsolakis (Ed.)

Surface Chemistry and Catalysis

MDPI

This book is a reprint of the Special Issue that appeared in the online, open access journal, *Catalysts* (ISSN 2073-4344) from 2015–2016, available at:

http://www.mdpi.com/journal/catalysts/special_issues/surface_chemistry

Guest Editor
Michalis Konsolakis
School of Production Engineering & Management
Technical University of Crete
Greece

Editorial Office
MDPI AG
St. Alban-Anlage 66
Basel, Switzerland

Publisher
Shu-Kun Lin

Managing Editor
Zu Qiu

1. Edition 2016

MDPI • Basel • Beijing • Wuhan • Barcelona • Belgrade

ISBN 978-3-03842-286-0 (Hbk)
ISBN 978-3-03842-287-7 (electronic)

Table of Contents

List of Contributors

R. Jürgen Behm Institute of Surface Chemistry and Catalysis, Ulm University, Albert-Einstein-Allee 47, D-89081 Ulm, Germany.

Kyriakos Bourikas School of Science and Technology, Hellenic Open University, GR-26222 Patras, Greece.

Magali Boutonnet KTH Royal Institute of Technology, Chemical Technology, Teknikringen 42, SE-100 44 Stockholm, Sweden.

Saul Cabrera UMSA Universidad Mayor de San Andrés, Instituto del Gas Natural, Campus Universitario, La Paz, Bolivia.

Gang Cao Environmental Science and Engineering Research Center, Shenzhen Graduate School, Harbin Institute of Technology, Shenzhen 518055, China.

Lidia Castoldi Laboratory of Catalysis and Catalytic Processes, Dipartimento di Energia, Politecnico di Milano, Via La Masa, 34, 20156 Milano, Italy.

Haiyang Cheng State Key Laboratory of Electroanalytical Chemistry, and Laboratory of Green Chemistry and Process, Changchun Institute of Applied Chemistry, Chinese Academy of Sciences, Changchun 130022, China.

Hui-Zhen Cui Key Laboratory for Colloid and Interface Chemistry, Key Laboratory of Special Aggregated Materials, School of Chemistry and Chemical Engineering, Shandong University, Jinan 250100, China.

Marco Daturi Laboratoire Catalyse et Spectrochimie, ENSICAEN, Université de Caen, CNRS, 6 Bd du Maréchal Juin, 14050 Caen, France.

Ana Raquel de la Osa Chemical Engineering Department, Faculty of Chemical Sciences and Technologies, University of Castilla La Mancha, Avda. Camilo José Cela 12, 13071 Ciudad Real, Spain.

Antonio de Lucas-Consuegra Department of Chemical Engineering, School of Chemical Sciences and Technologies, University of Castilla-La Mancha, Ave. Camilo José Cela 12, 13005 Ciudad Real, Spain.

Fernando Dorado Chemical Engineering Department, Faculty of Chemical Sciences and Technologies, University of Castilla La Mancha, Avda. Camilo José Cela 12, 13071 Ciudad Real, Spain.

Pio Forzatti Laboratory of Catalysis and Catalytic Processes, Dipartimento di Energia, Politecnico di Milano, Via La Masa, 34, 20156 Milano, Italy.

Jesús González-Cobos Department of Chemical Engineering, School of Chemical Sciences and Technologies, University of Castilla-La Mancha, Ave. Camilo José Cela 12, 13005 Ciudad Real, Spain.

Yu Guo Key Laboratory for Colloid and Interface Chemistry, Key Laboratory of Special Aggregated Materials, School of Chemistry and Chemical Engineering, Shandong University, Jinan 250100, China.

Yeusy Hartadi Institute of Surface Chemistry and Catalysis, Ulm University, Albert-Einstein-Allee 47, D-89081 Ulm, Germany.

Yuandan Huang The Key Laboratory of Food Colloids and Biotechnology, Ministry of Education, School of Chemical and Material Engineering, Jiangnan University, Wuxi 214122, China.

Zisis Ioakimidis Chemical Process & Energy Resources Institute, Centre for Research & Technology Hellas, 6th km. Charilaou—Thermi Rd., P.O. Box 60361, GR-57001 Thermi, Thessaloniki, Greece; Department of Mechanical Engineering, University of Western Macedonia, Bakola & Sialvera, GR-50100 Kozani, Greece.

Theophilos Ioannides Foundation for Research and Technology-Hellas (FORTH), Institute of Chemical Engineering Sciences (ICE-HT), Stadiou str., Platani, GR-26504 Patras, Greece.

Sven Järås KTH Royal Institute of Technology, Chemical Technology, Teknikringen 42, SE-100 44 Stockholm, Sweden.

Chun-Jiang Jia Key Laboratory for Colloid and Interface Chemistry, Key Laboratory of Special Aggregated Materials, School of Chemistry and Chemical Engineering, Shandong University, Jinan 250100, China.

Pingping Jiang The Key Laboratory of Food Colloids and Biotechnology, Ministry of Education, School of Chemical and Material Engineering, Jiangnan University, Wuxi 214122, China.

Stella Kennou Department of Chemical Engineering, University of Patras, GR-26504 Patras, Greece.

Dmitri Kilin Department of Chemistry, University of South Dakota, 414 E Clark Street, Vermillion, SD 57069, USA; Department of Chemistry and Biochemistry, North Dakota State University, Fargo, ND 58108, USA.

Michalis Konsolakis School of Production Engineering and Management, Technical University of Crete, 73100 Chania, Greece.

Ranjit Koodali Department of Chemistry, University of South Dakota, 414 E Clark Street, Vermillion, SD 57069, USA.

Christos Kordulis Department of Chemistry, University of Patras, GR-26504 Patras, Greece; Institute of Chemical Engineering Science (FORTH/ICE-HT), Stadiou Str. Platani, P.O. Box 1414, GR-26500 Patras, Greece.

Tzouliana Kraia Department of Mechanical Engineering, University of Western Macedonia, Bakola & Sialvera, GR-50100 Kozani, Greece; Chemical Process & Energy Resources Institute, Centre for Research & Technology Hellas, 6th km. Charilaou—Thermi Rd., P.O. Box 60361, GR-57001 Thermi, Thessaloniki, Greece.

Lukasz Kubiak Laboratory of Catalysis and Catalytic Processes, Dipartimento di Energia, Politecnico di Milano, Via La Masa, 34, 20156 Milano, Italy.

Spyridon Ladas Department of Chemical Engineering, University of Patras, GR-26504 Patras, Greece.

Xiaoting Li The Key Laboratory of Food Colloids and Biotechnology, Ministry of Education, School of Chemical and Material Engineering, Jiangnan University, Wuxi 214122, China.

Luca Lietti Laboratory of Catalysis and Catalytic Processes, Dipartimento di Energia, Politecnico di Milano, Via La Masa, 34, 20156 Milano, Italy.

Weiwei Lin Laboratory of Green Chemistry and Process, and State Key Laboratory of Electroanalytical Chemistry, Changchun Institute of Applied Chemistry, Chinese Academy of Sciences, Changchun 130022, China.

Tong Liu State Key Laboratory of Electroanalytical Chemistry, and Laboratory of Green Chemistry and Process, Changchun Institute of Applied Chemistry, Chinese Academy of Sciences, Changchun 130022, China;University of Chinese Academy of Sciences, Beijing 100049, China.

Luis Lopez KTH Royal Institute of Technology, Chemical Technology, Teknikringen 42, SE-100 44 Stockholm, Sweden; UMSA Universidad Mayor de San Andrés, Instituto del Gas Natural, Campus Universitario, La Paz, Bolivia.

Alexis Lycourghiotis Department of Chemistry, University of Patras, GR-26504 Patras, Greece.

Alberto Marinas Organic Chemistry Department, University of Córdoba, ceiA3, Marie Curie Building, E-14014 Córdoba, Spain.

George E. Marnellos Department of Environmental Engineering, and Department of Mechanical Engineering, University of Western Macedonia, Bakola & Sialvera, GR-50100 Kozani, Greece; Chemical Process & Energy Resources Institute, Centre for Research & Technology Hellas, 6th km. Charilaou—Thermi Rd., P.O. Box 60361, GR-57001 Thermi, Thessaloniki, Greece.

Roberto Matarrese Laboratory of Catalysis and Catalytic Processes, Dipartimento di Energia, Politecnico di Milano, Via La Masa, 34, 20156 Milano, Italy.

Vicente Montes Organic Chemistry Department, University of Córdoba, ceiA3, Marie Curie Building, E-14014 Córdoba, Spain.

Feng Ouyang Environmental Science and Engineering Research Center, Shenzhen Graduate School, Harbin Institute of Technology, Shenzhen 518055, China.

George D. Panagiotou Department of Chemistry, University of Patras, GR-26504 Patras, Greece.

Dandan Pang Environmental Science and Engineering Research Center, Shenzhen Graduate School, Harbin Institute of Technology, Shenzhen 518055, China.

Eftichia Papadopoulou Foundation for Research and Technology-Hellas (FORTH), Institute of Chemical Engineering Sciences (ICE-HT), Stadiou str., Platani, GR-26504 Patras, Greece.

Lu Qiu Environmental Science and Engineering Research Center, Shenzhen Graduate School, Harbin Institute of Technology, Shenzhen 518055, China.

Amaya Romero Chemical Engineering Department, Faculty of Chemical Sciences and Technologies, University of Castilla La Mancha, Avda. Camilo José Cela 12, 13071 Ciudad Real, Spain.

Paula Sánchez Chemical Engineering Department, Faculty of Chemical Sciences and Technologies, University of Castilla La Mancha, Avda. Camilo José Cela 12, 13071 Ciudad Real, Spain.

Wendi Sapp Department of Chemistry, University of South Dakota, 414 E Clark Street, Vermillion, SD 57069, USA.

Minhua Shao Department of Chemical and Biomolecular Engineering, The Hong Kong University of Science and Technology, Clear Water Bay, Kowloon, Hong Kong, China.

Rui Si Key Laboratory of Interfacial Physics and Technology, Shanghai Synchrotron Radiation Facility, Shanghai Institute of Applied Physics, Chinese Academy of Sciences, Shanghai 201204, China.

Labrini Sygellou Department of Chemical Engineering, University of Patras, GR-26504 Patras, Greece; Institute of Chemical Engineering Science (FORTH/ICE-HT), Stadiou Str. Platani, P.O. Box 1414, GR-26500 Patras, Greece.

Antonios Tribalis Department of Chemistry, University of Patras, GR-26504 Patras, Greece.

José Luis Valverde Chemical Engineering Department, Faculty of Chemical Sciences and Technologies, University of Castilla La Mancha, Avda. Camilo José Cela 12, 13071 Ciudad Real, Spain.

Jorge Velasco KTH Royal Institute of Technology, Chemical Technology, Teknikringen 42, SE-100 44 Stockholm, Sweden; UMSA Universidad Mayor de San Andrés, Instituto del Gas Natural, Campus Universitario, La Paz, Bolivia.

Guoping Wang School of Energy and Power Engineering, Nanjing University of Science and Technology, Nanjing 210094, China.

Wenju Wang Department of Chemical and Biomolecular Engineering, The Hong Kong University of Science and Technology, Clear Water Bay, Kowloon, Hong Kong, China; School of Energy and Power Engineering, Nanjing University of Science and Technology, Nanjing 210094, China.

Xu Wang Key Laboratory of Interfacial Physics and Technology, Shanghai Synchrotron Radiation Facility, Shanghai Institute of Applied Physics, Chinese Academy of Sciences, Shanghai 201204, China.

Yun Wang Environmental Science and Engineering Research Center, Shenzhen Graduate School, Harbin Institute of Technology, Shenzhen 518055, China.

Zhuangqing Wang The Key Laboratory of Food Colloids and Biotechnology, Ministry of Education, School of Chemical and Material Engineering, Jiangnan University, Wuxi 214122, China.

Daniel Widmann Institute of Surface Chemistry and Catalysis, Ulm University, Albert-Einstein-Allee 47, D-89081 Ulm, Germany.

Yancun Yu State Key Laboratory of Electroanalytical Chemistry, and Laboratory of Green Chemistry and Process, Changchun Institute of Applied Chemistry, Chinese Academy of Sciences, Changchun 130022, China.

Changliang Zhang Environmental Science and Engineering Research Center, Shenzhen Graduate School, Harbin Institute of Technology, Shenzhen 518055, China.

Chao Zhang State Key Laboratory of Electroanalytical Chemistry, and Laboratory of Green Chemistry and Process, Changchun Institute of Applied Chemistry, Chinese Academy of Sciences, Changchun 130022, China.

Fengyu Zhao State Key Laboratory of Electroanalytical Chemistry, and Laboratory of Green Chemistry and Process, Changchun Institute of Applied Chemistry, Chinese Academy of Sciences, Changchun 130022, China.

About the Guest Editor

Michalis Konsolakis graduated from the Department of Chemical Engineering of the University of Patras, Greece, in 1997 and obtained his PhD in 2001 from the same university. Part of his PhD studies was carried out in the Department of Chemistry, Cambridge University, within the framework of the Greek–British Joint Research and Technology Programme. He is currently an Associate Professor of "Surface Science and Heterogeneous Catalysis" at the School of Production Engineering and Management, Technical University of Crete, Greece, where he lectures on various topics related to chemistry, thermodynamics, materials science, surface science and heterogeneous catalysis. His research activities are mainly focused on materials science and heterogeneous catalysis with particular emphasis on surface and interface analyses. His publications include more than 140 papers in international journals and conference proceedings. He is a member of the Editorial Board of seven international journals related to catalysis, materials science and surface science. He also serves as a regular reviewer for more than 50 scientific journals and research funding agencies. (http://www.tuc.gr/konsolakis.html)

Preface to "Surface Chemistry and Catalysis"

Michalis Konsolakis

Reprinted from *Catalysts*. Cite as: Konsolakis, M. Surface Chemistry and Catalysis. *Catalysts* **2016**, *6*, 102.

1. Background

Nowadays, heterogeneous catalysis plays a prominent role. The majority of industrial chemical processes, involving the manufacturing of commodity chemicals, pharmaceuticals, clean fuels, etc., as well as pollution abatement technologies, have a common catalytic origin. As catalysis proceeds at the surface, it is of paramount importance to gain insight into the fundamental understanding of local surface chemistry, which in turn governs the catalytic performance. The deep understanding at the atomic level of a catalyst surface could pave the way towards the design of novel catalytic systems for real-life energy and environmental applications.

Thanks to surface science we can obtain profound insight into the structure of a surface, the chemical state of active sites, the interfacial reactivity, the way molecules bind and react, the role of surface defects and imperfections (e.g., surface oxygen vacancies), and the mode of action of various surface promoters/poisons. To elucidate the aforementioned surface phenomena, sophisticated techniques in combination with theoretical studies are necessary to reveal the composition and the structure/morphology of the surface as well as the chemical entity of adsorbed species. Moreover, time-resolved methods are required to investigate the dynamic phenomena occurring at the surface, such as adsorption/desorption, diffusion and chemical reactions. Under this perspective, it was clearly revealed, based on the recently published review articles by the Guest Editor, that the complete elucidation of a catalytic phenomenon (e.g., metal-support interactions [1]) or the fundamental understanding of a specific catalytic process (e.g., N_2O decomposition [2]) requires a holistic approach involving the combination of advanced ex situ experimental and theoretical studies with in situ operando studies.

2. This Special Issue

In light of the above aspects, the present themed issue aims to cover the recent advances in "surface chemistry and catalysis" that can be obtained by means of advanced characterization techniques, computational calculations and time-resolved methods, with particular emphasis on the structure-activity relationships (SARs). It consists of 14 high-quality papers, involving: a comprehensive review article on the surface analysis techniques that can be employed to elucidate the phenomenon of electrochemical promotion in catalysis [3]; two theoretical studies (Density

Functional Theory, DFT) on H_2O dissociation and its implications in catalysis [4,5]; two mechanistic studies by means of temperature-programmed desorption/surface reaction (TPD/TPSR) and/or operando spectroscopy on N_2O formation over NO_x storage-reduction (NSR) catalysts [6] and on methanol reforming over cobalt catalysts [7]; two articles on H_2 production by the steam reforming of ethanol [8] or diesel [9] over transition metal–based catalysts; two articles on the production of commercial fuels by Fisher-Tropsch synthesis [10,11]; two articles on Au-catalyzed CO oxidation [12] and preferential CO oxidation [13]; and three experimental investigations regarding the structure-activity correlation of NO oxidation to NO_2 over Mn-Co binary oxides [14], cyclohexene oxidation on TiZrCo mixed oxides [15] and alkene epoxidation on silica nanoparticles [16].

Contribution Highlights

The comprehensive review of González-Cobos J. and de Lucas-Consuegra A. [3] addresses the latest contributions made in the field of surface analysis towards the fundamental understanding of the Electrochemical Promotion of Catalysis (EPOC). The authors clearly revealed that the combination of in situ and ex situ surface analysis techniques, such as SEM, XPS, STM, SPEM, UPS, can provide the basis of a better understanding of the alkali-induced promotional effects. This multifunctional surface approach allows the authors to gain insight into the local surface structure of promoter species as well as of back-spillover phenomena taking place under electrochemical promotion conditions. Given the paramount importance of alkali promoters in catalysis, the results presented in this review can shed some light into the underlying mechanism of electrochemical promotion, also paving the way towards the design of highly active conventional heterogeneous catalysts.

Kilin D. and co-workers [4] investigated, by means of DFT calculations combined with density matrix equations of motion, the charge transfer mechanism involved in H_2O dissociation over titanium-doped microporous silica. The results revealed that silica substrates contain electrons and hole trap states, which could facilitate the water splitting. This provided strong evidence towards the key role of the substrate in the electron/hole dynamic processes involved in H_2O dissociation. In this regard, the fine-tuning of metal and/or substrate characteristics could lead to the optimization of (photo)catalytic efficiency.

In a similar manner, Wang W. et al. [5] explored, by means of DFT calculations, the adsorption of H_2O and its dissociation fragments (OH, H and O) on clean and O-pre-adsorbed Fe(100) surfaces. It was demonstrated that interactions between the different adsorbates and catalyst surface followed the order: H_2O < OH < H < O. More interestingly, it was revealed that both the H abstraction from the H_2O molecule and the subsequent OH dissociation are favored over O-pre-adsorbed Fe(100) surfaces. The results confirmed that the presence of pre-adsorbed oxygen on the catalyst surface can notably enhance the H_2O dissociation, opening new horizons towards the development of more efficient catalysts.

In a comprehensive mechanistic study by Forzatti P. and co-workers [6], the origin of N_2O formation over $Pt-BaO/Al_2O_3$ and $Rh-BaO/Al_2O_3$ model NSR catalysts was investigated by micro-reactor transient reactivity experiments and operando Fourier transform infrared (FT-IR) spectroscopy. It was clearly revealed that N_2O formation involves the coupling of undissociated NO molecules with N-adspecies formed via NO dissociation onto the reduced metal sites. In this regard, the N_2O formation is dependent on the oxidation state of the metal sites. At high temperatures, where the reductants effectively keep the metal sites in a fully reduced state, complete NO dissociation is achieved, thus hindering the N_2O formation. In contrast, at low temperatures, where the reductants start to reduce the active sites, the N_2O formation is favored.

The adsorption characteristics of methanol and its reforming products over Co-Mn catalysts were systematically investigated by Papadopoulou E. and Ioannides T. [7] through temperature-programmed desorption (TPD) studies. The influence of various parameters, in relation to the synthesis procedure and the Co/Mn ratio, into adsorptive properties was systematically explored in order to gain insight into the structure-activity relationships. The results indicated that the activity differences can be mainly related to the relative population and the nature of active sites. In particular, Co and Mn sites were considered to be responsible for H_2 desorption at low and high temperatures, respectively, whereas the interfacial sites can be considered for the intermediate temperature H_2 desorption.

Konsolakis et al. [8] reported on the Ethanol Steam Reforming (ESR) of various transition metals (Ni, Co, Cu, Fe) supported on CeO_2 with particular attention to surface chemistry aspects. A complementary surface characterization study was undertaken to reveal the impact of metal entity and/or metal-support interactions on the reforming activity. The results revealed the excellent reforming performance of Co/CeO_2 catalysts, both in terms of H_2 yield and life-time stability. The latter was attributed, inter alia, to the high oxygen mobility of cobalt-ceria binary oxides, mainly linked with the high population of lattice oxygen species. This factor can be considered responsible for the facile gasification of the carbonaceous species, thus preventing catalyst deactivation.

The impact of structural promoters (La, Ba, Ce) on the Diesel Steam Reforming (DSR) of Ni/Al_2O_3 catalysts was explored in detail by Tribalis A. et al. [9]. Incorporation of dopants into the Al_2O_3 carrier was found to be always beneficial, but to a different extent, depending on the nature of the promoter. The optimum performance, in terms of activity and stability, was obtained by simultaneously doping the Al_2O_3 support with Ba and La modifiers. On the basis of a complimentary surface characterization study, the latter was attributed to the increase of dispersion and reducibility of the Ni phase in conjunction with the decrease of the support acidity.

Lopez L. et al. [10] comparatively explored the catalytic performance of Rh/MCM-41 and Rh/SiO$_2$ catalysts for ethanol synthesis from syngas. The obtained differences in activity and selectivity were attributed, on the basis of complementary catalytic and surface characterization studies, to the different concentrations of water vapor in the pores of Rh/MCM-41. The latter was considered responsible for the enhanced formation of CO$_2$ and H$_2$ over Rh/MCM-41 catalysts through the water-gas-shift-reaction (WGSR).

De la Osa A.R. et al. [11] investigated Fisher-Tropsch synthesis (FTS) over Co/SiC catalysts with particular emphasis on the impact of cobalt precursor (nitrate, acetate, chloride, citrate) on FTS activity. Surface titration techniques along with an extended characterization by TPR and TEM were employed to gain insight into the structure-activity correlation. It was found that the nature of the precursor notably affects the acid/base properties as well as the metallic particle size, with great consequences on the FTS activity and chain growth probability. Cobalt nitrate provided the optimum activity and selectivity to C$_5^+$, which was attributed to the higher particle size, degree of reduction and basicity as compared to the other precursors.

Cui H.-Z. et al. [12] studied low-temperature CO oxidation over Au/FeO$_x$ catalysts, employing two different types of iron oxide supports, i.e., hydroxylated (Fe-OH) and dehydrated (Fe-O) iron oxides, and different preparation procedures (precipitation pH, calcination temperature). Surface characterization by a series of advanced characterization techniques, i.e., high-resolution transmission electron microscopy (HRTEM), X-ray photoelectron spectroscopy (XPS) and X-ray absorption near edge structure (XANES) spectroscopy, was carried out to explore the relationship between the nature of the oxide matrix and the catalytic activity. The results revealed that the surface chemistry of Au nanoparticles can be notably influenced by the nature of the support as well as by the preparation procedure following, with vast consequences on the catalytic activity. Metallic gold particles strongly interacting with the oxide carrier were determined as the active sites for CO oxidation.

Widmann D. and co-workers [13] explored, in a comprehensive manner, the underlying mechanism of the CO-PROX reaction over Au/TiO$_2$ catalysts by means of quantitative temporal analysis of products (TAP) reactor measurements. The authors concluded that CO and H$_2$ are oxidized by the same active oxygen species under PROX conditions, independently of the CO/H$_2$ ratio; both CO and H$_2$ compete for TiO$_2$ surface lattice oxygen located at the perimeter sites of the metal-support interface. In light of these findings, the authors suggested that the strategies for more selective Au catalysts should focus on the fine-tuning of the support material and/or the metal-support interface perimeter sites.

Qiu L. et al. [14] investigated the low-temperature oxidation of NO to NO_2 over Co-Mo/TiO_2 catalysts with particular attention to the impact of Co loading and the calcination temperature. Cobalt incorporation into Mo/TiO_2 catalysts in conjunction with calcination at moderate temperatures (300–400 °C) resulted in the optimum oxidation performance. By means of various surface characterization techniques, a close correlation between the redox characteristics (Mn oxidation state, amount of surface adsorbed oxygen) and the NO oxidation performance was revealed.

Liu T. et al. [15] investigated the aerobic oxidation of cyclohexene to value-added chemicals over TiZrCo metallic catalysts. High conversion of cyclohexene (>90%) accompanied by high selectivity to 2-cyclohexen-1-one (ca. 58%) was obtained by this particular catalyst, and these are among the highest values reported. Surface characterization revealed that CoO and Co_3O_4 are the active sites contributing to the superior performance of TiZrCo catalysts.

Finally, Li X. et al. [16] examined the cyclooctene epoxidation over phosphotungstate-based ionic silica nanoparticles. The novel synthesized silica network was extremely active for cyclooctene epoxidation, offering almost complete conversion and selectivity to epoxy-cyclooocten at 70 °C. The superior efficiency of the as-synthesized material as compared to bare SiO_2 nanoparticles was ascribed, based on an extended characterization study, to their optimal textural and structural/morphological characteristics.

In summary, the aforementioned special issue highlights the ongoing importance of the "surface chemistry approach", from both theoretical and experimental points of view, towards the fundamental understanding of catalytic phenomena. I am very pleased to serve as the Guest Editor of this thematic issue involving 14 high-quality studies on the interfacial discipline of "surface chemistry and catalysis". Firstly, I would like to express my gratitude to Professor Keith Hohn, Editor-in-chief of the *Catalysts* journal, for his kind invitation to organize this thematic issue. Special thanks to the editorial staff of *Catalysts*, particularly to Senior Assistant Editor Ms. Mary Fan, for their efforts and continuous support. Moreover, I am most appreciative to all authors for their contributions and hard work in revising them as well as to all reviewers for their valuable recommendations that assisted authors in upgrading their work to meet the high standards of *Catalysts*. I hope that this special issue will be a valuable resource for researchers, students and practitioners to promote and advance research and applications in the field of "surface chemistry and catalysis", since the fundamental understanding of catalysis will definitely be the vehicle towards the rational design of highly efficient and low-cost catalysts.

Conflicts of Interest: The author declares no conflict of interest.

References

1. Konsolakis, M. The role of Copper–Ceria interactions in catalysis science: Recent theoretical and experimental advances. *Appl. Catal. B* **2016**, *198*, 49–66.

2. Konsolakis, M. Recent advances on nitrous oxide (N_2O) decomposition over non-noble metal oxide catalysts: Catalytic performance, mechanistic considerations and surface chemistry aspects. *ACS Catal.* **2015**, *5*, 6397–6421.

3. González-Cobos, J.; de Lucas-Consuegra, A. A Review of Surface Analysis Techniques for the Investigation of the Phenomenon of Electrochemical Promotion of Catalysis with Alkaline Ionic Conductors. *Catalysts* **2016**, *6*, 15.

4. Sapp, W.; Koodali, R.; Kilin, D. Charge Transfer Mechanism in Titanium-Doped Microporous Silica for Photocatalytic Water-Splitting Applications. *Catalysts* **2016**, *6*.

5. Wang, W.; Wang, G.; Shao, M. First-Principles Modeling of Direct versus Oxygen-Assisted Water Dissociation on Fe(100) Surfaces. *Catalysts* **2016**, *6*, 29.

6. Kubiak, L.; Matarrese, R.; Castoldi, L.; Lietti, L.; Daturi, M.; Forzatti, P. Study of N_2O Formation over Rh- and Pt-Based LNT Catalysts. *Catalysts* **2016**, *6*, 36.

7. Papadopoulou, E.; Ioannides, T. Methanol Reforming over Cobalt Catalysts Prepared from Fumarate Precursors: TPD Investigation. *Catalysts* **2016**, *6*, 33.

8. Konsolakis, M.; Ioakimidis, Z.; Kraia, T.; Marnellos, G. Hydrogen Production by Ethanol Steam Reforming (ESR) over CeO_2 Supported Transition Metal (Fe, Co, Ni, Cu) Catalysts: Insight into the Structure-Activity Relationship. *Catalysts* **2016**, *6*, 39.

9. Tribalis, A.; Panagiotou, G.; Bourikas, K.; Sygellou, L.; Kennou, S.; Ladas, S.; Lycourghiotis, A.; Kordulis, C. Ni Catalysts Supported on Modified Alumina for Diesel Steam Reforming. *Catalysts* **2016**, *6*, 11.

10. Lopez, L.; Velasco, J.; Montes, V.; Marinas, A.; Cabrera, S.; Boutonnet, M.; Järås, S. Synthesis of Ethanol from Syngas over Rh/MCM-41 Catalyst: Effect of Water on Product Selectivity. *Catalysts* **2015**, *5*, 1737–1755.

11. De la Osa, A.; Romero, A.; Dorado, F.; Valverde, J.; Sánchez, P. Influence of Cobalt Precursor on Efficient Production of Commercial Fuels over FTS Co/SiC Catalyst. *Catalysts* **2016**, *6*, 98.

12. Cui, H.-Z.; Guo, Y.; Wang, X.; Jia, C.-J.; Si, R. Gold-Iron Oxide Catalyst for CO Oxidation: Effect of Support Structure. *Catalysts* **2016**, *6*, 37.

13. Hartadi, Y.; Behm, R.; Widmann, D. Competition of CO and H_2 for Active Oxygen Species during the Preferential CO Oxidation (PROX) on Au/TiO_2 Catalysts. *Catalysts* **2016**, *6*, 21.

14. Qiu, L.; Wang, Y.; Pang, D.; Ouyang, F.; Zhang, C.; Cao, G. Characterization and Catalytic Activity of Mn-Co/TiO_2 Catalysts for NO Oxidation to NO_2 at Low Temperature. *Catalysts* **2016**, *6*, 9.

15. Liu, T.; Cheng, H.; Lin, W.; Zhang, C.; Yu, Y.; Zhao, F. Aerobic Catalytic Oxidation of Cyclohexene over TiZrCo Catalysts. *Catalysts* **2016**, *6*, 24.

16. Li, X.; Jiang, P.; Wang, Z.; Huang, Y. Phosphotungstate-Based Ionic Silica Nanoparticles Network for Alkenes Epoxidation. *Catalysts* **2015**, *6*, 2.

A Review of Surface Analysis Techniques for the Investigation of the Phenomenon of Electrochemical Promotion of Catalysis with Alkaline Ionic Conductors

Jesús González-Cobos and Antonio de Lucas-Consuegra

Abstract: Electrochemical Promotion of Catalysis (EPOC) with alkali ionic conductors has been widely studied in literature due to its operational advantages *vs.* alkali classical promotion. This phenomenon allows to electrochemically control the alkali promoter coverage on a catalyst surface in the course of the catalytic reaction. Along the study of this phenomenon, a large variety of *in situ* and *ex situ* surface analysis techniques have been used to investigate the origin and mechanism of this kind of promotion. In this review, we analyze the most important contributions made on this field which have clearly evidenced the presence of adsorbed alkali surface species on the catalyst films deposited on alkaline solid electrolyte materials during EPOC experiments. Hence, the use of different surface analysis techniques such as scanning electron microscopy (SEM), energy-dispersive X-ray spectroscopy (EDX), X-ray diffraction (XRD), X-ray photoelectron spectroscopy (XPS), scanning photoelectron microscopy (SPEM), or scanning tunneling microscopy (STM), led to a better understanding of the alkali promoting effect, and served to confirm the theory of electrochemical promotion on this kind of catalytic systems. Given the functional similarities between alkali electrochemical and chemical promotion, this review aims to bring closer this phenomenon to the catalysis scientific community.

Reprinted from *Catalysts*. Cite as: González-Cobos, J.; de Lucas-Consuegra, A. A Review of Surface Analysis Techniques for the Investigation of the Phenomenon of Electrochemical Promotion of Catalysis with Alkaline Ionic Conductors. *Catalysts* **2016**, *6*, 15.

1. General Features of Alkaline Electrochemical Promotion

Promoters are widely used in the heterogeneous catalysis field [1,2]. Structural promoters improve the dispersion and stability of the active phase on the catalyst support, while electronic ones enhance the catalytic properties of the active phase itself. This latter kind of promoters can be added to the catalyst *ex situ*, *i.e.*, during the catalyst preparation step, or *in situ*, *i.e.*, in the course of the catalytic reaction, through the phenomenon of Electrochemical Promotion Of Catalysis (EPOC). This phenomenon, also known in literature as "Non-faradaic Electrochemical Modification of Catalytic Activity" (NEMCA effect) [3], is based on the modification

1

of the performance of a catalyst by the electrochemical pumping of promoter ions from an electro-active catalyst support, which is a solid electrolyte material, e.g., H^+, Na^+, K^+, O^{2-}, or F^- ionic conductors [4]. Although the first works on electrochemical promotion were carried out by using yttria-stabilized zirconia (YSZ), i.e., an O^{2-} ionic conductor material, as solid electrolyte [5], alkaline conductors such as the β-alumina family or NASICON-like compounds (e.g., $Na_3Zr_2Si_2PO_{12}$, $K_2YZr(PO_4)_3$, or $Li_{14}ZnGe_4O_{16}$), among others, have also been widely studied on the electrochemical promotion field [6,7].

In the case of electrochemical catalysts based on alkali (M^+)-conductors, the application of a cathodic polarization (i.e., negative current or overpotential) between the catalyst film-working electrode, which is deposited on one side of the electrolyte, and an inert counter electrode (typically gold) located at the opposite side (see Figure 1) leads to the migration of promoter (M^+) ions to the catalyst film, which is also called as back-spillover phenomenon. Once located over the catalyst surface, as in chemical (classical) promotion, these ionic species modify its chemisorption properties and hence its catalytic performance [3]. Thus, some of the main advantages of electrochemical promotion vs. classical promotion in heterogeneous catalysis are the capability of optimizing the amount of promoter coverage under changing reaction conditions and the possibility of in situ tuning the catalyst performance to maximize its activity and selectivity toward the desired product, even preventing catalyst deactivation or allowing its regeneration. These operational advantages vs. classical promotion, among others, have recently been revised by A. de Lucas-Consuegra [6]. It should also be noted that there are important differences in operating EPOC systems depending on the nature of the employed electro-active catalyst support. When using ionic conductor materials where the supplied ions can also participate in the catalytic reaction under study (e.g., O^{2-} ions in the case of catalytic oxidations or H^+ ions in catalytic hydrogenations) then these ions act as "sacrificial promoters" and present a finite mean residence time on the catalyst surface. In these cases, both galvanostatic and potentiostatic operations allow to obtain a steady-state catalytic reaction rate at each applied current or potential, respectively. However, when using solid electrolytes where the ionic conducting species are not involved in the catalytic reaction (e.g., Na^+ or K^+ ions) then only potentiostatic operation leads to steady-state reaction rates and, under galvanostatic operation, the coverage of the M^+ promoter, θ_{M^+}, increases with time as long as a constant electric current is maintained [3]. However, in these cases, as will be shown later, these alkali ions typically react with co-adsorbed reactant molecules on the gas-exposed catalyst surface leading to the formation of a large variety of promotional species.

ELECTROCHEMICAL PROMOTION (EPOC)

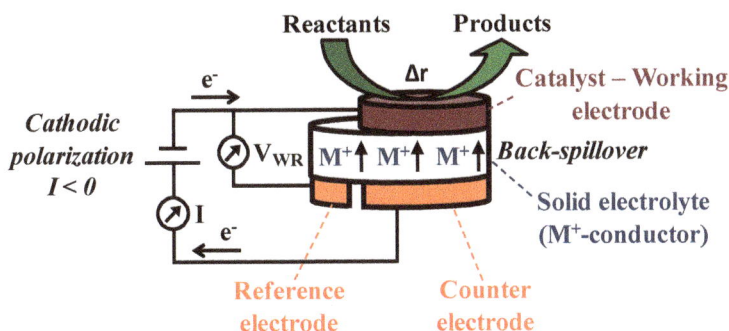

Figure 1. Scheme of the electrochemical cell used in electrochemical promotion studies with an alkaline ionic conductor (generally indicated as M^+-conductor) solid electrolytes.

Vayenas *et al.* performed the first electrochemical promotion study with alkaline solid electrolyte (Na-βAl$_2$O$_3$) in 1991 [8]. From this pioneer work, Na$^+$-conductors have been widely employed in many catalytic systems such as ethylene [9,10], CO [11], propane [12] and propylene oxidation [13], NO reduction [14–16], Fischer Tropsch synthesis [17], or hydrogenation of benzene [18] and CO$_2$ [19]. On the other hand, the first EPOC study using a K$^+$-conductor electrolyte (K$_2$YZr(PO$_4$)$_3$) dates from 1997 and addressed the Fe-catalyzed ammonia decomposition [20]. Urquhart *et al.* used other K$^+$-conductor solid electrolyte (K-βAl$_2$O$_3$) in Fischer-Tropsch reaction studies under both atmospheric [21] and high pressure [22], and de Lucas-Consuegra *et al.* introduced the use of this kind of ion-conducting catalyst support for the electrochemical promotion of Pt in CO [23] and propylene [24] oxidation, as well as in NO$_x$ reduction reactions [25,26]. More recent alkaline electrochemical promotion studies on CO$_2$ hydrogenation [27–30] and methanol conversion reactions [31–33] should also be highlighted. Additionally, in order to understand the mechanism of the phenomenon of electrochemical promotion of catalysis with both anionic and cationic conductors, a wide variety of characterization techniques have been used in the fields of catalysis (e.g., TPD, TPO, or work function measurement), electrochemistry (e.g., cyclic/linear sweep voltammetry or impedance spectroscopy), and surface science (e.g., XPS, UPS, SPEM, or STM) [3]. This paper aims to summarize the most relevant contributions relative to the latter techniques carried out in literature on electrochemical promotion with alkaline conductors. For this purpose, the surface characterization studies summarized in the next two sections have been divided into two categories, depending on whether they were performed under potentiostatic/galvanostatic control (*in situ* analysis) or not (*ex situ* analysis), as schematically shown in Figure 2.

All these techniques have contributed to the further understanding of the alkali electro-promotional effect in good agreement with the general rules of chemical and electrochemical promotion, valid for the different kinds of electronic promoters (both anionic and cationic ones) [3].

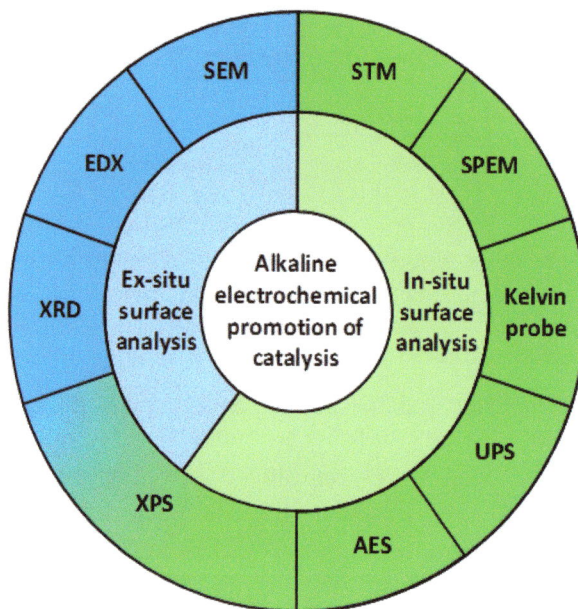

Figure 2. Scheme of the different surface analysis techniques used for investigating alkaline Electrochemical Promotion effect.

2. *Ex Situ* Characterization of Alkali-Promoted Catalyst Surfaces

The post-reaction characterization and analysis of the surface state of a catalyst film, previously subjected to given electrochemical promotion conditions, can be carried out *ex situ* by the following generic procedure. In the first place, the catalyst film is exposed to the reaction mixture while applying a certain positive or negative overpotential (for establishing an unpromoted or electropromoted state, respectively). After a given time, the reactor is cooled down to room temperature at the same applied overpotential. Then, the reactor and pipe lines are swept with inert gas and open-circuit conditions are established when the temperature is below 373 K (approximately), *i.e.*, when alkali ions mobility is too low. Finally, the electrochemical catalyst is transferred, under inert atmosphere, from the reactor to the characterization equipment. The aim of this procedure is to "freeze out" the catalyst surface state pertaining to the desired unpromoted/electropromoted state.

4

SEM and other microscopy techniques are widely employed in catalysis field to evaluate the structure and morphology of the catalysts. In EPOC studies, this technique, together with Energy-dispersive X-ray (EDX) spectroscopy, has also been employed to identify the arrangement and nature of the promoter phases present on the catalyst surface. For instance, Figure 3 shows the SEM micrograph, along with the corresponding elemental mapping and spectra by EDX, of a selected region of a Cu catalyst film deposited on a K-βAl$_2$O$_3$ pellet (K$^+$-conductor) used for the electrochemical promotion of the methanol partial oxidation reaction [32]. Prior to the surface analysis, the catalyst was subjected to certain reaction conditions and an applied potential, V_{WR} = -0.5 V, such that K$^+$ promoter ions were electrochemically supplied to the catalyst surface. As a consequence, in these micrographs, a large concentration of potassium (in blue) was found on the Cu catalyst surface (in green). This demonstrated that K$^+$ ions were able to migrate through the catalyst film and reached the gas-exposed catalyst surface. As also revealed by the EDX analysis taken from different areas of the micrograph (Figure 3c1,c2), oxygen- and potassium-containing surface compounds seemed to be formed on the metal catalyst film during the EPOC experiments, probably in form of some potassium oxides or carbonate molecules. In fact, an excess of these surface species (supplied under high cathodic polarization) could block the Cu active sites causing a decrease in the catalytic reaction rate in agreement with conventional chemical promotion [32]. Furthermore, some nitrogen (in red) was also noticed, homogeneously distributed on the catalyst surface, which was attributed to the K$^+$-promotional effect on the ammonia formation via reaction of hydrogen and nitrogen, both of them present in the gas reaction atmosphere.

The presence of potassium carbonate/bicarbonate species was also identified on the surface of a promoted Pt/K-βAl$_2$O$_3$ electrochemical catalyst employed in a methanol partial oxidation reaction [31]. In this study, these kinds of promoter-derived compounds were also detected *ex situ* by X-ray diffraction analysis after EPOC experiments, as shown in Figure 4. As in the case of the electropromoted Cu catalyst film, a poisoning effect derived from an excess of K$^+$-derived surface species was found on this Pt film [31]. Furthermore, in both mentioned studies [31,32], the obtained electropromotional effect was completely reversible since all the promoter phases were decomposed and the alkali ions were transferred from the catalyst back to the solid electrolyte after applying a positive enough potential at the end of the experiments. In this way, a clean, un-promoted catalyst surface, free of any promoter species, was achieved. These surface species were also observed by SEM-EDX analysis on Pt catalyst films deposited on K-βAl$_2$O$_3$ solid electrolyte for propene oxidation reaction [24]. In this case, the presence of potassium oxides and superoxides, along with carbon deposited fragments were observed after catalytic experiments. Moreover, these K$^+$-derived species seemed to induce a permanent

EPOC effect and showed to be more stable than potassium carbonates or bicarbonates, since the latter needed lower positive potentials to be decomposed. These results demonstrated that the nature of the final form of alkali promoter species and their chemical structure (e.g., oxides, superoxides, carbonates, *etc.*) strongly influence the final electropromotional behavior.

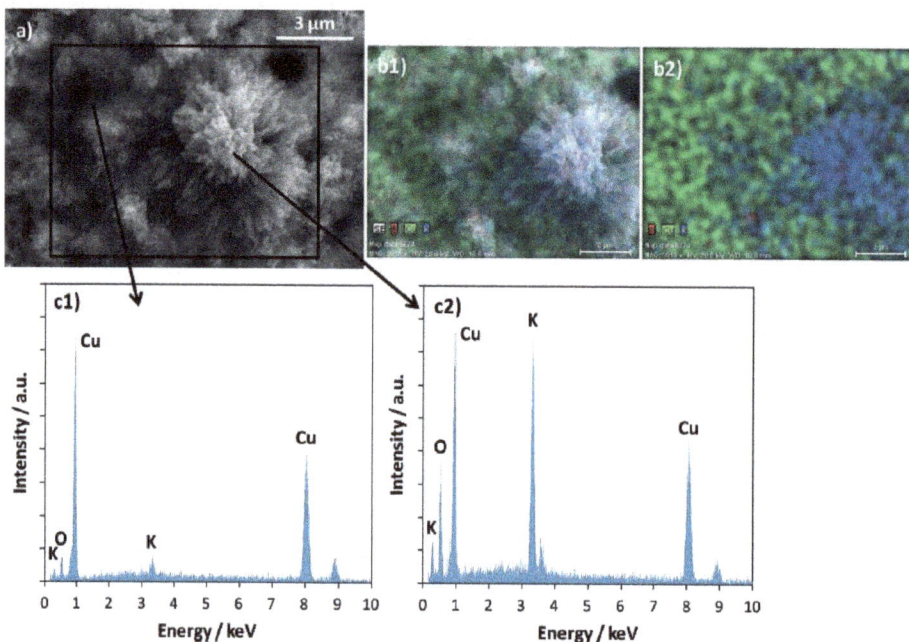

Figure 3. Top view SEM image of a selected area of a Cu/K-β Al$_2$O$_3$ electrochemical catalyst (**a**) after the EPOC experiments (593 °C, CH$_3$OH/O$_2$ = 4.4%/0.3%, V_{WR} = -0.5 V for 1 h), along with the corresponding elemental mapping (**b1** and **b2**) of Cu (green), K (blue), and N (red) and the EDX spectra from different regions (**c1** and **c2**). Reprinted with permission from Ref. [32].

It should be noted that other characterization techniques such as X-ray photoelectron (XPS) [32] and Fourier transform infrared (FTIR) [24] spectroscopies have also been *ex situ* employed in the past to study the chemical state of the different alkali promoter species formed on the catalyst surface under EPOC conditions, obtaining similar information about the nature of the different promotional species.

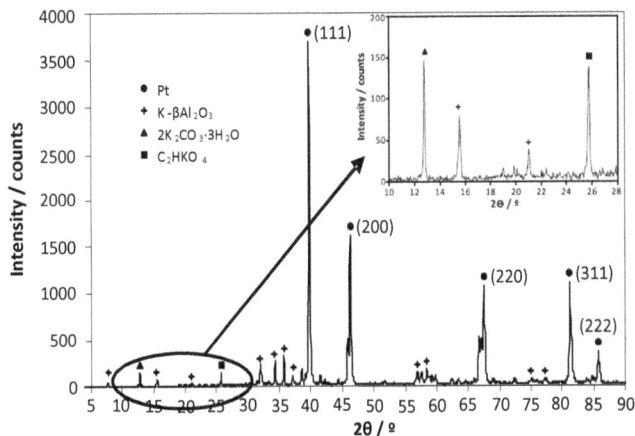

Figure 4. XRD spectra after EPOC experiments ($CH_3OH/O_2 = 7.2\%/4.6\%$, 593 K, overnight at -2 V) of a Pt film prepared by impregnation on a K-βAl_2O_3 solid electrolyte. Reprinted with permission from Ref. [31].

3. *In Situ* Characterization of Alkali-Promoted Catalyst Surfaces

During the alkaline electrochemical promotion of a catalyst film, the influence of the applied potential on the promoter coverage, as well as the nature and stability of the alkali-derived surface species, can be *in situ* evaluated by different techniques. Indeed, the progress made in the development and understanding of the phenomenon of electrochemical promotion over the years could not be conceived without the *in situ* spectroscopy studies performed by the group of professor Lambert and co-workers [34–45]. In all cases, the spectra were obtained immediately after exposing the appropriately polarized catalyst film (either unpromoted or electrochemically promoted) to conditions of temperature and reactant partial pressures similar to those encountered in the electrochemical promotion reactor, in order to simulate the different surface conditions of interest. For this purpose, spectrometers equipped with a reaction cell (under galvanostatic/potentiostatic control) and an ultra high vacuum chamber were used, in such a way that the electropromoted sample was mounted on a manipulator that allowed its translation between both chambers.

X-ray photoelectron spectroscopy (XPS) and Auger electron spectroscopy (AES) measurements have been carried out on Pt [34–38,40], Rh [39,41–43] and Cu [41,44,45] catalyst films used in alkaline electrochemical promotion studies for the oxidation of ethylene [34] or propene [37], hydrogenation of acetylene [38], and reduction of NO by CO [36,40–45] or by propene [35,39,42,43]. In first place, all these studies demonstrated that the mode of operation of the electrochemically promoted catalyst films involves reversible pumping (backspillover) of Na$^+$ or K$^+$ ions from the solid electrolyte in agreement with the theory and rules of Electrochemical Promotion of Catalysis. For instance, Figure 5a shows the Na 1s XPS spectra obtained on a Rh catalyst supported on Na-βAl$_2$O$_3$, at 580 K, as a function of catalyst potential (V_{WR}) under ultra-high vacuum conditions [41]. The +1 V spectrum corresponds to the clean (unpromoted) sample, while increasingly negative values of V_{WR} correspond to increasing amounts of electropumped Na$^+$ on the catalyst surface. As typically observed in these studies [35,38,40–42,44,46], the Na 1s emission comprises two components. The first one exhibits invariant binding energy (BE) and its intensity increases with decreasing V_{WR}, i.e., as Na$^+$ ions are electropumped to the catalyst. This is ascribed to sodium at the surface of the (grounded) Rh catalyst film. The second (shaded) component exhibits constant intensity and a systematic shift in apparent BE. This shift is numerically equal to the change in catalyst potential, strongly suggesting that this emission arises from the underlying solid electrolyte, whose electrostatic potential differs from that of the Rh film by the change in V_{WR}. This interpretation was confirmed by the experimental results obtained by grazing exit synchrotron photoemission, where the signal from the electrolyte vanished [46]. As also stated in the other referenced studies, the spectral behaviour was reversible and reproducible as a function of V_{WR}, consistent with the reversible and reproducible catalytic response observed during the electrochemical promotion of the Rh catalyst for the NO reduction reaction with both CO and C$_3$H$_6$ [41]. Hence, as it can be drawn from Figure 5b, the decrease in the catalyst potential to −1 V leads to a linear increase in the Na$^+$ coverage (θ_{Na^+}), up to around 0.025 monolayers, which may be estimated from the integrated Na 1s intensity of the component associated with the metal surface [47]. Moreover, the catalyst work function (ϕ) also shows to vary linearly with V_{WR} in such a low θ_{Na^+} range. In this case, work function changes were determined by ultraviolet photoelectron spectroscopy (UPS), by measuring the change in secondary electron cutoff in the spectrum relative to the Fermi edge [41], although it can also be in situ measured with a Kelvin probe [48]. Very interestingly, XPS data also allowed verifying that electrochemically pumped sodium is identical in behavior and in chemical state with Na supplied by vacuum deposition from a Na evaporation source [35], which is in very good agreement with the close similarities found between electrochemically-promoted catalysts and conventionally-promoted ones [35,49].

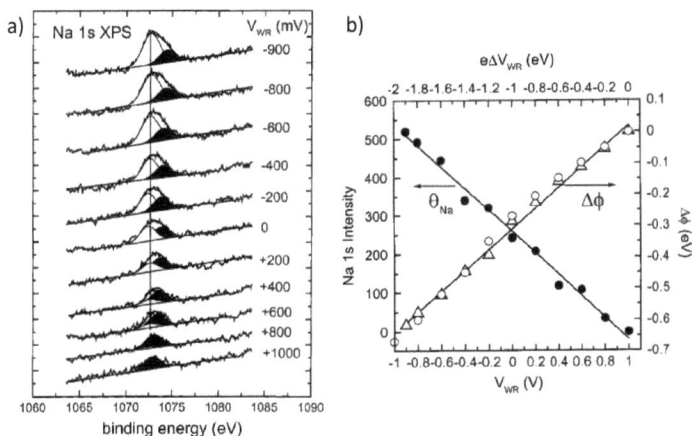

Figure 5. Na 1s XPS spectra taken on a Rh/Na-βAl$_2$O$_3$ electrochemical catalyst (580 K, UHV conditions), showing the effect of catalyst potential (V_{WR}) on sodium coverage (**a**). Invariant component due to Na on Rh; shifting component due to Na in solid electrolyte. Influence of V_{WR} on the integrated Na 1s XPS intensity due to Na on the Rh surface and associated work function change of the catalyst film (**b**). Reprinted with permission from Ref. [41]. Copyright 2000 American Chemical Society.

XPS is also a very useful technique for the in-situ identification of alkali-derived surface compounds and for the study of their formation/decomposition on the catalyst surface during the electrochemical promotion experiments. In this way, depending on the reaction atmosphere, NaNO$_2$ [40], NaNO$_3$ [40,43–45] and Na$_2$CO$_3$ [34,37,39,43] were found on the electropromoted catalyst surface which, in excess, induced a poisoning effect on the catalytic activity [34,37,40]. A very illustrative example is the work carried out by Filkin *et al.* on propene oxidation reaction with a Pt/Na-βAl$_2$O$_3$ catalyst, where not only XPS but also X-ray excited AES and X-ray absorption near edge structure (XANES) were performed [37]. Figure 6a,b show the Na KLL Auger and Na 1s XPS spectra obtained after exposure of the catalyst to the reaction mixture and to an applied potential (V_{WR}) such that the Pt film was either poisoned (V_{WR} = −0.6 V, spectrum 1), promoted (V_{WR} = −0.1 V, spectrum 2) or electrochemically clean (V_{WR} = +0.5 V, spectrum 3). The Na KLL auger data show that the catalyhst surface promoted at slightly negative potential presents a lower amount of Na-containing compounds than the poisoned surface (*i.e.*, at higher negative potential), and that these promoter phases are stable at reaction temperature (588 K) but decomposed by pumping the Na$^+$ ions back to the solid electrolyte. Moreover, as the authors stated, during the positive polarization, the pressure in the vacuum system increased, denoting that gaseous molecules released upon decomposition of the alkali-derived compounds. The Na 1s XPS spectra

9

confirm the stability of the Na$^+$-derived compounds at reaction conditions and their decomposition under positive polarization. However, this kind of spectrometry technique does not allow distinguishing the Na loading obtained under promoted and poisoned conditions, in contrast to Auger electron spectroscopy (Figure 6a). This feature can be explained according to the different electron escape depths related to the AE and XPS spectra (with sampling depths of around 25 and 8 Å, respectively [37]). On the other hand, the carbon 1s XPS spectra (Figure 6c) not only confirm that the poisoned surface presents a much higher amount of Na$^+$-derived compounds formed under reaction conditions, but also provide some proof concerning the chemical nature of these species, which likely consist of sodium carbonates, as also verified by the XANES results (Figure 6d).

In two of the previously mentioned studies [38,46], scanning photoelectron microscopy (SPEM) was also used to *in situ* analyze the surface of Cu [46] and Pt [38] catalysts promoted by Na$^+$ and K$^+$, respectively. For instance, Figure 7 shows the results obtained on this latter work for the alkali-promoted acetylene hydrogenation reaction. Specifically, 6.4 μm × 6.4 μm Pt 4f$_{7/2}$ (raw data), K 2p$_{3/2}$ (raw data) and topography corrected [50] K 2p$_{3/2}$ intensity maps were taken under different applied potentials (V_{WR}) from +0.4 V (unpromoted state) to −0.8 V (electropromoted state). These micrographs show that the Pt signal does not attenuate as K$^+$ ions are electrochemically supplied to the catalyst film and that a very small thermal drift takes place during the experiment. On the other hand, after correcting the intensity modulations derived from topography, the K 2p$_{3/2}$ maps show that the alkali promoter is relatively uniformly distributed on the Pt surface, and that its concentration clearly increases upon decreasing V_{WR}.

Furthermore, it should be mentioned that the back-spillover phenomenon with alkaline ions has also been confirmed by scanning tunneling microscopy (STM) over Pt catalysts [51–53]. In these studies, unfiltered STM images were firstly obtained from an air-exposed Pt(111) catalyst film deposited on a Na-βAl$_2$O$_3$ solid electrolyte, under open circuit conditions, *i.e.*, before applying any electric current or potential. In this way, a Pt(111)-(2×2)-O adlattice (interatomic distance of 5.6 Å) was found along with an overlapping Pt(111)-(12×12)-Na adlattice (interatomic distance of 33.2 Å) in some regions, the latter being attributed to thermal diffusion of sodium from the Na-βAl$_2$O$_3$ during the deposition of the Pt(111) film [51–53]. Then, STM was carried out under both unpromoted and electropromoted established conditions. In the first case, after the application of a positive current or potential, only the Pt(111)-(2×2)-O adlattice remained on the micrograph, thus denoting a Na-free catalyst surface. This demonstrates that an unpromoted (reference) state can be defined by applying a positive enough potential on the catalyst film supported on an alkaline solid electrolyte. In the second case, after applying a negative current (of the order of −1 μA) for a few minutes, the Pt(111)-(12×12)-Na adlattice reappeared [51–53]. Then, all these studies show that the origin of alkali-EPOC is clearly due to the reversible migration (back-spillover) of alkali ions to the catalyst surface. These ions may interact with co-adsorbed reactant molecules depending on the reaction conditions forming a wide variety of surface compounds and altering the chemisorption properties of the catalyst in a pronounce and controllable way.

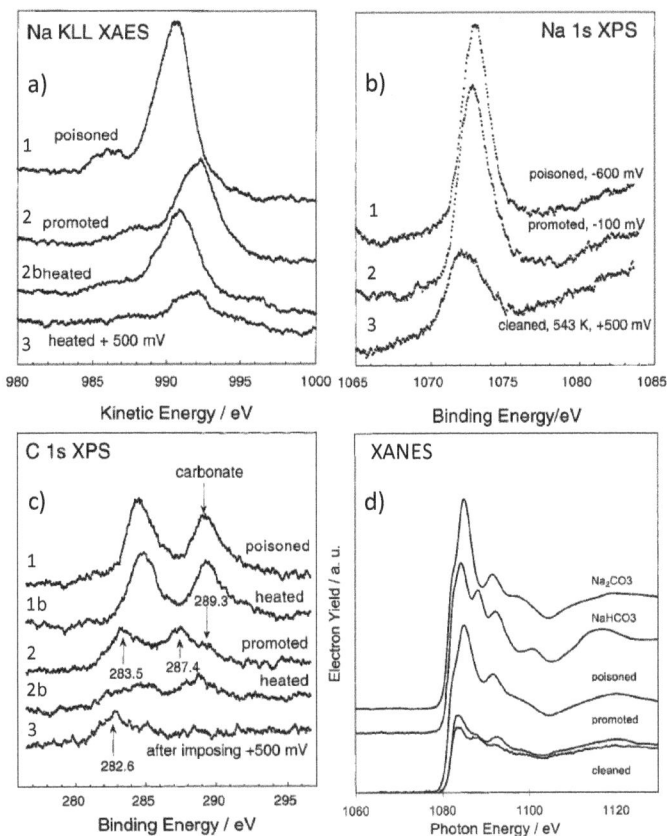

Figure 6. (a) Sodium KLL XAES, (b) Sodium 1s XPS and (c) Carbon 1s XPS analysis of a Pt/Na-β Al$_2$O$_3$ electrochemical catalyst under propene combustion atmosphere (0.6 kPa propene, 2.5 kPa oxygen). All spectra acquired at room temperature, open circuit, after exposure to different un-promoted and electropromoted conditions: *1*, poisoned conditions, V_{WR} = −600 mV; *1b*, as *1* but after heating to 588 K; *2*, promoted conditions, V_{WR} = −100 mV; *2b*, as *2* but after heating to 588 K; *3*, as *2* but after imposing V_{WR} = +500 mV. (d) XANES spectra after exposure to propene and oxygen at 500 K under promoted and poisoned conditions (See Ref. [37] for temperature and partial pressures conditions). Reference XANES spectra for the cleaned surface, Na$_2$CO$_3$ and NaHCO$_3$ are also shown. Reprinted with permission from Ref. [37].

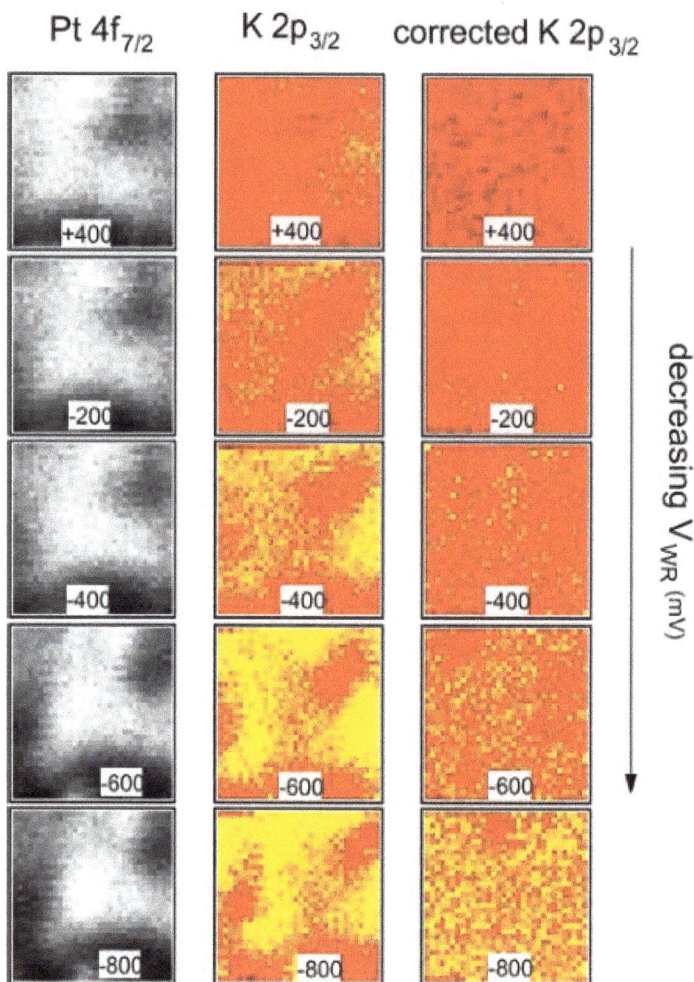

Figure 7. Pt 4f7/2 and K 2p3/2 photoelectron intensity maps taken on a Pt/K-βAl₂O₃ catalyst as a function of V_{WR} at 473 K during the selective hydrogenation of acetylene. Reprinted with permission from Ref. [38].

Finally, it should be noted that there is a kind of *in situ* characterization methodology called operando which is acquiring increasing interest in the catalysis scientific community and could be also very helpful in the study of the alkaline electrochemical promotion. This methodology is based on the *in situ* characterization of the catalyst surface while the catalytic activity is simultaneously measured, under real working conditions. However, the perfect correlation between the catalytic activity measurement and the surface analysis is very difficult mainly due to problems arising from the operando cell reactor design. This reactor cell must operate under specific pressure and temperature conditions while keeping the three-electrode configuration and allowing the spectra collection under such reaction conditions. Moreover, the presence of spectator species in the reaction and void volumes in the reactor, which are typically high for gas phase reactions, may affect the obtained spectra. Thus, new insights in the *in situ* characterization of the EPOC phenomenon could be opened in so far as the operando reactor design is improved and such limitations are overcome.

4. Conclusions and Prospects

It is well known that alkali promoters play a key role in heterogeneous catalysis with special emphasis on catalytic reactions of large scale industrial application such as ammonia or hydrocarbons synthesis. As classical alkali promotion, the phenomenon of electrochemical promotion of catalysis (EPOC) is based on the addition of promoting species to the catalyst active surface in order to modify its chemisorptive properties and, hence, its activity and selectivity. However, in the case of EPOC, the electrically induced back-spillover of the promoter species enables the straightforward study of the promoting role of certain alkali coverage and the *in situ*, controlled, enhancement of the catalytic performance under dynamic reaction conditions. For this reason, it becomes essential the employment of proper surface analysis techniques which allow monitoring the amount and state of the promoter phases as well as a deep understanding of the back-spillover phenomenon. In this sense, several *ex situ* and *in situ* characterization techniques have been carried out for the last two decades that have been described on this paper.

Although the *ex situ* surface analysis techniques involve a series of inaccuracies derived from the necessary handling and transfer of the electrochemical catalyst to the characterization equipment, they may constitute a useful and easy tool to determine the stability of the electrochemical catalyst and to obtain qualitative information about the different species adsorbed on the catalyst surface. On the other hand, the *in situ* characterization techniques provide detailed information about the mechanism of alkali ions backspillover, the nature of the promoter-derived surface compounds and their influence on the catalytic properties, as a function of both the catalyst potential variation and the reaction conditions. Hence, the implementation of surface analysis techniques herein mentioned and other possible such as fourier transform infrared (FTIR) spectroscopy, photoelectron emission microscopy (PEEM), or atomic force microscopy (AFM), in conjunction with the development of alkaline electrochemical promotion experiments, is of paramount importance not only for better understanding of this phenomenon, but also for the design of more efficient and competitive conventional heterogeneous catalysts.

Acknowledgments: The financial support of the Spanish Government and European Union is gratefully acknowledged. The authors also thank all nice scientist, colleagues and co-workers in the EPOC field.

Author Contributions: Jesús Gonzalez Cobos was the primary author of this review, selecting and discussing the most important results published on the field. Antonio de Lucas Consuegra provided assistance in writing, revising the case studies and updating the article in response to the reviewers.

Conflicts of Interest: The authors declare no conflict of interest.

References

1. Mross, W.D. Alkali doping in heterogeneous catalysis. *Catal. Rev.* **1983**, *25*, 591–637.
2. Farrauto, R.J.; Bartholomew, C.H. *Fundamentals of Industrial Catalytic Processes*; Chapman & Hall: London, UK, 1997.
3. Vayenas, C.G.; Bebelis, S.; Pliangos, C.S.; Brosda, D. *Tsiplakides, Electrochemical Activation of Catalysis: Promotion, Electrochemical Promotion and Metal-Support Interactions*; Kluwer Academic Publishers/Plenum Press: New York, NY, USA, 2001.
4. Gellings, P.J.; Bouwmeester, H.J.M. *The CRC Handbook of Solid State Electrochemistry*; CRC Press: Boca Raton, FL, USA, 1997.
5. Stoukides, M.; Vayenas, C.G. The effect of electrochemical oxygen pumping on the rate and selectivity of ethylene oxidation on polycrystalline silver. *J. Catal.* **1981**, *70*, 137–146.
6. De Lucas-Consuegra, A. New trends of Alkali Promotion in Heterogeneous Catalysis: Electrochemical Promotion with Alkaline Ionic Conductors. *Catal. Surv. Asia* **2015**, *19*, 25–37.
7. Vernoux, P.; Lizarraga, L.; Tsampas, M.N.; Sapountzi, F.M.; de Lucas-Consuegra, A.; Valverde, J.L.; Souentie, S.; Vayenas, C.G.; Tsiplakides, D.; Balomenou, S.; Baranova, E.A. Ionically conducting ceramics as active catalyst supports. *Chem. Rev.* **2013**, *113*, 8192–8260.
8. Vayenas, C.G.; Bebelis, S.; Despotopoulou, M. Non-faradaic electrochemical modification of catalytic activity 4. The use of β''-Al_2O_3 as the solid electrolyte. *J. Catal.* **1991**, *128*, 415–435.
9. Karavasilis, C.; Bebelis, S.; Vayenas, C.G. *In situ* controlled promotion of catalyst surfaces via NEMCA: The effect of Na on the Ag-catalyzed ethylene epoxidation in the presence of chlorine moderators. *J. Catal.* **1996**, *160*, 205–213.
10. Petrolekas, P.D.; Brosda, S.; Vayenas, C.G. Electrochemical promotion of Pt catalyst electrodes deposited on $Na_3Zr_2Si_2PO_{12}$ during ethylene oxidation. *J. Electrochem. Soc.* **1998**, *145*, 1469–1477.
11. Yentekakis, I.V.; Moggridge, G.; Vayenas, C.G.; Lambert, R.M. *In Situ* controlled promotion of catalyst surfaces via NEMCA: The effect of Na on the Pt-catalyzed CO oxidation. *J. Catal.* **1994**, *146*, 292–305.
12. Kotsionopoulos, N.; Bebelis, S. *In situ* electrochemical modification of catalytic activity for propane combustion of Pt/β-Al_2O_3 catalyst-electrodes. *Top. Catal.* **2007**, *44*, 379–389.

13. Vernoux, P.; Gaillard, F.; Lopez, C.; Siebert, E. *In-situ* electrochemical control of the catalytic activity of platinum for the propene oxidation. *Solid State Ionics* **2004**, *175*, 609–613.

14. Dorado, F.; de Lucas-Consuegra, A.; Vernoux, P.; Valverde, J.L. Electrochemical promotion of platinum impregnated catalyst for the selective catalytic reduction of NO by propene in presence of oxygen. *Appl. Catal. B* **2007**, *73*, 42–50.

15. Palermo, A.; Lambert, R.M.; Harkness, I.R.; Yentekakis, I.V.; Mar'ina, O.; Vayenas, C.G. Electrochemical promotion by Na of the platinum-catalyzed reaction between CO and NO. *J. Catal.* **1996**, *161*, 471–479.

16. Vernoux, P.; Gaillard, F.; Lopez, C.; Siebert, E. Coupling catalysis to electrochemistry: A solution to selective reduction of nitrogen oxides in lean-burn engine exhausts? *J. Catal.* **2003**, *217*, 203–208.

17. Williams, F.J.; Lambert, R.M. A study of sodium promotion in Fischer-Tropsch synthesis: Electrochemical control of a ruthenium model catalyst. *Catal. Lett.* **2000**, *70*, 9–14.

18. Cavalca, C.A.; Haller, G.L. Solid electrolytes as active catalyst supports: Electrochemical modification of benzene hydrogenation activity on Pt/β"(Na)Al$_2$O$_3$. *J. Catal.* **1998**, *177*, 389–395.

19. Bebelis, S.; Karasali, H.; Vayenas, C.G. Electrochemical promotion of the CO$_2$ hydrogenation on Pd/YSZ and Pd/β"-Al$_2$O$_3$ catalyst-electrodes. *Solid State Ionics* **2008**, *179*, 1391–1395.

20. Pitselis, G.E.; Petrolekas, P.D.; Vayenas, C.G. Electrochemical promotion of ammonia decomposition over Fe catalyst films interfaced with K$^+$- & H$^+$- conductors. *Ionics* **1997**, *3*, 110–116.

21. Urquhart, A.J.; Keel, J.M.; Williams, F.J.; Lambert, R.M. Electrochemical Promotion by Potassium of Rhodium-Catalyzed Fischer-Tropsch Synthesis: XP Spectroscopy and Reaction Studies. *J. Phys. Chem. B* **2003**, *107*, 10591–10597.

22. Urquhart, A.J.; Williams, F.J.; Lambert, R.M. Electrochemical promotion by potassium of Rh-catalysed fischer-tropsch synthesis at high pressure. *Catal. Lett.* **2005**, *103*, 137–141.

23. De Lucas-Consuegra, A.; Dorado, F.; Valverde, J.L.; Karoum, R.; Vernoux, P. Electrochemical activation of Pt catalyst by potassium for low temperature CO deep oxidation. *Catal. Commun.* **2008**, *9*, 17–20.

24. De Lucas-Consuegra, A.; Dorado, F.; Valverde, J.L.; Karoum, R.; Vernoux, P. Low-temperature propene combustion over Pt/K-βAl$_2$O$_3$ electrochemical catalyst: Characterization, catalytic activity measurements, and investigation of the NEMCA effect. *J. Catal.* **2007**, *251*, 474–484.

25. De Lucas-Consuegra, A.; Caravaca, A.; Dorado, F.; Valverde, J.L. Pt/K-βAl$_2$O$_3$ solid electrolyte cell as a "smart electrochemical catalyst" for the effective removal of NO$_x$ under wet reaction conditions. *Catal. Today* **2009**, *146*, 330–335.

26. De Lucas-Consuegra, A.; Dorado, F.; Jiménez-Borja, C.; Valverde, J.L. Influence of the reaction conditions on the electrochemical promotion by potassium for the selective catalytic reduction of N$_2$O by C$_3$H$_6$ on platinum. *Appl. Catal. B* **2008**, *78*, 222–231.

27. Ruiz, E.; Cillero, D.; Martínez, P.J.; Morales, Á.; Vicente, G.S.; De Diego, G.; Sánchez, J.M. Bench scale study of electrochemically promoted catalytic CO_2 hydrogenation to renewable fuels. *Catal. Today* **2013**, *210*, 55–66.

28. Ruiz, E.; Cillero, D.; Martínez, P.J.; Morales, Á.; Vicente, G.S.; de Diego, G.; Sánchez, J.M. Electrochemical synthesis of fuels by CO_2 hydrogenation on Cu in a potassium ion conducting membrane reactor at bench scale. *Catal. Today* **2014**, *236*, 108–120.

29. Makri, M.; Katsaounis, A.; Vayenas, C.G. Electrochemical promotion of CO_2 hydrogenation on Ru catalyst-electrodes supported on a K-β''-Al_2O_3 solid electrolyte. *Electrochim. Acta* **2015**, *179*, 556–564.

30. Gutiérrez-Guerra, N.; González-Cobos, J.; Serrano-Ruiz, J.C.; Valverde, J.L.; de Lucas-Consuegra, A. Electrochemical Activation of Ni Catalysts with Potassium Ionic Conductors for CO_2 Hydrogenation. *Top. Catal.* **2015**, *58*, 1256–1269.

31. De Lucas-Consuegra, A.; González-Cobos, J.; García-Rodríguez, Y.; Mosquera, A.; Endrino, J.L.; Valverde, J.L. Enhancing the catalytic activity and selectivity of the partial oxidation of methanol by electrochemical promotion. *J. Catal.* **2012**, *293*, 149–157.

32. González-Cobos, J.; Rico, V.J.; González-Elipe, A.R.; Valverde, J.L.; de Lucas-Consuegra, A. Electrochemical activation of an oblique angle deposited Cu catalyst film for H_2 production. *Catal. Sci. Tec.* **2015**, *5*, 2203–2214.

33. González-Cobos, J.; López-Pedrajas, D.; Ruiz-López, E.; Valverde, J.L.; de Lucas-Consuegra, A. Applications of the Electrochemical Promotion of Catalysis in Methanol Conversion Processes. *Top. Catal.* **2015**, *58*, 1290–1302.

34. Harkness, I.R.; Hardacre, C.; Lambert, R.M.; Yentekakis, I.V.; Vayenas, C.G. Ethylene oxidation over platinum: In situ electrochemically controlled promotion using Na-β'' alumina and studies with a Pt(111)/Na model catalyst. *J. Catal.* **1996**, *160*, 19–26.

35. Konsolakis, M.; Palermo, A.; Tikhov, M.; Lambert, R.M.; Yentekakis, I.V. Electrochemical *vs.* conventional promotion: A new tool to design effective, highly dispersed conventional catalysts. *Ionics* **1998**, *4*, 148–156.

36. Lambert, R.M.; Tikhov, M.; Palermo, A.; Yentekakis, I.V.; Vayenas, C.G. Electrochemical promotion of environmentally important catalytic reactions. *Ionics* **1995**, *1*, 366–376.

37. Filkin, N.C.; Tikhov, M.S.; Palermo, A.; Lambert, R.M. A Kinetic and Spectroscopic Study of the in Situ Electrochemical Promotion by Sodium of the Platinum-Catalyzed Combustion of Propene. *J. Phys. Chem. A* **1999**, *103*, 2680–2687.

38. Williams, F.J.; Palermo, A.; Tracey, S.; Tikhov, M.S.; Lambert, R.M. Electrochemical promotion by potassium of the selective hydrogenation of acetylene on platinum: Reaction studies and XP spectroscopy. *J. Phys. Chem. B* **2002**, *106*, 5668–5672.

39. Williams, F.J.; Palermo, A.; Tikhov, M.S.; Lambert, R.M. Electrochemical promotion by sodium of the rhodium-catalyzed reduction of NO by propene: Kinetics and spectroscopy. *J. Phys. Chem. B* **2001**, *105*, 1381–1388.

40. Yentekakis, I.V.; Palermo, A.; Filkin, N.C.; Tikhov, M.S.; Lambert, R.M. *In situ* electrochemical promotion by sodium of the platinum-catalyzed reduction of NO by propene. *J. Phys. Chem. B* **1997**, *101*, 3759–3768.

41. Williams, F.J.; Palermo, A.; Tikhov, M.S.; Lambert, R.M. Electrochemical promotion by sodium of the rhodium-catalyzed NO + CO reaction. *J. Phys. Chem. B* **2000**, *104*, 11883–11890.

42. Williams, F.J.; Palermo, A.; Tikhov, M.S.; Lambert, R.M. Mechanism of alkali promotion in heterogeneous catalysis under realistic conditions: Application of electron spectroscopy and electrochemical promotion to the reduction of NO by CO and by propene over rhodium. *Surf. Sci.* **2001**, *482–485*, 177–182.

43. Williams, F.J.; Tikhov, M.S.; Palermo, A.; Macleod, N.; Lambert, R.M. Electrochemical promotion of rhodium-catalyzed NO reduction by CO and by propene in the presence of oxygen. *J. Phys. Chem. B* **2002**, *105*, 2800–2808.

44. Lambert, R.M.; Williams, F.; Palermo, A.; Tikhov, M.S. Modelling alkali promotion in heterogeneous catalysis: *In situ* electrochemical control of catalytic reactions. *Top. Catal.* **2000**, *13*, 91–98.

45. Williams, F.J.; Palermo, A.; Tikhov, M.S.; Lambert, R.M. First Demonstration of in Situ Electrochemical Control of a Base Metal Catalyst: Spectroscopic and Kinetic Study of the CO + NO Reaction over Na-Promoted Cu. *J.Phys. Chem. B* **1999**, *103*, 9960–9966.

46. Williams, F.J.; Palermo, A.; Tikhov, M.S.; Lambert, R.M. The Origin of Electrochemical Promotion in Heterogeneous Catalysis: Photoelectron Spectroscopy of Solid State Electrochemical Cells. *J. Phys. Chem. B* **2000**, *104*, 615–621.

47. Carley, A.F.; Roberts, M.W. X-ray photoelectron spectroscopic study of the interaction of oxygen and nitric oxide with aluminium. *Proc. R. Soc. London Ser. A* **1978**, *363*, 403–424.

48. Vayenas, C.G.; Bebelis, S.; Ladas, S. Dependence of catalytic rates on catalyst work function. *Nature* **1990**, *343*, 625–627.

49. Yentekakis, I.V.; Lambert, R.M.; Tikhov, M.S.; Konsolakis, M.; Kiousis, V. Promotion by sodium in emission control catalysis: A kinetic and spectroscopic study of the Pd-catalyzed reduction of NO by propene. *J. Catal.* **1998**, *176*, 82–92.

50. Günther, S.; Kolmakov, A.; Kovac, J.; Kiskinova, M. Artefact formation in scanning photoelectron emission microscopy. *Ultramicroscopy* **1998**, *75*, 35–51.

51. Makri, M.; Vayenas, G.G.; Bebelis, S.; Besocke, K.H.; Cavalca, C. Atomic resolution STM imaging of electrochemically controlled reversible promoter dosing of catalysts. *Surf. Sci.* **1996**, *369*, 351–359.

52. Makri, M.; Vayenas, C.G.; Bebelis, S.; Besocke, K.H.; Cavalca, C. Atomic resolution scanning tunneling microscopy imaging of Pt electrodes interfaced with β''-Al_2O_3. *Ionics* **1996**, *2*, 248–253.

53. Archonta, D.; Frantzis, A.; Tsiplakides, D.; Vayenas, C.G. STM observation of the origin of electrochemical promotion on metal catalyst-electrodes interfaced with YSZ and β''-Al_2O_3. *Solid State Ionics* **2006**, *177*, 2221–2225.

17

Charge Transfer Mechanism in Titanium-Doped Microporous Silica for Photocatalytic Water-Splitting Applications

Wendi Sapp, Ranjit Koodali and Dmitri Kilin

Abstract: Solar energy conversion into chemical form is possible using artificial means. One example of a highly-efficient fuel is solar energy used to split water into oxygen and hydrogen. Efficient photocatalytic water-splitting remains an open challenge for researchers across the globe. Despite significant progress, several aspects of the reaction, including the charge transfer mechanism, are not fully clear. Density functional theory combined with density matrix equations of motion were used to identify and characterize the charge transfer mechanism involved in the dissociation of water. A simulated porous silica substrate, using periodic boundary conditions, with Ti^{4+} ions embedded on the inner pore wall was found to contain electron and hole trap states that could facilitate a chemical reaction. A trap state was located within the silica substrate that lengthened relaxation time, which may favor a chemical reaction. A chemical reaction would have to occur within the window of photoexcitation; therefore, the existence of a trapping state may encourage a chemical reaction. This provides evidence that the silica substrate plays an integral part in the electron/hole dynamics of the system, leading to the conclusion that both components (photoactive materials and support) of heterogeneous catalytic systems are important in optimization of catalytic efficiency.

Reprinted from *Catalysts*. Cite as: Sapp, W.; Koodali, R.; Kilin, D. Charge Transfer Mechanism in Titanium-Doped Microporous Silica for Photocatalytic Water-Splitting Applications. *Catalysts* **2016**, *6*, 34.

1. Introduction

Economical water splitting technology for the development of low-cost and efficient fuels motivates research in many fields. This particular application of the conversion of solar energy into chemical reaction can be accomplished using a catalytic site mounted on a substrate. The periodic pore system that is characteristic of silica substrates is considered to be one of the best options for use in heterogeneous catalysis research, mostly due to its high surface-to-volume ratio [1]. Micro- and mesoporous materials have a wide range of functionality, some of which are still being discovered [2]. Solar energy conversion [3], energy storage, and photodynamic therapy [4] are just a few areas in which these materials have found wide-spread utility.

The addition of titanium atoms to the inner surface of the silica pore walls provides a photoactive material [5] and introduces numerous charge transfer states, which can be analyzed by atomistic computational modeling [6–8]. It has been known since the 1970s that titanium dioxide (TiO_2) can be used in water-splitting applications [9]. Since then, numerous researchers have been trying to improve conditions for photocatalysis using TiO_2 in conjunction with various substrates [10,11], dopants [12–14], dyes [15,16], *etc.* Titanium's distribution onto a periodic and porous substrate has been shown to increase the effectiveness of hydrogen evolution when compared to bulk TiO_2 [17]. Evaluation of these parameters is, by no means, a simple task. It is a goal of this research to evaluate one of these parameters and its role in catalysis: a silica substrate. At present, many materials have been reported as photocatalysts for water splitting. However, titanium-doped microporous silica demonstrates H_2 production reaction using sacrificial reagents such as triethanolamine (TEOA) or methanol. This material, thus, is able to perform a significant part of the water-splitting mechanism. By simulation of this model, we learn more details of the role of charge transfer for photocatalysis. Additionally, while it is important to evaluate the other aspects of catalysis, which are known to improve photocatalytic efficiency (*i.e.*, co-catalysts, alternative dopants, dyes, and the use of sacrificial agents), we ventured to simplify the model in order to eliminate parameters that might obscure the role of the silica substrate.

Charge transfer dynamics that occur prior to the dissociation of water molecules near the titanium atoms can be scrutinized using non-adiabatic molecular dynamics [18]. In order to further understand the process by which charge transfer favors water-splitting, electron dynamics may be employed. Ideally, one would analyze bandgap excitation, electron relaxation and dynamics of ions all at once. Since this is not possible with current molecular simulation techniques, a separation of the governing aspects of the reaction must occur. To make this investigation more practical, ground state electronic structure is explored first. This information helps to identify orbital transitions due to excitations, which are of interest. Molecular dynamics [19] are simulated, which leads to couplings [20] and eventually to electron dynamics of photoexcitation [21] that then provides excitation lifetimes [22] and details of the charge transfer meachanism [6–8]. These lifetimes are an integral part of making conclusions about a material and its photocatalytic properties. If electron relaxation rates are low enough, catalytic reactions may occur.

The ability to use photoexcitations as initial conditions for electron dynamics allows for investigations into the dynamics of the formation of possible charge transfer states, which could lead to chemical reactions. A range of the most important orbitals can be narrowed down by visualizing them and identifying orbitals of interest based on their locations on the model. These orbitals can be used as an input to electron dynamics to compute relaxation dynamics and rates for electrons and holes. These data (on dynamics and rates) allow for the identification of so-called "trap states" [23]. A trap state is defined as an energy state in which the electron or hole becomes trapped for a relatively long time when compared to the overall time of relaxation. The basic idea is that the existence of trapping states could facilitate charge transfer because it encourages charge separation, which can then induce a chemical reaction.

2. Results and Discussion

An alpha-quartz silica slab, of the dimensions $5.4 \times 14.7 \times 17.0$ Å^3 (cubic Angstroms) and composition $Si_{36}O_{72}$, was the basis for this model. Atoms were systematically removed until a ~1 nm cylindrical pore was created, leading to a modified composition of $Si_{24}O_{48}$. A tetrahedrally-coordinated Ti atom was added to the inner surface of the pore wall and capped, as well as the addition of H caps onto the exposed O atoms. Water molecules were added to the inside of the pore, giving an overall chemical composition of $Si_{24}O_{48} + TiO_2H_2 + 11H_2O$. A fraction of water molecules dissociate and adsorb onto uncoordinated ions from the inner surface of the pore: H^+ to uncoordinated O, and OH^- to uncoordinated Si or Ti atoms. Thus, the model contains intact and dissociated water molecules. Figure 1 displays a diagram of the model, indicating the position of the Ti atom in relation to the placement of water molecules within the pore. Oxygen atoms represent the inner surface of the pore. Utilizing periodic boundary conditions in three dimensions, this model simulates a periodic and tubular pore system, which is similar to forms of silica substrates used in experimental photocatalysis research. We compare four models: silica slab with an empty pore, water alone (single and multiple molecules), silica with Ti and a water-filled pore.

In catalytic simulations, it is important to investigate basic electronic properties as a starting point for further inquiries into charge transfer and molecular dynamics. In Figure 2, densities of states and absorbance spectra are computed according to Equations (5)–(8), which are simple ways to visualize observables related to basic electronic properties. Comparing the isolated water molecule and a cluster of H_2O in Figure 2A,B, it is seen that a cluster of H_2O molecules provides an occupied orbital in the valence band that does not exist in a single water molecule. Perhaps this is indicative of water's ability to ionize $(H_2O + H_2O \rightleftharpoons OH^- + H_3O^+)$. A close comparison of the densities of states of all models under consideration (top row, Figure 2) reveals similarities. One notable similarity can be seen between the silica slab only (Figure 2C) and the Ti-functionalized pore (Figure 2D). There is a similar pattern in the valence bands from -10 to ~0 eV and conduction band in the range of 0 to 10 eV in these two models. However, the addition of a Ti atom provides more states in the ~0 to 4 eV range in Figure 2D. We expect this to occur with the addition of a transition metal, as these states are Ti d-orbitals. The density of states for this primary model (Figure 2D) shows a small band gap of ~0.032 eV. Additionally, one can see electron occupation in the valance band of the water cluster, in the water-filled pore, and also in the Ti-functionalized pore (see Figure 2B–D), potentially indicating a more facile charge transfer.

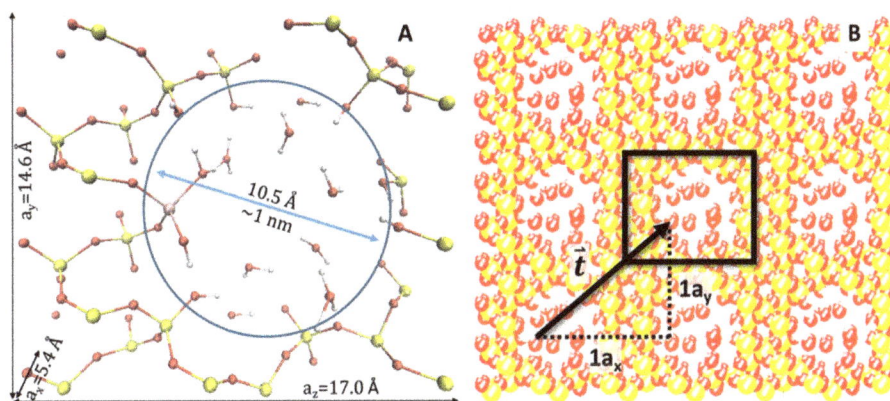

Figure 1. (A) A silicon dioxide slab was created, measuring 5.4 × 14.7 × 17.0 Å³ (cubic Angstroms). Si and O atoms were removed from bulk silica in a deliberate fashion in order to produce a cylindrical pore with a diameter of 10.5 Å On the inner surface of the pore, a Ti atom was added and capped with two hydroxyl groups, giving a tetrahedrally-coordinated Ti; **(B)** Several units are repeated in the x, y, and z directions to demonstrate periodicity.

Absorbance spectra were analyzed with two main goals in mind: first, to gain an understanding of the basic electronic properties of the model, and, second, to obtain the initial conditions for future electron dynamics simulations. With the first goal in mind, the components of the system are investigated separately in order to determine how each contributes to the overall character. The multi-molecular water (Figure 2F) shows an absorbance peak at about 380 nm, which is outside of the visible spectrum. A comparison of the absorbance spectra is made between the components of the model (as displayed in Figure 2E–G), and the complete model's absorbance spectrum (Figure 2H). The complete model has its peaks labeled based on the analysis of orbital transitions and comparison to the spectra of the components of the model. The model absorbs in the UV region of the electromagnetic spectrum, which is consistent with known absorption of silica.

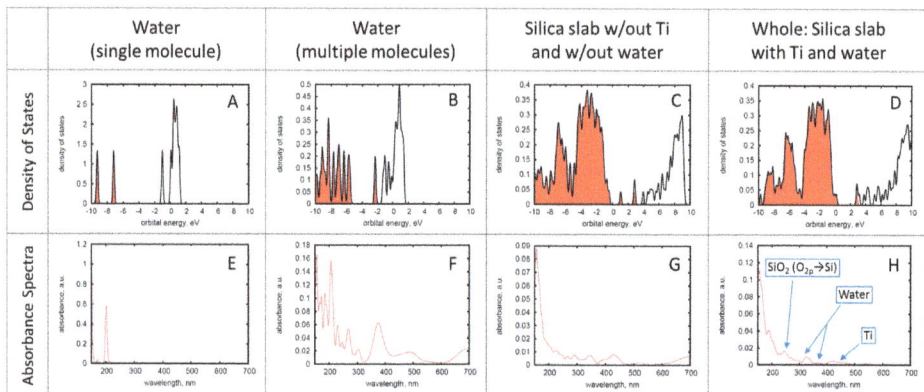

Figure 2. Signatures of electronic structure: densities of states and absorption spectra for the explored model and its partial components. (**A–D**) Densities of states for individual and combinations of components of the model and (**E–H**) absorbance spectra of individual and combinations of components of the model: (**A,E**) H_2O; (**B,F**) 10 H_2O; (**C,G**) SiO_2 pore; and (**D,H**) SiO_2 pore with Ti and 10 H_2O.

The inclusion of a Ti atom on the inner pore wall has contributed transition metal d-orbitals to the electronic structure of the model. The titanium exists in a +4 oxidation state, which means all of its d-orbitals are empty, with an electron configuration of [Ar] $3d^0$ $4s^0$. Since the Ti is tetrahedrally-coordinated, it is expected to see tetrahedral crystal field splitting among the orbital energies, ε_i [24,25]. In fact, this was found as one of the results of this research. These orbitals were characterized as d-orbitals by plotting isosurfaces and identifying the characteristic shapes of d-orbitals. LU+5 and LU+4 orbitals have d-character with some p-orbital shape contribution, which is why there are six d-orbitals reported, instead of five. In Figure 3, there exists a gap between the two lowest-energy d-orbitals and the energies of the remaining d-orbitals. Tetrahedral crystal field splitting of Ti's d-orbitals exists within a reasonably-sized sub-gap of energy between energies of the two groups of nearly-degenerate orbitals. This sub-gap energy is about 1.2 eV, which is of realistic magnitude to confirm the accuracy of the model.

The model can exist in several excited states. Some of the excitations have neutral character, others exhibit charge transfer character. These charge transfer states have been included due to the trapping states that are contained in each of them. Furthermore, HO-11, HO-6, LU+7, and LU+8 orbitals exist on water molecules within the pore, as displayed in Figure 4. In a photoexcited state, one can observe the non-radiative transition from one excitation to another due to non-adiabatic couplings. If there is a possibility of a chemical reaction occurring, the focus should be on those molecules and orbitals involved in such a reaction. In this model, the dynamics of charge transfer states formation have been identified. The dynamics of charge transfer are triggered by the photoexcitation to a neutral state (HO-6, LU+7) and (HO-11, LU+8). The dynamics induced by different photoexcitations are compared in Figure 5.

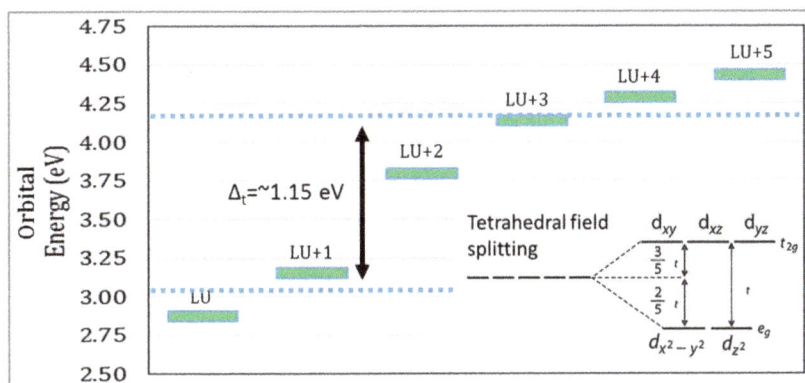

Figure 3. (**Left**) Energies of selected orbitals and (**Right**) orbital energies for LU through LU+5. Plotting these orbitals indicated d-orbital character. LU+4 and LU+5 maintain d character but show a signature of orbital mixing, which is why there are a total of six d-like orbitals. A gap in energy exists between LU+1 and LU+2, indicating tetrahedral field splitting. Right inset: expected tetrahedral crystal field d-orbital splitting.

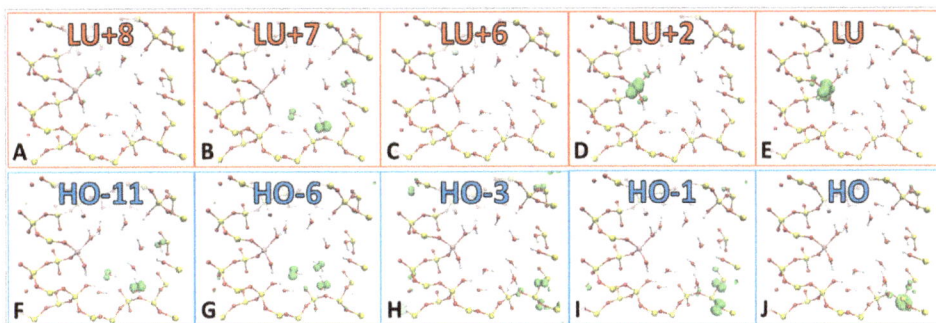

Figure 4. Green clouds represents isosurfaces for orbitals of interest, which participate in the charge transfer dynamics of the system: (**A–E**) unoccupied orbitals and (**F–J**) occupied orbitals. These pairs of occupied and unoccupied orbitals represent different photoexcitations.

23

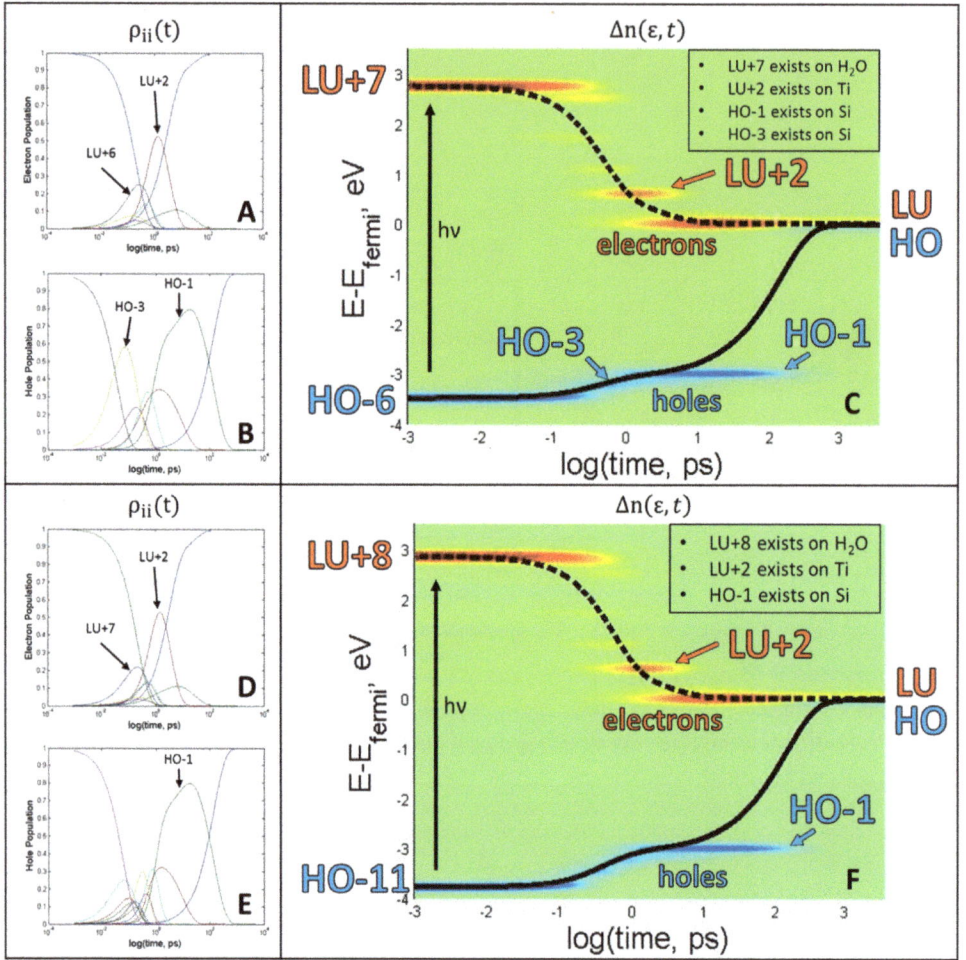

Figure 5. Photo-induced excited state dynamics for two typical excitations: (A–C) (HO-6, LU+7) and (D–F) (HO-11, LU+8). (**A,D**) Electron population over actual time; (**B,E**) hole populations over actual time; and (**C,F**) distribution of electron and hole relaxations over time and computed according to Equation (18). The solid line represents the movement of the hole through orbitals over time, while the dashed line represents electron movement. Time is displayed in a logarithmic scale.

The pathways of occupation dynamics are determined by the initial excitation A → B, see Equation (14). Solutions of Equations (13)–(16) first provide occupations as functions of time, ρ_{ii} (t), which are displayed in Figure 5A,B, for one excitation and another in Figure 5D,E. In Figure 5A, electron occupation relaxes from LU+7, but not before transitioning through several short-lived states and one longer-lived state at LU+2, which exists on the Ti atom (Figure 4D). LU+2 behaves as an electron trap, which slows its relaxation rate. Ultimately, LU+7 → LU relaxation occurs at a rate of 0.8141 ps^{-1}, which is computed according to Equation (22). A summary of relaxation rates can be found in Table 1. In contrast, the HO-6 → HO relaxation (Figure 5B) occurs at a much slower rate of 0.0301 ps^{-1}. This slower rate is due to stable hole trapping states, which occur at HO-3 and HO-1 and are located within the silica substrate (Figure 4H,I). Viewing the larger peaks contained in Figure 5A helps to identify more influential trap states. The observation that LU+6 and LU+2 peaks have higher amplitude than other states indicates that more time is spent in those states. The same can be said for Figure 5B, containing larger peaks at HO-3 and HO-1. These states are occupied for a longer time. All of the above information can be included in one figure (Figure 5C and are derived using Equations (17) and (18)). Figure 5C,F is helpful in visualizing the movement of electrons and holes in both orbitals and energy over time.

Table 1. Relaxation rates, K, and duration, τ, of electrons and holes for A (HO-6, LU+7) and B (HO-11, LU+8) excitations.

Transition	K_e, ps^{-1}	τ^e, ps	K_h, ps^{-1}	τ^h, ps
A	0.8141	1.2284	0.0301	33.2226
B	0.7919	1.2628	0.0341	29.3255

Figure 5D,E tells similar stories, with meta-stable electron traps occurring at LU+7 and LU+2. The rate of electron relaxation is calculated to be 0.7919 ps^{-1}. A stable hole trap exists at HO-1, which lengthens the hole's relaxation rate compared to the electron, and can be visualized more clearly in Figure 5E. The more interesting aspect of this story is that the hole traps exist on Si orbitals, which could be a testament to why silica is such an effective substrate for heterogeneous catalysis, as it facilitates charge separation. Until the hole relaxes, the model resides in the oxidation state $O^- + Ti^{3+}$. The lowest occupied orbital is located on Ti and is occupied by the electron, which means that the titanium is in the Ti^{3+} state rather than Ti^{4+}, which is specific for the ground state. The titanium in a photo-induced metastable Ti^{3+} state is expected to be a strong reducing agent. Thus, in this excited state, Ti^{3+} is able to promote reducing half reactions $2H^+ + 2Ti^{3+} \rightarrow H_2 + 2Ti^{4+}$ contributing to water splitting and hydrogen evolution. Additionally, the hole trap occurring at HO-1 is a consequence of the energy gap law [26], because there is a sizable energy gap between HO and HO-1 (~3 eV), which makes it difficult for the energy to dissipate into vibrations during the relaxation process, as all vibrational frequencies are below this sub-gap.

The electron traps for both excitations are located on one of the Ti atom's orbitals, which seems logical since the Ti is known to be the catalytic site of the water-splitting reaction. The rates of electron relaxation differ by 0.0222 ps^{-1}. The difference of the rates of hole relaxation, however, is only 0.0040 ps^{-1}. The hole relaxation rates are less dependent upon initial excitation than the electrons. The excitations of HO-6 → LU+7 and HO-11 → LU+8 have features in common, such as the existence of the same electron trap state of LU+2 and hole trap state HO-1. This indicates that these two trap states exist, independent of the excitation. Additionally, the rates of hole relaxation are very similar, indicating that the existence of the hole trap is the determining factor in the magnitude of the rate. This feature of dynamics is dictated by electronic structure with an occupied orbital in the middle of the gap.

3. Experimental Section

Atomic structure determines initial positions of each ion, $\vec{R_I}$, in the model. Density-Functional Theory (DFT) was chosen because it allows one to analyze electronic structure and spectra [27]. The Vienna *Ab-Initio* Simulation Package (VASP v5.4, Computational Materials Physics, Vienna, Austria, 2015) was used for calculations, with Projector Augmented Wave (PAW) pseudopotentials and Generalized Gradient Approximation (GGA) Perdew-Burke-Ernzerhof (PBE) functionals [28]. Periodic boundary conditions are used in order to simulate the periodicity of typical silica substrate material. Figure 1A provides a visualization of the model as a unit cell. As portrayed in Figure 1B, the unit cell may be replicated and then translated in the x, y, or z directions. The potential is not affected by this replication and subsequent movement along a translation vector, $\vec{t}$, and therefore, the potential experienced by the electrons [29,30] of the original unit cell is equivalent to any replica as given by Equation (1)

$$u\left(\vec{r}\right) = u\left(\vec{r} + k\cdot\vec{a_x} + l\cdot\vec{a_y} + m\cdot\vec{a_z}\right) = u\left(\vec{r} + \vec{t}\right) \tag{1}$$

with component vectors $\vec{a_x}$, $\vec{a_y}$, and $\vec{a_z}$, and integers k, l, and m.

The Kohn-Sham equation is the Schrödinger equation of a fictional system of particles which do not interact and that generate the same density as any system of interacting particles. This is given by Equation (2)

$$\left(\frac{-\hbar^2}{2m}\nabla^2 + v\left[\vec{r}, \rho\left(\vec{r}\right)\right]\right)\psi_i^{KS}\left(\vec{r}\right) = \varepsilon_i\psi_i^{KS}\left(\vec{r}\right) \tag{2}$$

where the first term corresponds to kinetic energy, T, and includes a symbol for the gradient $\nabla = \left(\frac{\partial}{\partial x}, \frac{\partial}{\partial y}, \frac{\partial}{\partial z}\right)$. In solving Equation (1), a set of single-electron orbitals $\Psi_i^{KS}\left(\vec{r}\right)$ and their energies ε_i is determined. These orbitals are then combined with an orbital occupation function, f_i, for assembling the overall density of electrons, as given by Equation (3)

$$\rho\left(\vec{r}\right) = \sum_{i \leqslant HO} f_i \psi_i^{KS*}\left(\vec{r}\right) \psi_i^{KS}\left(\vec{r}\right)$$ (3)

In addition, the total density defines the potential, is defined by Equation (4)

$$v\left[\vec{r}, \rho\left(\vec{r}\right)\right] = \frac{\partial}{\partial\rho}\left(E^{tot}\left[\rho\right] - T\right)$$ (4)

which is defined as functional derivative of the total energy with respect to variation of the total density and includes the interactions of electrons with ions, and three electron iterations: Coulomb, correlation, and exchange. Rectangular brackets symbolize a functional. Equations (2)–(4) are solved in a self-consistent and iterative fashion. Electrostatic potential introduced in Equation (4) demonstrates the interaction of a valence electron with the remaining charge of electrons and ions. This electrostatic potential is a consequence computed in DFT.

The electronic density of state (DOS) is defined as the number of states per interval of energy that are capable of electron population [29]. DOS are used to describe electronic structure of the model. DOS is given by Equation (5)

$$n\left(\varepsilon\right) = \sum_i \delta\left(\varepsilon - \varepsilon_i\right)$$ (5)

where the index, i, runs over all orbitals calculated and ε_i is the energy of a given orbital (calculated using DFT). The Dirac delta function is approximated by a finite width Gaussian function.

The optical absorption spectrum is computed based on the transition dipole moment, by Equation (6)

$$\vec{D}_{ij} = \int \psi_i^* \vec{r} \psi_j d\vec{r}$$ (6)

which calculates transition dipoles from initial ψ_j to final ψ_i and is directly related to the oscillator strength, which is then given by Equation (7)

$$f_{ij} = \left|\vec{D}_{ij}\right|^2 \frac{4\pi m_e v_{ij}}{3\hbar e^2}$$ (7)

Here, v_{ij} is the angular frequency of the excitation from state i to state j, D_{ij} is the transition dipole from state i to state j, with $\hbar$ and e as fundamental constants.

The oscillator strength contributes a weight in the absorption spectrum, leading to Equation (8)

$$\alpha(\varepsilon) = \sum_{ij} f_{ij}\delta(\varepsilon - \Delta\varepsilon_{ij}) \tag{8}$$

in order to apply intensities to the Eigen energy differences which are most probable to absorb light in specific energy ranges.

Molecular dynamics (MD) [30] was used to simulate the change of the positions of the atoms after the cluster is "heated" at a given temperature for a given amount of time. The kinetic energy of the model should remain constant, so the velocities of all the atoms were rescaled to mimic a constant temperature during the MD process according to Equation (9), where M_I and $\dfrac{d\vec{R_I}}{dt}$ denote the mass and initial velocity of Ith nucleus, respectively, N is the number of nuclei, k_B is the Boltzmann constant, and T is the temperature. The Hellman-Feynman forces, $\vec{F_I}$, act on each atom to determine the acceleration that is to be used in the Newton equation of motion, Equation (10),

$$\sum_{I=1}^{N} \frac{M_I\left(\frac{d\vec{R_I}}{dt}\Big|_{t=0}\right)}{2} = \frac{3}{2}Nk_BT \tag{9}$$

$$M_I\frac{d^2\vec{R_I}(t)}{dt^2} = \vec{F_I}(t) \tag{10}$$

by iterative solving of the Kohn–Sham equation and then updating the ionic positions to generate a single example of nuclear positions trajectory $\{\vec{R_I}(t)\}$ for a micro-canonical ensemble. The time dependent adjustments of KS orbitals ψ_i^{KS} and orbital energies ε_i can also be monitored as a function of time. On-the-fly non-adiabatic couplings [20,31] were computed along the nuclear trajectory $\{\vec{R_I}(t)\}$ by Equation (11).

$$
\begin{aligned}
\omega_{ij}(t) &= -i\hbar\langle\psi_i^{KS}\left|\frac{d}{dt}\right|\psi_j^{KS}\rangle \\
&= \frac{-i\hbar}{2\Delta t}\int dr[\psi_i^{KS*}(\{\vec{R_I}(t)\},r)\psi_j^{KS}(\{\vec{R_I}(t+\Delta t)\},r) \\
&\quad -\psi_j^{KS*}(\{\vec{R_I}(t+\Delta t)\},r)\psi_i^{KS}(\{\vec{R_I}(t)\},r)]
\end{aligned} \tag{11}
$$

Next, the autocorrelation functions of these couplings were processed with Equation (12).

$$M_{ijkl}(\tau) = \frac{1}{T}\int_0^T dt\omega_{ij}(t+\tau)\omega_{kl}(t) \tag{12}$$

According to Redfield theory [32], the Fourier transform of $M(\tau)$ provides parameters R_{jklm} for the equation of motion of electronic degrees of freedom, as per Equation (13)

$$\frac{d\rho_{ij}}{dt} = -\frac{i}{\hbar} \sum_{k} \left(F_{ik}\rho_{kj} - \rho_{ik}F_{ki} \right) + \left(\frac{d\rho_{ij}}{dt} \right)_{diss} = (\check{\zeta} + \hat{R})\rho \tag{13}$$

where $\check{\zeta}\rho$ and $\hat{R}\rho = \left(\frac{d\rho_{ij}}{dt} \right)_{diss} = \sum_{lm} R_{jklm}\rho_{lm}$ represent the Liouville and Redfield superoperators acting on the density operator.

The initial values of the density matrix at time $t = 0$, were symbolized by $\rho_{ij}^{(A,B)}(0)$ and served as the initial condition for a system of differential equations, as per Equation (14).

$$\rho_{ij}^{(A,B)}(0) = \delta_{ij}\left(f_i - \delta_{iA} + \delta_{iB} \right) \tag{14}$$

Here, indices A and B label specific initial conditions, the Dirac delta function δ_{ij} reflects the fact that off-diagonal elements are set to zero, f_i stands for the initial occupation of ith Kohn–Sham orbital so that $f_i = 1, i \leqslant$ LUMO and $f_i = 0, i \geqslant$ LUMO. A Dirac delta with a negative sign $-\delta_{ia}$ corresponds to the removal of an electron from the Ath Kohn–Sham orbital, Dirac delta with a positive sign $+\delta_{iB}$ corresponds to the addition of an electron to the Bth Kohn–Sham orbital. Equation (14) summarizes the following situation: an electron-hole pair A $\rightarrow$ B is excited, with A and B being indices of orbitals corresponding to the valence band, where an electron is promoted from, and the conduction band, where the electron is promoted to. The pairs of orbitals, (A,B), were chosen according the maximum values of oscillator strengths for the corresponding excitation f_{AB}.

The time evolution of the electronic state was calculated by solving the equation of motion as follows:

First, Liouville–Redfield superoperator is diagonalized, providing a set of eigenvalues $\{\Omega_\xi\}$ and eigenvectors, $\{\rho^\xi\}$, according to Equation (15)

$$(\check{\zeta} + \hat{R})\rho^\xi = \Omega_\xi \rho^\xi \tag{15}$$

Here solutions of Equation 15 are labeled by index ξ. Then, the superposition of eigenvectors, given by Equation (16),

$$\rho_{ij}(t) = \left\langle \rho_{ij}(0) \,|\, \rho_{ij}^\xi \right\rangle \rho_{ij}^\xi e^{\Omega_\xi t} \tag{16}$$

were used to compose the density matrix for each instant of time in such a way that it matches the initial conditions Equation (14) [6–8,22]. The electron in the conduction band and the hole in the valence band are indicated by e and h, respectively. In addition, index e was used with the appreciation that equations for holes are the same as for electrons.

Utilizing the above information, the charge density distribution, rate of energy dissipation, and rate of charge transfer were calculated as follows [6–8,22,33].

The distribution of a carrier (electron or hole) in space Equation (17) and in energy Equation (18) are as follows:

$$n(z,t) = \sum \rho_{ii}(t)\,\rho_i(z) \tag{17}$$

$$n(\varepsilon,t) = \sum \rho_{ii}(t)\,\delta(\varepsilon - \varepsilon_i) \tag{18}$$

Here, $\rho_{ii}(t)$ represents the occupation of the ith orbital, $\rho_i(z) = \iint dxdy\psi_i^*(x,y,z)\,\psi_i(x,y,z)$. The expectation value of energy of a carrier is given by Equation (19)

$$\langle \Delta\varepsilon_e\rangle(t) = \sum_i \rho_{ii}(t)\,\varepsilon_i(t) \tag{19}$$

Equation (18) can be rewritten to have a dimensionless energy in this manner:

$$\langle E_e\rangle = \frac{\langle\Delta\varepsilon_e\rangle(t) - \langle\Delta\varepsilon_e\rangle(\infty)}{\langle\Delta\varepsilon_e\rangle(0) - \langle\Delta\varepsilon_e\rangle(\infty)} \tag{20}$$

If one assumes a single exponential fit of the carrier energy dissipation, then as per Equation (21)

$$\langle E_e\rangle(t) = \exp\{-k_e t\} \tag{21}$$

then the energy dissipation rate can be expressed by Equation (22):

$$k_e = \{\tau^e\}^{-1} = \left\{\int_0^\infty \langle E_e\rangle(t)\,dt\right\}^{-1} \tag{22}$$

The constant $\tau^{e(h)}$ represents the average relaxation time for the electron (or hole), and therefore the dynamics of the electronic relaxation.

A photoexcitation from ground state to excited state (A,B) results in a sequence of elementary relaxation events (refer to Scheme 1) with the electron and hole subsequently moving one step towards lower excitation, however, the time needed for each step is different, this relaxation dynamics may often exhibit so-called "trapping" when a charge carrier gets trapped in a specific orbital and requires substantial time for the next relaxation step.

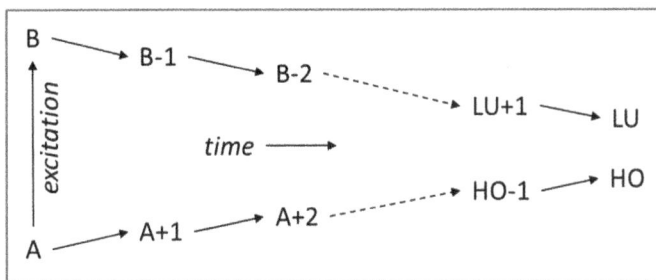

Scheme 1. Description of the initial excitations of electron and hole and their paths to relax down to a neutral state over time. A denotes occupied orbitals and B denotes unoccupied ones.

4. Conclusions

Heterogeneous catalysis, as modeled here, with a silica substrate and titanium atom, has been evaluated using DFT. Since the formation of charge transfer states is a prerequisite for a chemical reaction, it was necessary to identify such states in this model, which was accomplished in this work. This is another successful application of the density matrix method for electron dynamics. A survey of DOS and absorbance spectra of individual components and of the whole system concludes that a composite of the individual components of the system combine to form the whole density of states.

The calculated data reveal a much longer relaxation time for holes relative to that for electrons, which is dictated by energy offsets. It may seem intuitive that only electron trap states would facilitate a catalytic reaction, but hole trap states would also, as this study demonstrates: either would prevent recombination of the electron–hole pair. Holes affect electrons insofar as that the electron cannot recombine if the hole is still in an excited state. In fact, a prolonged separation of charge was found, which may indicate a higher probability of reaction since there is more electron movement within the system's components. In order for water splitting to occur, the reaction must happen while the charge transfer state exists, before it recombines. Interestingly, an important trap state was found to be located within the silica substrate. This trap state prolongs relaxation, widening the window available for a chemical reaction to occur.

The presence and location of the trap state indicates that such heterogeneous substrates may play a greater role in catalysis than previously envisaged. This allows one to modify the system to provide better electronic characteristics for more efficient water-splitting, *i.e.*, tuning the band-gap. In other words, in the future, one may be able to choose substrates and catalytic sites to correspond to desired electronic properties in order to optimize catalytic reactions. This research will continue by applying this methodology to different situations (such as introduction of oxygen vacancy and other metal oxide materials) to clarify the charge transfer mechanism in various photocatalytic materials.

Acknowledgments: This research was supported by NSF awards EPS-0903804 and CHE-1413614 for methods development and by DOE, BES Chemical Sciences, NERSC Contract No. DE-AC02-05CH11231, allocation Awards 86898, 88777, and 89959 "*Computational Modeling of Photo-catalysis and Photo-induced Charge Transfer Dynamics on Surfaces*". Authors acknowledge the use of computational resources of USD HPC cluster managed by Douglas Jennewein. W.S. thanks DGE-0903685 South Dakota IGERT support.

Author Contributions: Wendi K. Sapp performed *ab initio* computations, data collection, figure preparation, and manuscript writing. Ranjit T. Koodali performed introduction of the open problems in mesoporous materials for catalysis, provided intellectual contribution in choice of atomistic models and interpretation of results, and edited the manuscript. Dmitri S. Kilin coordinated this research effort, executed choice of appropriate methodology for charge transfer dynamics, and contributed to planning and editing the manuscript.

Conflicts of Interest: The authors declare no conflict of interest.

References

1. Gies, H.; Grabowski, S.; Bandyopadhyay, M.; Grunert, W.; Tkachenko, O.P.; Klementiev, K.V.; Birkner, A. Synthesis and characterization of silica MCM-48 as carrier of size-confined nanocrystalline metal oxides particles inside the pore system. *Microporous Mesoporous Mater.* **2003**, *60*, 31–42.

2. Kudo, A.; Miseki, Y. Heterogeneous photocatalyst materials for water splitting. *Chem. Soc. Rev.* **2009**, *38*, 253–278.

3. Li, Y.; Fu, Z.Y.; Su, B.L. Hierarchically structured porous materials for energy conversion and storage. *Adv. Funct. Mater.* **2012**, *22*, 4634–4667.

4. Jeong, E.Y.; Park, S.E. Synthesis of porphyrin-bridged periodic mesoporous organosilica and their catalytic applications. *Res. Chem. Intermed.* **2012**, *38*, 1237–1248.

5. Horiuchi, Y.; Toyao, T.; Takeuchi, M.; Matsuoka, M.; Anpo, M. Recent advances in visible-light-responsive photocatalysts for hydrogen production and solar energy conversion–from semiconducting TiO_2 to MOF/PCP photocatalysts. *Phys. Chem. Chem. Phys.* **2013**, *15*, 13243–13253.

6. Inerbaev, T.M.; Hoefelmeyer, J.D.; Kilin, D.S. Photoinduced Charge Transfer from Titania to Surface Doping Site. *J. Phys. Chem. C* **2013**, *117*, 9673–9692.

7. Jensen, S.J.; Inerbaev, T.M.; Kilin, D.S. Spin Unrestricted Excited State Relaxation Study of Vanadium (IV) Doped Anatase. *J. Phys. Chem. C* **2016**.

8. Huang, S.; Kilin, D.S. Charge Transfer, Luminescence, and Phonon Bottleneck in TiO_2 Nanowires Computed by Eigenvectors of Liouville Superoperator. *J. Chem. Theory Comput.* **2014**, *10*, 3996–4005.

9. Fujishima, A. Electrochemical photolysis of water at a semiconductor electrode. *Nature* **1972**, *238*, 37–38.

10. Park, J.H.; Kim, S.; Bard, A.J. Novel carbon-doped TiO_2 nanotube arrays with high aspect ratios for efficient solar water splitting. *Nano Lett.* **2006**, *6*, 24–28.

11. Zhang, X.-Y.; Li, H.-P.; Cui, X.-L.; Lin, Y. Graphene/TiO_2 nanocomposites: Synthesis, characterization and application in hydrogen evolution from water photocatalytic splitting. *J. Mater. Chem.* **2010**, *20*, 2801–2806.

12. Kitano, M.; Takeuchi, M.; Matsuoka, M.; Thomas, J.M.; Anpo, M. Photocatalytic water splitting using Pt-loaded visible light-responsive TiO_2 thin film photocatalysts. *Catal. Today* **2007**, *120*, 133–138.

13. Burda, C.; Lou, Y.; Chen, X.; Samia, A.C.; Stout, J.; Gole, J.L. Enhanced nitrogen doping in TiO_2 nanoparticles. *Nano Lett.* **2003**, *3*, 1049–1051.

14. Park, J.H.; Park, O.O.; Kim, S. Photoelectrochemical water splitting at titanium dioxide nanotubes coated with tungsten trioxide. *Appl. Phys. Lett.* **2006**, *89*, 163106.

15. Youngblood, W.J.; Lee, S.-H.A.; Maeda, K.; Mallouk, T.E. Visible light water splitting using dye-sensitized oxide semiconductors. *Acc. Chem. Res.* **2009**, *42*, 1966–1973.

16. Rasalingam, S.; Peng, R.; Wu, C.-M.; Mariappan, K.; Koodali, R.T. Robust and effective Ru-bipyridyl dye sensitized Ti-MCM-48 cubic mesoporous materials for photocatalytic hydrogen evolution under visible light illumination. *Catal. Commun.* **2015**, *65*, 14–19.

17. Anpo, M.; Yamashita, H.; Ikeue, K.; Fujii, Y.; Zhang, S.G.; Ichihashi, Y.; Park, D.R.; Suzuki, Y.; Koyano, K.; Tatsumi, T. Photocatalytic reduction of CO_2 with H_2O on Ti-MCM-41 and Ti-MCM-48 mesoporous zeolite catalysts. *Catal. Today* **1998**, *44*, 327–332.

18. Akimov, A.V.; Muckerman, J.T.; Prezhdo, O.V. Nonadiabatic Dynamics of Positive Charge during Photocatalytic Water Splitting on GaN(10-10) Surface: Charge Localization Governs Splitting Efficiency. *J. Am. Chem. Soc.* **2013**, *135*, 8682–8691.

19. Marx, D.; Hutter, J. Ab Initio Molecular Dynamics: Basic Theory and Advanced Methods. In *Ab Initio Molecular Dynamics: Basic Theory and Advanced Methods*; Cambridge University Press: London, UK, 2012.

20. Hammes-Schiffer, S.; Tully, J.C. Vibrationally enhanced proton transfer. *J. Phys. Chem.* **1995**, *99*, 5793–5797.

21. Nelson, T.; Fernandez-Alberti, S.; Chernyak, V.; Roitberg, A.E.; Tretiak, S. Nonadiabatic excited-state molecular dynamics modeling of photoinduced dynamics in conjugated molecules. *J. Phys. Chem. B* **2011**, *115*, 5402–5414.

22. Kilin, D.S.; Micha, D.A. Relaxation of Photoexcited Electrons at a Nanostructured Si(111) Surface. *J. Phys. Chem. Lett.* **2010**, *1*, 1073–1077.

23. Liu, I.-S.; Lo, H.-H.; Chien, C.-T.; Lin, Y.-Y.; Chen, C.-W.; Chen, Y.-F.; Su, W.-F.; Liou, S.-C. Enhancing photoluminescence quenching and photoelectric properties of CdSe quantum dots with hole accepting ligands. *J. Mater. Chem.* **2008**, *18*, 675–682.

24. Van Vleck, J.H. *The Theory of Electric and Magnetic Susceptibilities*; Oxford At The Clarendon Press: Gloucestershire, UK, 1932.

25. Vincent, A. *Molecular Symmetry and Group Theory: A Programmed Introduction to Chemical Applications*; John Wiley & Sons: New York, NY, USA, 2013.

26. Englman, R.; Jortner, J. The energy gap law for radiationless transitions in large molecules. *Mol. Phys.* **1970**, *18*, 145–164.

27. Kohn, W.; Sham, L.J. Self-consistent equations including exchange and correlation effects. *Phys. Rev.* **1965**, *140*, A1133.

28. Kresse, G.; Furthmüller, J. Efficient iterative schemes for ab initio total-energy calculations using a plane-wave basis set. *Phys. Rev. B* **1996**, *54*, 169.

29. Parr, R.G.; Yang, W. *Density-Functional Theory of Atoms and Molecules*; Oxford University Press: New York, NY, USA, 1989; Volume 16.

30. Johnson, B.G.; Gill, P.M.; Pople, J.A.; Fox, D.J. Computing molecular electrostatic potentials with the PRISM algorithm. *Chem. Phys. Lett.* **1993**, *206*, 239–246.

31. Marx, D.; Hutter, J. Ab initio molecular dynamics: Theory and implementation. *Mod. Methods Algorithms Quantum Chem.* **2000**, *1*, 301–449.

32. Egorova, D.; Gelin, M.F.; Thoss, M.; Wang, H.; Domcke, W. Effects of intense femtosecond pumping on ultrafast electronic-vibrational dynamics in molecular systems with relaxation. *J. Chem. Phys.* **2008**, *129*, 214303–214311.

33. Chen, J.; Schmitz, A.; Kilin, D.S. Computational simulation of the p-n doped silicon quantum dot. *Int. J. Quantum Chem.* **2012**, *112*, 3879–3888.

First-Principles Modeling of Direct *versus* Oxygen-Assisted Water Dissociation on Fe(100) Surfaces

Wenju Wang, Guoping Wang and Minhua Shao

Abstract: The O–H bond breaking in H_2O molecules on metal surfaces covered with pre-adsorbed oxygen atoms is an important topic in heterogeneous catalysis. The adsorption configurations of H_2O and relevant dissociation species on clean and O-pre-adsorbed Fe(100) surfaces were investigated by density functional theory (DFT). The preferential sites for H_2O, HO, O, and H were investigated on both surfaces. Both the first H abstraction from adsorbed H_2O and the subsequent OH dissociation are exothermic on the O-pre-adsorbed Fe(100) surface. However, the pre-adsorbed O significantly reduces the kinetics energy barriers for both reactions. Our results confirmed that the presence of pre-adsorbed oxygen species could significantly promote H_2O dissociation.

Reprinted from *Catalysts*. Cite as: Wang, W.; Wang, G.; Shao, M. First-Principles Modeling of Direct *versus* Oxygen-Assisted Water Dissociation on Fe(100) Surfaces. *Catalysts* **2016**, *6*, 29.

1. Introduction

Water (H_2O) is one of the most widespread resources involved in many chemical processes [1]. It has also been proposed as the significant feedstock to produce H_2 in steam and oxidative reforming technologies [2–8], which involve adsorption and dissociation of H_2O molecules. A fundamental study of water-metal surface interactions will assist in understanding the reaction mechanisms of this heterogeneous catalytic reaction. Noble metals such as Ru [9,10], Rh [10,11], Pd [10], Ir [10–13], and Pt [10,14] are active for steam and oxidative reforming but are not preferred in conventional industrial reformers due to their high costs. This motivates studies on non-precious metal catalysts, such as Ni- [15–17], Cu- [18,19], Co- [20,21], and Fe- [22–27] based materials. Due to extremely low cost and high catalytic activity, Fe-based catalysts have caught great attentions.

Adsorbed hydroxyl, atomic oxygen and hydrogen are the simplest dissociation products of water on metal surfaces. Also, hydroxyl can further dissociate, giving rise to adsorbed atomic hydrogen and oxygen. Some experimental studies and theoretical calculations on adsorption/dissociation of water on the Fe(100) surface have been reported [28–37]. Eder and Terakura [33] demonstrated that water preferred to adsorb at the bridge site via an upright molecular adsorption configuration. However, it was unstable leading to a spontaneous dissociation into H and OH species. The same conclusion was achieved by Jung and Kang [34] using density functional theory (DFT) calculations. The proposed upright H_2O configuration at the bridge site by these authors, however, is not in line with the flat-lying configuration on the top site on other metals, such as Ru, Rh, Pd, Pt, Al, Cu, Ag, and Au [38–41].

The promotion effect of pre-adsorbed O atoms on the surfaces of the catalysts has been discovered and investigated by both experimental and theoretical approaches. The pre-adsorbed O atoms have various degrees of effect towards H_2O dissociation on different metals. A comprehensive review on this topic has been provided by Thiel *et al.* [42] and Henderson [43]. The positive effect of pre-adsorbed O atoms on water dissociation on Pd single crystals and thin films has been reported [44–48]. On the other hand, the water dissociation on the clean Pd surface has not been observed, in consistent with the DFT calculations [49]. Shavorskiy *et al.* [50] also found the formation of OH when water reacted with co-adsorbed O on the missing-row reconstructed Pt{110}(1 × 2) surface. A similar promotion effect on the Fe(111) surface has been reported [51]. Hung *et al.* [52] found that adsorbed oxygen facilitated water dissociation via the hydrogen transfer process on the pre-oxidized Fe(100) surface. Liu *et al.* [37] studied the effect of the O atom which came from the dissociation of the first H_2O molecule on the consequent dissolution of the second H_2O molecule. The O atom assisted H_2O dissociation (O + H_2O = 2OH) was kinetically favorable, and further OH dissociation was roughly thermo-neutral. In that case, the co-adsorption of H coming from the dissociation of the first H_2O molecule on the Fe(100) might have an effect on the reaction pathway. In our manuscript, we studied the effect of two O atoms which came from the dissociation of O_2 molecule on the adsorption and dissociation of water molecule.

Herein, we report a systematic DFT study of the adsorption of H_2O and its dissociation fragments (OH, H and O) on clean and O-pre-adsorbed Fe(100) surfaces. The effect of two O atoms which came from the dissociation of O_2 molecule on the adsorption and dissociation of water molecule has not been reported previously. The dissociation pathways on both surfaces were also discussed based on the calculation results to understand the effect from the pre-adsorbed O atoms.

2. Results and Discussion

2.1. Adsorption of H_2O, OH, O and H on the Clean Fe(100) Surface

The adsorption properties of H_2O and the dissociation products H, O, and OH species on a clean Fe(100) surface will be discussed first. The most stable adsorption configurations of these species are presented in Figure 1. The corresponding adsorption energies and structural details are summarized in Tables 1 and 2. On the

Fe(100) surface, H_2O was found to be preferentially adsorbed on a top site via the oxygen atom with the H_2O plane being almost parallel to the Fe surface (Figure 1a). The interaction between the Fe(100) surface and H_2O molecule is weak, evidenced by a small adsorption energy (-0.65 eV). The O–H bond length of 0.98 Å and the $\angle$HOH bond angle of 105.4°, as listed in Table 2, are almost identical to those of a free H_2O molecule (0.98 Å and 104.4°, respectively). Similar adsorption behaviors of H_2O on Ni(111) [53–55], Ni(100) [55], Ni(110) [55], Cu(111) [56], Cu(100) [57] and Cu(110) [58,59] were observed.

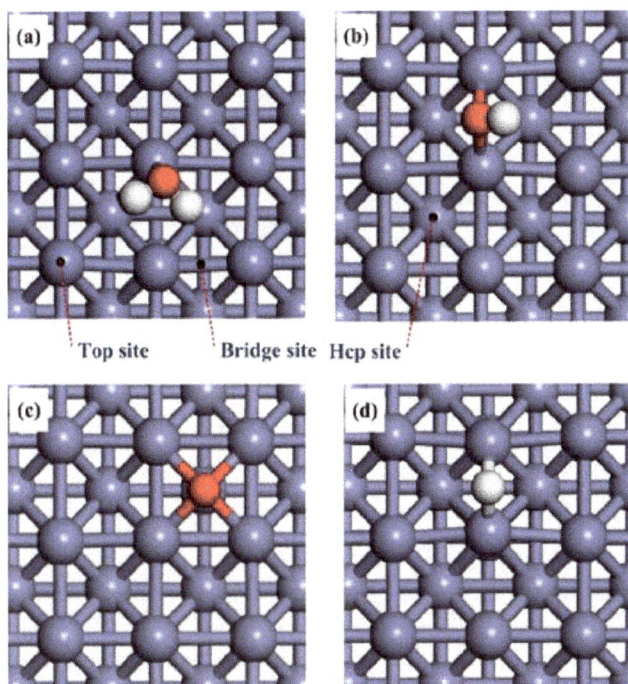

Figure 1. Top views of the adsorption configurations of (a) H_2O; (b) OH; (c) O; and (d) Hon clean Fe(100) surfaces. The slate blue, red and white balls stand for Fe, O and H atoms, respectively.

The adsorption energies of the OH species at the bridge and hcp sites are -3.93 eV and -3.84 eV, respectively. The adsorption at the hollow hcp site is locally unstable. These results suggest that the most stable adsorption site for OH on Fe(100) is at the bridge, consistent with the previous study [28,33]. The O–Fe bond length is1.99 Å, which is shorter than that of H_2O adsorbed on clean Fe (100) surface (2.17 Å). Based on these results, one may conclude that the OH can be strongly adsorbed on the Fe(100) surface.

For the O species, Govender *et al.* [35] have found that the hcp and bridge sites were equally stable. On the other hand, Błoński *et al.* [60] and Lu *et al.* [30] suggested that O atom only occupied the hollow hcp sites. We found that the adsorption energy of O atom at the hollow hcp is higher than that at the bridge site (-3.67 eV at the former *vs.* -3.00 eV at the latter). On contrast, the top site is not stable for O adsorption. Thus, our results echo that reported by Błoński *et al.* [60] and Lu *et al.* [30].

Table 1. Adsorption energies (E_{ads}, in eV) of H_2O, OH, O and H on clean and O-pre-adsorbed Fe(100) surfaces.

Species	Clean Fe(100)			O-pre-adsorbed Fe(100)	
	Top	Bridge	Hcp	Top	Bridge
H_2O	-0.65	-	-	-1.13	-
OH	-	-3.93	-3.84	-	-4.02
O	-	-3.00	-3.67	-	-
H	-	-3.99	-3.86	-	-

Table 2. Geometrical parameters for H_2O, OH, O, and H adsorbed on clean and O-pre-adsorbed Fe(100) surfaces.

Species	Clean Fe(100)				O-pre-adsorbed Fe(100)		
	d_{O-H} (Å)	A_{H-O-H} (°)	$d_{O(/H)-Fe}$ (Å) [a]	d_{O-H} (Å)	A_{H-O-H} (°)	d_{O-Fe} (Å) [a]
H_2O	0.99/0.98 (0.98/0.98) [b]	105.4 (104.4) [b]	2.17	0.99/0.98	106.4	2.20
OH	0.98 (0.99) [b]	-	1.99/1.99	0.98	-	1.98/1.98
O	-	-	2.04/2.04/2.04/2.04	-	-	-
H	-	-	1.70/1.70	-	-	-

[a] $d_{O(/H)-Fe}$ (Å) is the distance to the first Fe neighbor; [b] Values in parentheses correspond to gas-phase species or free radicals.

Finally, for the H atom, the adsorption energies at the bridge and hollow hcp sites are -3.99 eV and -3.86 eV, respectively. The top site is not stable for H adsorption. This result suggested that the bridge is the most stable adsorption site for H atoms, which is in good agreement with previous DFT calculation results [61].

Based on the calculated adsorption energies in Table 1, it can be concluded that the interaction with the Fe(100) surface increases in the order of H_2O < O < OH <H. In the following sections, we will discuss the dissociation of H_2O on the Fe(100) surface based on these results.

2.2. Adsorption of H_2O and OH on the O-Pre-adsorbed Fe(100) Surface

As mentioned in the Introduction, the pre-adsorbed oxygen atoms could play an important role in the activation of O–H bond. Therefore, it is of interest to investigate the adsorption behavior of H_2O on O-pre-adsorbed Fe(100) surface. We found that the H_2O molecule could also be adsorbed on the top site on the O-pre-adsorbed Fe surface, with a higher interaction energy (-1.13 eV) than that on a clean surface (-0.65 eV). The adsorption configuration was listed in Table 2. The pre-adsorbed electronegative O atom may increase the acidity of neighboring Fe atoms [62], and the adsorption energy of H_2O via through-space electronic interaction. Another interesting finding is that the pre-adsorbed O atom and the H atom in the H_2O may form a hydrogen bond with a bond distance of 2.04 Å in the co-adsorption configuration, as shown in Figure 2a.

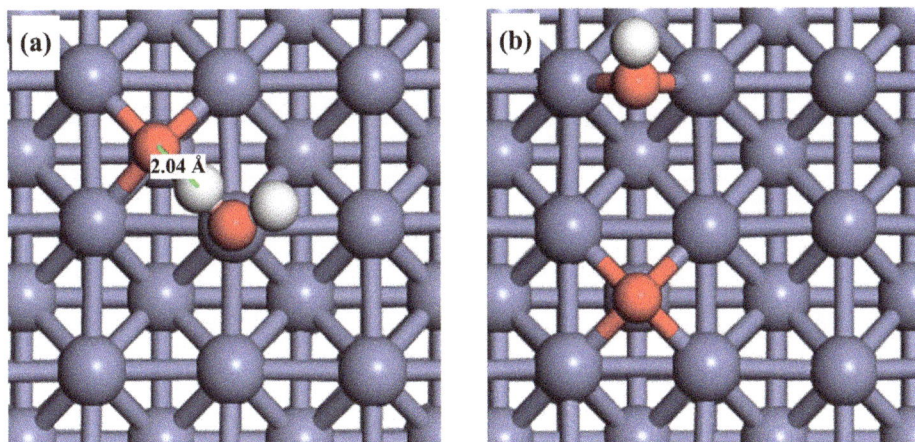

Figure 2. Top views of the adsorption configurations of (**a**) H_2O + O; and (**b**) OH+ O on O-pre-adsorbed Fe(100) surfaces.

As for the adsorption of OH on O-pre-adsorbed Fe(100) surface, both OH and O are placed on their favorable bridge sites. The most stable structure of OH co-adsorbed with O atom is shown in Figure 2b. The distance between the O atom of hydroxyl group and the nearest Fe atom remains 1.98 Å. However, the adsorption energy changes from -3.93 eV to -4.02 eV, suggesting a stronger interaction with the O-pre-adsorbed Fe(100) surface than with a clean surface. The bond distance of O–H is 0.98 Å, identical to that on the clean surface. Similar effects from the pre-adsorbed-O atom were also found on an Au(100) surface [63].

2.3. Reaction Mechanisms

2.3.1. H_2O Dissociation on the Clean Fe(100) Surface

Water may partially dissociate to produce OH_{ad} and H_{ad} (Equation (1)), and follow the dissociation of OH_{ad} to form H_{ad} and O_{ad} (Equation (2)). The calculated reaction energies and activation barriers for these two reaction steps are shown in Table 3. The associated transition states (TSs) are displayed in Figure 3. In the first step (Equation (1)), molecular water is adsorbed on the clean Fe(100) surface at the top site as discussed in Section 2.1 (Figure 3a). At TS1, one of the H atoms was stripped from the H_2O molecule (Figure 3b). The bond distance between this H atom and O atom in the OH_{ad} is 1.57 Å. The energy barrier and reaction energy for this step are 1.45 eV and −0.91 eV (Figure 4), respectively. At the equilibrium state, both the H and OH radicals are adsorbed at the bridge sites (Figure 3c).

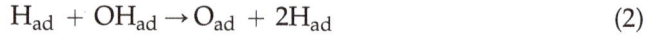

$$H_2O_{ad} \rightarrow H_{ad} + OH_{ad} \tag{1}$$

$$H_{ad} + OH_{ad} \rightarrow O_{ad} + 2H_{ad} \tag{2}$$

Table 3. Energy of reaction (ΔE) and energy of barrier (E_a) for the dissociation of H_2O on clean and O-pre-adsorbed Fe(100) surfaces.

Surfaces	Reactions	ΔE (eV)	E_a (eV)
Clean Fe(100)	Equation (1)	−1.02	1.45 (TS1)
	Equation (2)	−0.58	2.12 (TS2)
O-pre-adsorbed Fe(100)	Equation (3)	−0.53	0.92 (TS1′)
	Equation (4)	−0.10	2.02 (TS2′)

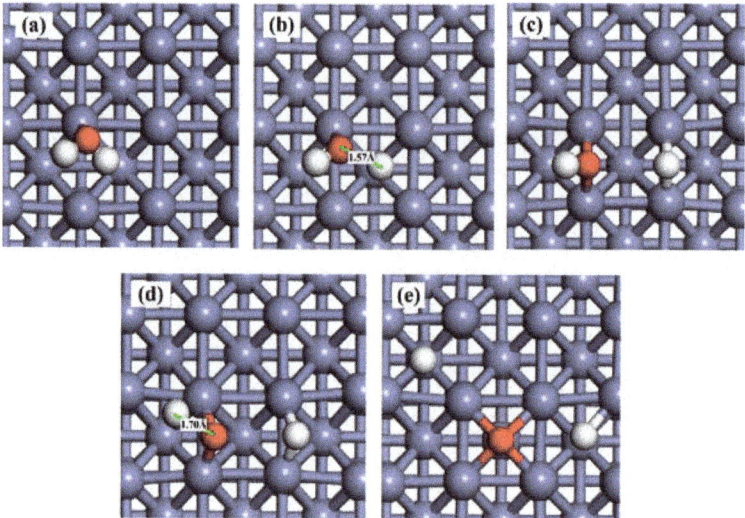

Figure 3. Snapshots of H_2O dissociation process on a clean Fe(100) surface. (a) H_2O_{ad}; (b) TS1; (c) $H_{ad} + OH_{ad}$; (d) OH_{ad}; (e) TS2; and (f) $O_{ad} + H_{ad}$.

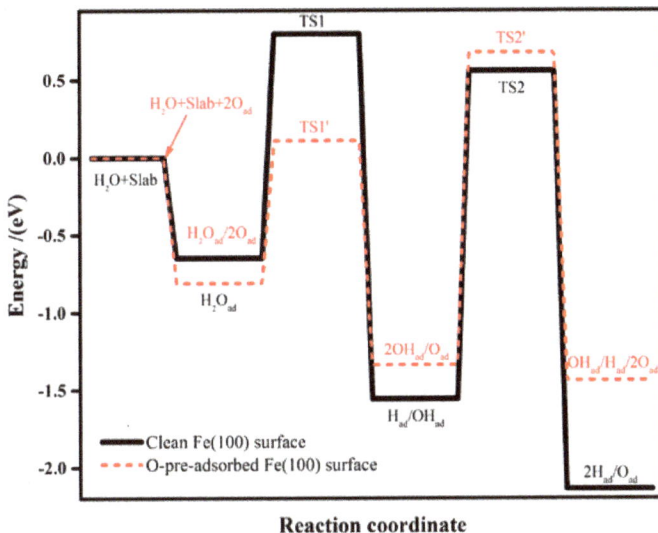

Figure 4. Schematic energy diagram for the dissociation of H_2O on clean (black lines) and O-pre-adsorbed (red lines) Fe(100) surfaces.

In the second step of dissociation (Equation (2)), the O atom remains at the bridge site while the H atom diffuses away and forms a new Fe–H bond (TS2 in Figure 3d). The distance between this H atom and the O atom is 1.70 Å, which is much longer than that of the initial value (0.98 Å). The Fe–H bond is not stable and H continues to move towards the nearby bridge site (Figure 3e). The energy barrier of the second step is 2.12 eV, which is almost 1.5 times of that of the first step. The reaction energy is −0.58 eV, indicating that the second step is an exothermic reaction. According to the activation energies for TS1 and TS2, both steps require significant energies to overcome the energy barriers.

2.3.2. H_2O Dissociation on the O-Preadsorbed Fe(100) Surface

The dissociation mechanisms of H_2O on the O-pre-adsorbed metal surfaces are significantly different from those on a clean surface. The process of H_2O dissociation on an iron surface, promoted by adsorbed oxygen, can be expressed as Equations (3) and (4).

$$H_2O_{ad} + 2O_{ad} \rightarrow 2OH_{ad} + O_{ad} \tag{3}$$

$$2OH_{ad} + O_{ad} \rightarrow OH_{ad} + 2O_{ad} + H_{ad} \tag{4}$$

The pre-adsorbed O atom may act as a reactant. The reaction energies and activation barriers for Equation (3) and the consequent Equation (4) are also calculated and shown in Table 3. The corresponding reaction steps are illustrated in Figure 4. On the O-pre-adsorbed Fe(100) surface, molecular water is adsorbed at the top site, similar to that on the clean surface (2.20 Å on the O-covered surface *vs.* 2.17 Å

41

on the clean surface). The pre-adsorbed O atom forms a hydrogen bond with one of the H atom in the adsorbed H_2O with a bond distance of 2.04 Å (Figure 2a). Due to the interaction with the pre-adsorbed O atom, the adsorption of H_2O molecule is slightly away from the top site (Figure 5a). Next, the pre-adsorbed O pulls away the hydrogen atom to form a new OH_{ad} species adsorbed at the short bridge site, leaving the remaining OH at another bridge site (Figure 5b,c). The energy barrier and heat energy for this step were 0.92 eV and -0.53 eV, respectively. Therefore, the energy barrier of H_2O splitting in the presence of O atom on the Fe(100) surface is much lower than that on the clean surface. The strong promotion effect of pre-adsorbed-O atoms on the Fe(100) surface may help design more active catalyst by considering partially oxidizing the metal surfaces. The decrease of the bond length of O–H from 1.57 Å in TS1 to 1.42 Å in TS1' leads to the reduction of the water dissociation barrier from 1.45 eV on a clean surface to 0.92 eV on an O-pre-adsorbed one. We may conclude that the pre-adsorbed O can significantly promote H_2O dissociation.

The energy barrier for Equation (4) reported in our work (0.92 eV) is significantly higher than that reported by Liu *et al.* (0.18 eV) [37]. One of the reasons is the different computation methods used. In the current study, GGA-PBE functional in CASTEP code was used, rather than GGA-PBE-D2 functional in VASP code that includes the long-range dispersion correction for van der Waals (vdW) interactions [37].

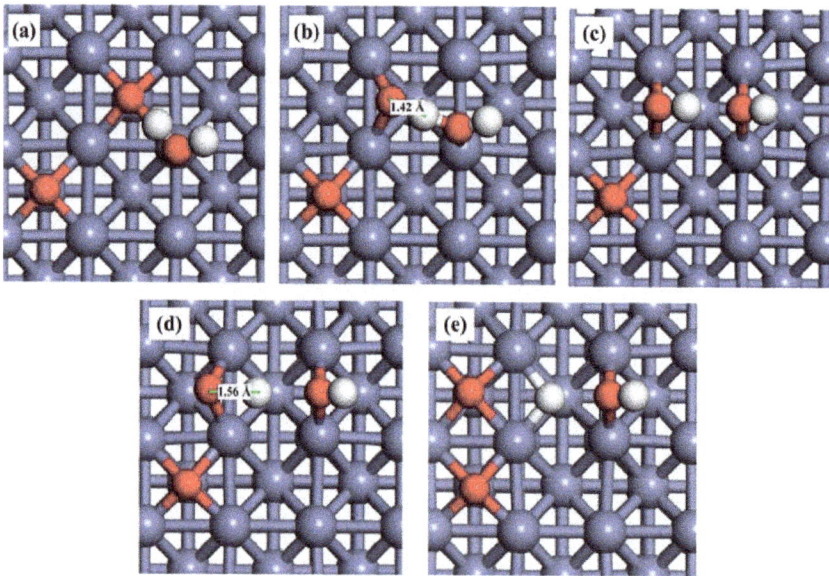

Figure 5. Snapshots of H_2O dissociation process on a O-pre-adsorbed Fe(100) surface. (a) $O_{ad} + H_2O_{ad}$; (b) TS1'; (c) $2OH_{ad}$; (d) $O_{ad} + OH_{ad}$; (e) TS2'; and (f) $2O_{ad} + H_{ad}$.

OH is also difficult to dissociate on clean Fe surfaces due to the high energy barrier (2.12 eV). In the presence of a pre-adsorbed O atom, the energy barrier is reduced to 2.02 eV. According to Figure 5d, the bond length of O–H at TS2′ (1.56 Å) is shorter than that on clean Fe surfaces (1.70 Å), suggesting that the O atom promotes the dissociation of hydroxyl. These results are also elucidated on Au(100) [61] and Pd (111) [64] surfaces. The high energy barriers of OH dissociation on the clean and O-pre-adsorbed Fe(100) surfaces make the occurrence of this reaction kinetically difficult. Consequently, the decomposition of the O-H bond in the OH_{ad} (Equations (2) and (4)) are the rate-determining steps of the whole dissociation reaction of H_2O.

3. Computational Methods

Spin-polarized periodic DFT calculations were performed using the Cambridge Sequential Total Energy Package (CASTEP) program in the Materials Studio 6.0 package (Accelrys Software Inc., San Diego, CA, USA). All calculations were performed using ultrasoft pseudopotentials, with kinetic energy cutoff of 400 eV. Electronic exchange and correlation effects were described within the generalized gradient approximation (GGA) [65] using Perdew-Burke-Ernzerhof (PBE) [66] functionals.

The optimized lattice parameter of bulk Fe was calculated using a bcc unit cell sampled with a 15 × 15 × 15 Monkhorst-Pack k-point grid. The calculated value was 2.816 Å, which was comparable to the most accepted experimental value of 2.87 Å [67], with a difference of less than 2.0%.

A (3 × 3) surface unit cell with a slab of four layers' thickness was selected as the model. This slab was repeated periodically with a 15 Å of vacuum region between the slabs. Only one H_2O molecule per super cell was adsorbed on one side of the slab to reduce lateral interactions between adsorbates. The geometry optimization including all degrees of freedom of the adsorbates and the two topmost metal layers were considered. The total energy calculation and the surface structural relaxation were performed by sampling the Brillouin zone with a 6 × 6 × 1 Monkhorst-Pack grid. All of the considered geometries were fully relaxed so that the forces became smaller than 0.01 eV· Å^{-1}. A 0.1 eV Fermi smearing was used and the convergence criteria for the geometry optimizations were 10^{-5} eV for the total energy. The computed lattice constant (2.643 Å) and magnetic moment (2.22 μB) are close to the experimental values (2.866 Å [68] and 2.22 μB [69]).

As usually defined [70], the adsorption energy (E_{ad}) was calculated by Equation (5):

$$E_{ad} = E_T - E_{Fe} - E_S \qquad (5)$$

where E_T, E_{Fe} and E_S are the total energy of the system, the energy associated to the isolated surface, and the energy of the isolated species (H, O, OH and H_2O in the present study), respectively. A negative value of E_{ad} indicates an exothermic chemisorption process. The linear synchronous transit (LST) calculation [71] combined with a quadratic synchronous transit (QST) calculation and conjugate gradient refinements [72] were used to obtain the transition state (TS). A LST optimization was performed in the calculations of LST/QST, while QST maximization was used to obtain the TS approximation. The same conjugate gradient minimization was repeated until a stationary point was obtained.

We carried the Hubbard correction for H_2O adsorption on the Fe(110) surface using the DFT+U method. DFT + U implementation in the code was based on the formalism summarized in Ref. [73]. Only on-site Coulomb repulsion was used and all higher-order multipolar terms were neglected. The effect of the Hubbard correction was found to be negligible for adsorption energies of H_2O, O, OH and H on the Fe(100) surface (Table 4).

Table 4. Comparison of adsorption energies of H_2O, OH, O and H on Fe(110) with and without Hubbard correction.

Species	DFT(GGA-PBE)	DFT + U
H_2O (top)	−0.65	−0.67
OH (brg)	−3.93	−3.89
O (hcp)	−3.67	−3.54
H (brg)	−3.99	−3.66

4. Conclusions

The adsorption configurations of H_2O and relevant dissociation species on clean and O-pre-adsorbed Fe(100) surfaces were investigated by DFT. It was found that H_2O is preferably adsorbed on the top site, O is absorbed on the bridge and hollow hcp site, while OH and H are adsorbed on the bridge site. The calculated adsorption energies revealed that the interactions between the adsorbates and the Fe surface increase in the order of H_2O < OH < H < O. In addition, we found that both the first H abstraction from adsorbed H_2O and the subsequent OH dissociation are exothermic on both clean Fe(100) and O-pre-adsorbed Fe(100) surfaces. However, the pre-adsorbed O significantly reduces the kinetics energy barriers for both reactions. Our results confirmed that the presence of pre-adsorbed oxygen species significantly promotes H_2O dissociation and will help in the design of better catalysts for water dissociation.

Acknowledgments: This work is financially supported by National Natural Science Foundation of China with Grant No. 21206074 and 61304137, Natural Science Foundation of Jiangsu province with Grant No. BK2012406, China Postdoctoral Science Foundation Funded Project with Grant No. 2014M561649, Research Fund for the Doctoral Program of Higher Education of China with Grant No. 20123219120033, the Fundamental Research Funds for the Central Universities with Grant No. 30920140112004, and Zijin Intelligent Program of Nanjing University of Science and Technology with Grant No. 2013-ZJ-0103. MS acknowledges a startup fund from the Hong Kong University of Science and Technology.

Author Contributions: W.J. Wang did the simulation work, wrote the manuscript; G.P. Wang revised the first version of the manuscript; M.H. Shao is a supervisor, made editing and improvement corrections of manuscript, and worked on all topics listed above.

Conflicts of Interest: The authors declare no conflict of interest.

References

1. Henniker, J.C. The depth of the surface zone of a liquid. *Rev. Mod. Phys.* **1949**, *21*, 322–341.
2. Bockris, J.O.M.; Veziroglu, T.N. Estimates of the price of hydrogen as a medium for wind and solar sources. *Int. J. Hydrogen Energy* **2007**, *32*, 1605–1610.
3. Suh, M.P.; Park, H.J.; Prasad, T.K.; Lim, D.W. Hydrogen storage in metal-organic frameworks. *Chem. Rev.* **2012**, *112*, 782–835.
4. Eberle, U.; Felderhoff, M.; Schüth, F. Chemical and physical solutions for hydrogen storage. *Angew. Chem. Int. Ed.* **2009**, *48*, 6608–6630.
5. Dicks, A.L. Hydrogen generation from natural gas for the fuel cell systems of tomorrow. *J. Power Sources* **1996**, *61*, 113–124.
6. Dybkjaer, I. Tubular reforming and autothermal reforming of natural gas—An overview of available processes. *Fuel Process. Technol.* **1995**, *42*, 85–107.
7. Takezawa, N.; Iwasa, N. Steam reforming and dehydrogenation of methanol: Difference in the catalytic functions of copper and group VIII metals. *Catal. Today* **1997**, *36*, 45–56.
8. Haryanto, A.; Fernando, S.; Murali, N.; Adhikari, S. Current status of hydrogen production techniques by steam reforming of ethanol: A review. *Energy Fuels* **2005**, *19*, 2098–2106.
9. Wei, J.M.; Iglesia, E. Reaction pathways and site requirements for the activation and chemical conversion of methane on Ru-based catalysts. *J. Phys. Chem. B* **2004**, *108*, 7253–7262.
10. Rostrupnielsen, J.R.; Hansen, J.H.B. CO_2-reforming of methane over transition metal. *J. Catal.* **1993**, *144*, 38–49.
11. Wei, J.M.; Iglesia, E. Structural requirements and reaction pathways in methane activation and chemical conversion catalyzed by rhodium. *J. Catal.* **2004**, *225*, 116–127.
12. Wei, J.M.; Iglesia, E. Isotopic and kinetic assessment of the mechanism of methane reforming and decomposition reactions on supported iridium catalysts. *Phys. Chem. Chem. Phys.* **2004**, *6*, 3754–3759.

13. Wei, J.M.; Iglesia, E. Structural and mechanistic requirements for methane activation and chemical conversion on supported iridium clusters. *Angew. Chem. Int. Ed.* **2004**, *43*, 3685–3688.

14. Wei, J.M.; Iglesia, E. Mechanism and site requirements for activation and chemical conversion of methane on supported Pt clusters and turnover rate comparisons among noble metals. *J. Phys. Chem. B* **2004**, *108*, 4094–4103.

15. Sehested, J. Four challenges for nickel steam-reforming catalysts. *Catal. Today* **2006**, *111*, 103–110.

16. Murakhtina, T.; Site, L.D.; Sebastiani, D. Vibrational frequencies of water adsorbed on (111) and (221) nickel surfaces from first principle calculations. *Chem. Phys. Chem.* **2006**, *7*, 1215–1219.

17. Catapan, R.C.; Oliveira, A.A.M.; Chen, Y.; Vlachos, D.G. DFT study of the water–gas shift reaction and coke formation on Ni(111) and Ni(211) surfaces. *J. Phys. Chem. C* **2012**, *116*, 20281–20291.

18. Lindström, B.; Pettersson, L.J. Hydrogen generation by steam reforming of methanol over copper-based catalysts for fuel cell applications. *Int. J. Hydrogen Energy* **2001**, *26*, 923–933.

19. Andersson, K.; Ketteler, G.; Bluhm, H.; Yamamoto, S.; Ogasawara, H.; Pettersson, L.G.M.; Salmeron, M.; Nilsson, A. Autocatalytic water dissociation on Cu(110) at near ambient conditions. *J. Am. Chem. Soc.* **2008**, *130*, 2793–2797.

20. Llorca, J.; Homs, N.; Sales, J.; de la Piscina, P.R. Efficient production of hydrogen over supported cobalt catalysts from ethanol steam reforming. *J. Catal.* **2002**, *209*, 306–317.

21. Xu, L.S.; Ma, Y.S.; Zhang, Y.L.; Chen, B.H.; Wu, Z.F.; Jiang, Z.Q.; Huang, W.X. Water adsorption on a Co(0001) surface. *J. Phys. Chem. C* **2010**, *114*, 17023–17029.

22. Murata, K.; Wang, L.S.; Saito, M.; Inaba, M.; Takahara, I.; Mimura, N. Hydrogen production from steam reforming of hydrocarbons over alkaline-earth metal-modified Fe- or Ni-based catalysts. *Energy Fuels* **2004**, *18*, 122–126.

23. Polychronopoulou, K.; Bakandritsos, A.; Tzitzios, V.; Fierro, J.L.G.; Efstathiou, A.M. Absorption-enhanced reforming of phenol by steam over supported Fe catalysts. *J. Catal.* **2006**, *241*, 132–148.

24. Liang, C.H.; Ma, Z.Q.; Lin, H.Y.; Ding, L.; Qiu, J.S.; Frandsen, W.; Su, D.S. Template preparation of nanoscale $Ce_xFe_{1-x}O_2$ solid solutions and their catalytic properties for ethanol steam reforming. *J. Mater. Chem.* **2009**, *19*, 1417–1424.

25. Noichi, H.; Uddin, A.; Sasaoka, E. Steam reforming of naphthalene as model biomass tar over iron-aluminum and iron-zirconium oxide catalyst catalysts. *Fuel Process. Technol.* **2010**, *91*, 1609–1616.

26. Mawdsley, J.R.; Krause, T.R. Rare earth-first-row transition metal perovskites as catalysts for the autothermal reforming of hydrocarbon fuels to generate hydrogen. *Appl. Catal. A* **2008**, *334*, 311–320.

27. Sato, K.; Nagaoka, K.; Nishiguchi, H.; Takita, Y. n-C_4H_{10} autothermal reforming over MgO-supported base metal catalysts. *Int. J. Hydrogen Energy* **2009**, *34*, 333–342.

28. Anderson, A.B. Reactions and structures of water on clean and oxygen covered Pt(111) and Fe(100). *Surf. Sci.* **1981**, *105*, 159–176.

29. Baró1, A.M.; Erley1, W. The adsorption of H_2O on Fe(100) studied by EELS. *J. Vac. Sci. Technol.* **1982**, *20*, 580–583.

30. Lu, J.P.; Albert, M.R.; Bernasek, S.L. The adsorption of oxygen on the Fe(100) surface. *Surf. Sci.* **1989**, *215*, 348–362.

31. Hung, W.H.; Schwartz, J.; Bernasek, S.L. Sequential oxidation of Fe(100) by water adsorption: formation of an ordered hydroxylated surface. *Surf. Sci.* **1991**, *248*, 332–342.

32. Freitas, R.R.Q.; Rivelino, R.; de Brito Mota, F.; de Castilho, C.M.C. Dissociative adsorption and aggregation of water on the Fe(100) surface: A DFT study. *J. Phys. Chem. C* **2012**, *116*, 20306–20314.

33. Eder, M.; Terakura, K. Initial stages of oxidation of (100) and (110) surfaces of iron caused by water. *Phys. Rev. B* **2001**, *64*, 115426:1–115426:7.

34. Jung, S.C.; Kang, M.H. Adsorption of a water molecule on Fe(100): Density-functional calculations. *Phys. Rev. B* **2010**, *81*, 115460:1–115460:7.

35. Govender, A.; Ferré, D.C.; Niemantsverdriet, J.W. The surface chemistry of water on Fe(100): A density functional theory study. *Chem. Phys. Chem.* **2012**, *13*, 1583–1590.

36. Lazar, P.; Otyepka, M. Dissociation of water at iron surfaces: Generalized gradient functional and range-separated hybrid functional study. *J. Phys. Chem. C* **2012**, *116*, 25470–25477.

37. Liu, S.; Tian, X.; Wang, T.; Wen, X.; Li, Y.W.; Wang, J.; Jiao, H. High coverage water aggregation and dissociation on Fe(100): A computational analysis. *J. Phys. Chem. C* **2014**, *118*, 26139–26154.

38. Michaelides, A.; Ranea, V.A.; de Andres, P.L.; King, D.A. General model for water monomer adsorption on close-packed transition and noble metal surfaces. *Phys. Rev. Lett.* **2003**, *90*, 216102:1–216102:4.

39. Michaelides, A.; Ranea, V.A.; de Andres, P.L.; King, D.A. First-principles study of H_2O diffusion on a metal surface: H_2O on Al{100}. *Phys. Rev. B* **2004**, *69*, 075409:1–075409:4.

40. Wang, S.; Cao, Y.; Rikvold, P.A. First-principles calculations for the adsorption of water molecules on the Cu(100) surface. *Phys. Rev. B* **2004**, *70*, 205410:1–205410:4.

41. Li, J.B.; Zhu, S.L.; Li, Y.; Wang, F.H. Water adsorption on Pd {100} from first principles. *Phys. Rev. B* **2007**, *76*, 235433:1–235433:8.

42. Thiel, P.A.; Madey, T.E. The interaction of water with solid surfaces: fundamental aspects. *Surf. Sci. Rep.* **1987**, *7*, 211–385.

43. Henderson, M.A. The interaction of water with solid surfaces: fundamental aspects revisited. *Surf. Sci. Rep.* **2002**, *46*, 1–308.

44. Heras, J.M.; Estiu, G.; Viscido, L. The interaction of water with clean palladium films: A thermal desorption and work function study. *Appl. Surf. Sci.* **1997**, *108*, 455–464.

45. Nyberg, C.; Tengstal, C.G. Adsorption and reaction of water, oxygen, and hydrogen on Pd(100): Identification of adsorbed hydroxyl and implications for the catalytic H_2–O_2 reaction. *J. Chem. Phys.* **1984**, *80*, 3463–3468.

46. Wolf, M.; Nettesheim, S.; White, J.M.; Hasselbrink, E.; Ertl, G. Ultraviolet-laser induced dissociation and desorption of water adsorbed on Pd(111). *J. Chem. Phys.* **1990**, *92*, 1509–1510.

47. Wolf, M.; Nettesheim, S.; White, J.M.; Hasselbrink, E.; Ertl, G. Dynamics of the ultraviolet photochemistry of water adsorbed on Pd(111). *J. Chem. Phys.* **1991**, *94*, 4609–4619.

48. Zhu, X.Y.; White, J.M.; Wolf, M.; Hasselbrink, E.; Ertl, G. Photochemical pathways of water on palladium (111) at 6.4 eV. *J. Phys. Chem.* **1991**, *95*, 8393–8402.

49. Cao, Y.L.; Chen, Z.X. Theoretical studies on the adsorption and decomposition of H_2O on Pd(111) surface. *Surf. Sci.* **2006**, *600*, 4572–4583.

50. Shavorskiy, A.; Eralp, T.; Gladys, M.J.; Held, G. A stable pure hydroxyl layer on Pt{110}-(1×2). *J. Phys. Chem. C* **2009**, *113*, 21755–21764.

51. Liu, S.L.; Tian, X.X.; Wang, T.; Wen, X.D.; Li, Y.W.; Wang, J.G.; Jiao, H.J. Coverage dependent water dissociative adsorption on the cleanand O-precovered Fe(111) surfaces. *J. Phys. Chem. C* **2015**, *119*, 11714–11724.

52. Hung, W.H.; Schwartz, J.; Bernasek, S.L. Adsorption of H_2O on oxidized Fe(100) surfaces: comparison between the oxidation of iron by H_2O and O_2. *Surf. Sci.* **1993**, *294*, 21–32.

53. Pozzo, M.; Carlini, G.; Rosei, R.; Alfè, D. Comparative study of water dissociation on Rh(111) and Ni(111) studied with first principles calculations. *J. Chem. Phys.* **2007**, *126*, 164706:1–164706:11.

54. Wang, W.J.; Wang, G.P. A theoretical study of water adsorption and dissociation on Ni(111) surface during oxidative steam reforming and water gas shift processes. *J. Energy Inst.* **2015**, *88*, 112–117.

55. Mohsenzadeh, A.; Bolton, K.; Richards, T. DFT study of the adsorption and dissociation of water on Ni(111), Ni(110) and Ni(100) surfaces. *Surf. Sci.* **2014**, *627*, 1–10.

56. Tang, Q.L.; Chen, Z.X.; He, X. A theoretical study of the water gas shift reaction mechanism on Cu(111) model system. *Surf. Sci.* **2009**, *603*, 2138–2144.

57. Wang, W.J.; Wang, G.P. Theoretical study of direct *versus* oxygen-assisted water dissociation on the Cu(110) surface. *Appl. Surf. Sci.* **2015**, *351*, 846–852.

58. Fajín, J.L.C.; Cordeiro, M.N.D.S.; Illas, F.; Gomes, J.R.B. Descriptors controlling the catalytic activity of metallic surfaces towardwater splitting. *J. Catal.* **2010**, *276*, 92–100.

59. Fajín, J.L.C.; Cordeiro, M.N.D.S.; Illas, F.; Gomes, J.R.B. Generalized Brønsted-Evans-Polanyi relationships and descriptors for O–H bond cleavage of organic molecules on transition metal surfaces. *J. Catal.* **2014**, *313*, 24–33.

60. Błoński, P.; Kiejna, A.; Hafner, J. Theoretical study of oxygen adsorption at the Fe(110) and (100) surfaces. *Surf. Sci.* **2005**, *590*, 88–100.

61. Sorescu, D.C. First principles calculations of the adsorption and diffusion of hydrogen on Fe(100) surface and in the bulk. *Catal. Today* **2005**, *105*, 44–65.

62. Wang, Y.Q.; Yan, L.F.; Wang, G.C. Oxygen-assisted water partial dissociation on copper: a model study. *Phys. Chem. Chem. Phys.* **2015**, *17*, 8231–8238.

63. Jiang, Z.; Li, M.M.; Yan, T.; Fang, T. Decomposition of H_2O on clean and oxygen-covered Au(100) surface: A DFT study. *Appl. Surf. Sci.* **2014**, *315*, 16–21.

64. Cao, Y.; Chen, Z. Slab model studies of water adsorption and decomposition on clean and *X*- (*X* = C, N and O) contaminated Pd(111) surfaces. *Phys. Chem. Chem. Phys.* **2007**, *9*, 739–746.

65. Kurth, S.; Perdew, J.P.; Blaha, P. Molecular and solid-state tests of density functional approximations: LSD, GGAs, and meta-GGAs. *Int. J. Quantum Chem.* **1999**, *75*, 889–909.

66. Perdew, J.P.; Burke, K.; Ernzerhof, M. Generalized gradient approximation made simple. *Phys. Rev. Lett.* **1996**, *77*, 3865–3868.

67. Kittel, C. *Introduction to Solid State Physics*, 8th ed.; Wiley: New York, NY, USA, 2005.

68. Kohlhaas, R.; Donner, P.; Schmitz-Pranghe, N. The temperature-dependence of the lattice parameters of iron, cobalt, and nickel in the high temperature range. *Z. Angew. Phys.* **1967**, *23*, 245.

69. Kittel, C. *Introduction to Solid State Physics*, 7th ed.; Wiley: New York, NY, USA, 1996.

70. Jiang, D.E.; Carter, E.A. Diffusion of interstitial hydrogen into and through bcc Fe from first principles. *Phys. Rev. B* **2004**, *70*, 064102:1–064102:9.

71. Halgren, T.A. The synchronous-transit method for determining reaction pathways and locating molecular transition states. *Chem. Phys. Lett.* **1977**, *49*, 225–232.

72. Govind, N.; Petersen, M.; Fitzgerald, G.; King-Smith, D.; Andzelm, J. A generalized synchronous transit method for transition state location. *Comp. Mater. Sci.* **2003**, *28*, 250–258.

73. Cococcioni, M.; de Gironcoli, S. Linear response approach to the calculation of the effective interaction parameters in the LDA+U method. *Phys. Rev. B* **2005**, *71*, 035105.

Study of N$_2$O Formation over Rh- and Pt-Based LNT Catalysts

Lukasz Kubiak, Roberto Matarrese, Lidia Castoldi, Luca Lietti, Marco Daturi and Pio Forzatti

Abstract: In this paper, mechanistic aspects involved in the formation of N$_2$O over Pt-BaO/Al$_2$O$_3$ and Rh-BaO/Al$_2$O$_3$ model NO$_x$ Storage-Reduction (NSR) catalysts are discussed. The reactivity of both gas-phase NO and stored nitrates was investigated by using H$_2$ and NH$_3$ as reductants. It was found that N$_2$O formation involves the presence of gas-phase NO, since no N$_2$O is observed upon the reduction of nitrates stored over both Pt- and Rh-BaO/Al$_2$O$_3$ catalyst samples. In particular, N$_2$O formation involves the coupling of undissociated NO molecules with N-adspecies formed upon NO dissociation onto reduced Platinum-Group-Metal (PGM) sites. Accordingly, N$_2$O formation is observed at low temperatures, when PGM sites start to be reduced, and disappears at high temperatures where PGM sites are fully reduced and complete NO dissociation takes place. Besides, N$_2$O formation is observed at lower temperatures with H$_2$ than with NH$_3$ in view of the higher reactivity of hydrogen in the reduction of the PGM sites and onto Pt-containing catalyst due to the higher reducibility of Pt $vs.$ Rh.

Reprinted from *Catalysts*. Cite as: Kubiak, L.; Matarrese, R.; Castoldi, L.; Lietti, L.; Daturi, M.; Forzatti, P. Study of N$_2$O Formation over Rh- and Pt-Based LNT Catalysts. *Catalysts* **2016**, *6*, 36.

1. Introduction

The removal of NO$_x$ under lean conditions according to the NO$_x$ Storage-Reduction (NSR) technique is a viable approach for both diesel and lean gasoline fueled engines [1]. In this process NO$_x$ conversion is based on a cyclic engine operation alternating between fuel lean and rich conditions: NO$_x$ (NO+NO$_2$) are adsorbed on the catalyst surface during lean operations, while the surface of the trap is regenerated under the fuel rich environment. The NO$_x$ adsorption step involves oxidation sites (generally provided by noble metals like Pt, Pd or Rh) and alkali or alkaline-earth oxides storage sites (Ba, K, Sr), whereas the reduction stage involves reduction sites provided by the noble metals. All these components are deposited over a high surface area support like γ-Al$_2$O$_3$.

The operation of NSR catalysts (also known as Lean NO$_x$ Traps, LNTs) reduces NO$_x$ mainly to N$_2$ although undesired by-products like NH$_3$ or N$_2$O may also be generated. NH$_3$ is not generally a concern since an ammonia Selective Catalytic

Reaction (SCR) catalyst can be placed downstream of the NSR system, such as in the case of combined NO_x storage and NH_3-SCR catalytic systems [2]. At variance, N_2O is a highly undesired by-product in view of its very high global warming potential (nearly 300 times that of CO_2). Over fully formulated catalysts, it has been shown that N_2O formation is apparent both upon switch from lean to rich mode but also during the alternation from rich to lean regime [3–8]. It has been suggested that N_2O formation at the lean to rich transition occurs at the regeneration front, upon reduction of the stored NO_x over not fully reduced Platinum-Group-Metal (PGM) sites. On the other hand, N_2O formation upon the rich to lean transition originates from reaction between residual surface NO_x with reductive species (like NCO, CO or NH_3) in an adsorbed state [3,4,7]. This N_2O peak can be reduced if a neutral or slightly lean phase is inserted between the rich and the lean phase, since a more complete regeneration of the catalyst can be attained [9].

In order to provide further insights on the pathway involved in N_2O formation, a fundamental study was carried out on a model Pt-BaO/Al_2O_3 catalyst under simplified and controlled reaction conditions [10]. Accordingly the reactivity of gaseous NO alone and in the presence of various reducing agents like H_2, NH_3, and CO was analyzed by micro-reactor transient reactivity experiments and *operando* Fourier transform infrared (FT-IR) spectroscopy, to provide a complementary overview on the pathway of N_2O evolution. It has been shown that N_2O is formed from gaseous NO provided that a proper oxidation degree of the Pt sites is attained [10]. Accordingly the N_2O evolution was observed only in a narrow temperature range corresponding to the PtO to Pt transformation, determined by the reactivity of the reducing agents ($H_2 > NH_3 > CO$).

In this study the investigation previously carried out over the model Pt-BaO/Al_2O_3 sample has been extended to a Rh-containing sample. Since Rh is generally considered to play an important role in the reduction of NO_x [11], it is of interest to analyze the role of this component in the pathways involved in N_2O formation. For this purpose the reduction of gaseous NO in the presence of various reductants (H_2, NH_3) was investigated by transient microreactor experiments and *operando* FT-IR spectroscopy to combine gas-phase data with surface species analysis. To point out the role of NO_x surface species (nitrates) in the formation of N_2O, experiments were also performed with the same reductants after adsorption of NO_x on the catalyst surface.

2. Results and Discussion

2.1. NO-Temperature Programmed Reaction (NO-TPR)

The reactivity of NO over the reduced Pt-BaO/Al_2O_3 and Rh-BaO/Al_2O_3 was investigated at first. The results of the gas phase analysis are shown in Figure 1A,B for the Rh-BaO/Al$_2$O$_3$ and Pt-BaO/Al$_2$O$_3$ catalyst, respectively. In both cases a NO consumption peak is observed, with a minimum in NO concentration centered near 250 and 200 °C for Rh-BaO/Al$_2$O$_3$ and Pt-BaO/Al$_2$O$_3$, respectively. The NO consumption is more evident in the case of the Pt-BaO/Al$_2$O$_3$ catalyst sample; in both cases it is accompanied by evolution of N$_2$O and of minor amounts of N$_2$. Notably, the NO consumption is not balanced by the evolution of N$_2$O (and of N$_2$), indicating the build up of N-containing species on the catalyst surface.

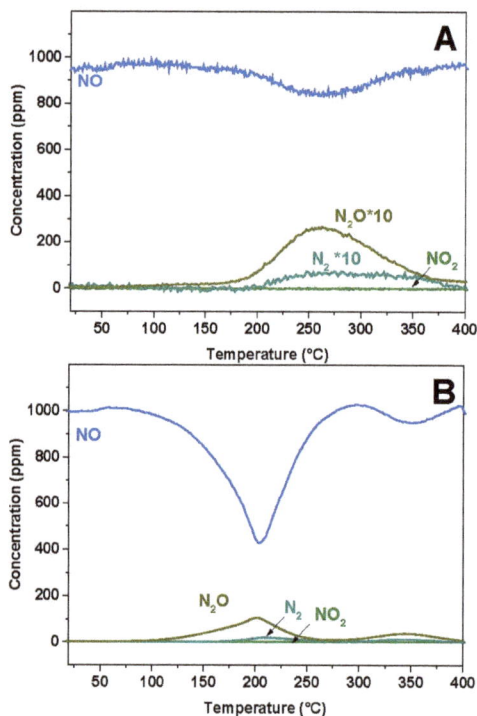

Figure 1. NO-TPR experiments with NO (1000 ppm) in He from 40–400 °C (10 °C/min) over pre-reduced Rh-BaO/Al$_2$O$_3$ (**A**) and Pt-BaO/Al$_2$O$_3$ (**B**) catalysts (rig #1).

The results of surface analysis are shown in Figure 2. In the case of Rh-BaO/Al$_2$O$_3$ catalyst (Figure 2A) upon exposure of NO at 50 °C (spectrum a) a band appears at 1210 cm^{-1} corresponding to ν_{asym}(NO$_2$) vibration of chelating nitrites on the barium phase. The presence of the ν_{sym}(NO$_2$) mode of nitrites may also be observed near 1365 cm^{-1}. Upon heating, the band located at 1210 cm^{-1} increases, reaches a maximum intensity around 200–250 °C (spectra e–g) and then decreases with temperature. The decrease of the bands of nitrites is accompanied by an increase in the bands located near 1327 and 1405 cm^{-1} associated with ionic nitrates ($\nu_{asym}(NO_3)$ at 1327 and 1405 cm^{-1}). The presence of an isosbestic point suggests the occurrence of the transformation of nitrites into nitrates on the catalyst surface. The appearance of nitrate species is also confirmed by the presence of a low intensity peak at 1030 cm^{-1} ($\nu_{sym}(NO_3)$). Furthermore, a shoulder near 1545 cm^{-1}, associated with bidentate nitrates ($\nu(N=O)$) is also detected. At the end of the temperature ramp (350 °C) the surface of the catalyst is dominated by the presence of nitrate species, however nitrites are still present (band at 1210 cm^{-1}).

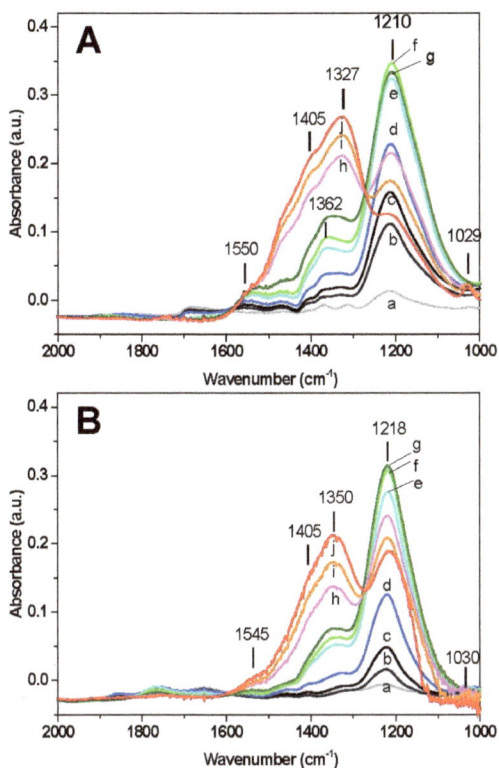

Figure 2. Operando FT-IR spectra recorded during NO-TPR experiments with NO (1000 ppm) in He over Rh-BaO/Al$_2$O$_3$ (**A**) and Pt-BaO/Al$_2$O$_3$ (**B**) from 50 to 350 °C (10 °C/min), spectra recorded every 1 min, displayed at selected temperatures: (**a**) 50 °C; (**b**) 100 °C; (**c**) 120 °C; (**d**) 150 °C; (**e**) 200 °C; (**f**) 220 °C; (**g**) 250 °C; (**h**) 300 °C; (**i**) 320 °C; (**l**) 350 °C (rig#2).

A similar picture was observed over the Pt-BaO/Al$_2$O$_3$ catalyst (Figure 2B), with only minor differences. Also in this case the formation of chelating nitrites (1218 cm^{-1}) is seen at the initial stage, reaching a maximum in the same temperature range (220–250 °C) observed for the Rh-based catalyst. Then the bands decrease in favor of nitrate formation, as pointed out by the isosbestic point and by the growth of the bands near 1350 and 1405 cm^{-1}. At the end of the run (350 °C), the final amount of nitrites on the Pt containing catalyst is higher if compared to the Rh-based sample, as pointed out by the relative intensity of the bands at 1218 cm^{-1} (nitrites) and at 1350–1405 cm^{-1} (nitrates).

The above results clearly show that upon contacting NO with the reduced catalyst surface, NO is consumed leading to N_2O and nitrites adsorbed over the barium surface. In particular the results are in line with the occurrence, below 250 °C, of the following NO disproportion reaction Equation (1):

$$4\,NO + BaO \rightarrow N_2O + Ba(NO_2)_2 \tag{1}$$

In fact the ratio between NO consumption and N_2O formation is not very far from the theoretical value 1:4 expected from the stoichiometry of the overall reaction Equation (1). At higher temperature, above 250 °C, nitrites are transformed into nitrates, in line with previous results [10].

The overall stoichiometry of the NO disproportion reaction Equation (1) can be explained by considering the occurrence of the following reactions Equations (2)–(4):

$$NO + 2\,Me \rightarrow Me\text{-}N + Me\text{-}O \tag{2}$$

$$NO + Me\text{-}N \rightarrow N_2O + Me \tag{3}$$

$$2\,NO + Me\text{-}O + BaO \rightarrow Me + Ba(NO_2)_2 \tag{4}$$

where Me is the metal site (Pt or Rh). According to this scheme, the reaction is initiated at low temperature by the decomposition of NO over reduced Me sites, leading to the formation of N- and O-adspecies (reaction Equation (2)). The formation of oxidized metal sites (Me-O species) prevents the dissociation of other NO molecules and this results in the formation of N_2O (reaction Equation (3)) and of nitrites on Ba (reaction 4)). The occurrence of reaction Equation (4) removes O-adatoms on metal sites leaving vacant metal sites accessible for further NO dissociation. Reactions Equations (2)–(4) lead to the overall stoichiometry of reaction Equation (1).

Worth noting is that reactions Equations (2)–(4) require the presence of reduced metal sites, to initiate the NO dissociation reaction. In fact the NO disproportion reaction is observed to a much lower extent over an oxidized Pt-based catalyst sample (data not reported). Along similar lines, the storage component BaO is also important in that it provides the adsorption sites for the nitrite (and nitrate) surface species. In fact, experiments carried out over a bare Pt/Al_2O_3 sample pointed out a much lower N_2O (and nitrite) formation. However, the presence of BaO may also increase the activity of the metal sites in the dissociative chemisorption of NO, as reported elsewhere [12].

2.2. NO/H$_2$-TPR

To investigate the effect of the presence of a reductant in N$_2$O formation, experiments were carried out in the presence of H$_2$ and results are shown in Figure 3A,B in the case of Rh-BaO/Al$_2$O$_3$ and Pt-BaO/Al$_2$O$_3$ catalysts, respectively.

In the case of the Rh-based catalyst, NO starts to be converted near 150 °C, with formation of N$_2$O and N$_2$ in nearly equal amounts. Then, above 200 °C, NO is fully consumed and N$_2$O formation rapidly drops. Accordingly above 200 °C only N$_2$ and NH$_3$ are seen in the reaction products.

Very similar results are observed in the case of the Pt-BaO/Al$_2$O$_3$ catalyst (Figure 3B). However, in this case NO consumption is observed at much lower temperatures (onset temperature near 50 °C), and the NH$_3$ selectivity at high temperatures is much higher if compared to the Rh-BaO/Al$_2$O$_3$ sample.

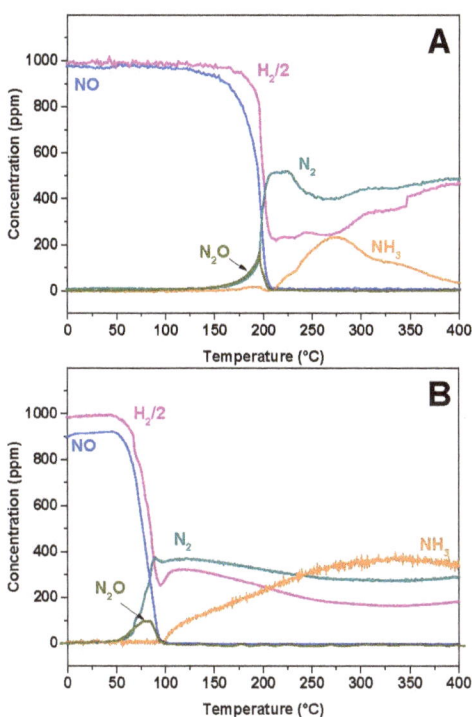

Figure 3. NO/H$_2$-TPR experiments with NO (1000 ppm) + H$_2$ (2000 ppm) over (**A**) Rh-BaO/Al$_2$O$_3$ and (**B**) Pt-BaO/Al$_2$O$_3$ from 40–400 °C (4 °C/min), (rig#3).

The correspondent surface analysis is shown for the Pt-BaO/Al$_2$O$_3$ sample in Figure 4. Upon NO admission, the formation of a very weak band at 1218 cm^{-1} is seen, related to the formation of chelating nitrites on the Ba phase. This band slightly increases with temperature, reaches a weak maximum near 85 °C and then disappears at higher temperatures, along with N_2O. The spectra also show complex bands in the range 1300–1600 cm^{-1}, with positive and negative contributions. These bands are related to modifications of a covalent carbonate species, initially present on the surface, into more symmetric, ionic species upon interaction with water formed during the reduction of NO with H_2 and upon the temperature increase.

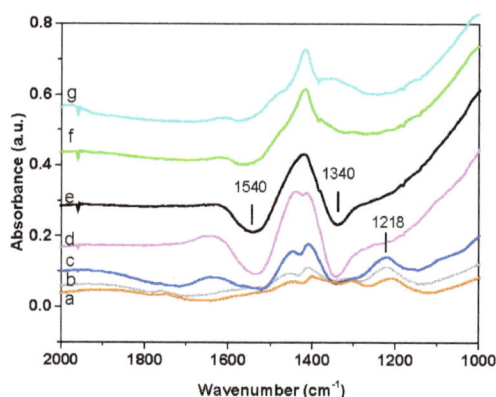

Figure 4. FT-IR spectra obtained during NO/H_2–TPR with NO (1000 ppm) + H_2 (2000 ppm) over Pt-BaO/Al$_2$O$_3$ from 40–400 °C (4 °C/min) at: (**a**) 40 °C; (**b**) 77 °C; (**c**) 85 °C; (**d**) 152 °C; (**e**) 252 °C; (**f**) 354 °C; (**g**) 400 °C (rig#3).

Very similar results were obtained in the case of the Rh-based sample (not shown): also in this case we observed the formation of a weak nitrite band in the temperature range where N_2O is formed.

Upon comparing these results with those obtained in the absence of H_2 (*i.e.*, the NO-TPR runs, Figures 1 and 2), it clearly appears that N_2O formation is seen at much lower temperatures in the case of the Pt-BaO/Al$_2$O$_3$ sample, and at slightly higher temperatures in the case of the Rh-based sample. In both cases N_2O formation is accompanied by nitrite evolution, although the band of nitrites is much lower in the presence of H_2.

As suggested above (see reactions Equations (2)–(4)), N_2O (and nitrite) formation requires the presence of reduced Me (Pt, Rh) sites. Hence the reaction is initiated by the reduction of the PGM sites by H_2: the lower onset temperature for NO reduction illustrates the much higher reducibility of the Pt-based sample if compared to the Rh-BaO/Al$_2$O$_3$ catalyst.

As shown in Figure 3, the NO reduction reaction leads at low temperatures to the formation of N_2 and N_2O. Then, at higher temperatures, N_2O formation becomes nil and the selectivity shifts towards N_2 and NH_3. It is speculated that upon reduction of the PGM sites by H_2, N_2O (and nitrite) formation occurs according to reactions

Equations (2)–(4). The presence of H_2 leads also to the formation of Rh-H and Pt-H species, according to the reaction Equation (5).

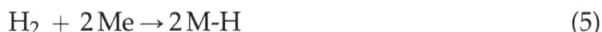

$$H_2 + 2\,Me \rightarrow 2\,M\text{-}H \tag{5}$$

where Me is the metal site (Pt or Rh). The so formed Me-H species scavenge O-adatoms, reaction Equation (6):

$$2\,Me\text{-}H + Me\text{-}O \rightarrow 3\,Me + H_2O \tag{6}$$

The scavenging of O-adatoms according to reaction Equation (6) limits the formation of nitrites according to reaction Equation (4), as indeed is observed in the presence of H_2.

Besides, N_2 formation is also observed along with N_2O, coming from the coupling of N-adatoms (reactions Equations (2) and (7)):

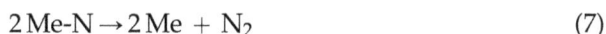

$$2\,Me\text{-}N \rightarrow 2\,Me + N_2 \tag{7}$$

Since the O-removal efficiency by H_2 increases with temperature, upon increasing the temperature NO completely dissociates and N-adatoms lead to the formation of N_2 (reaction Equation (7)) or of NH_3 (reaction Equation (8)), thus shifting the selectivity towards N_2 and NH_3:

$$M\text{-}N + 3\,M\text{-}H \rightarrow 4\,M + NH_3 \tag{8}$$

Notably, in the case of the Rh-containing catalyst, the NH_3 concentration decreases above 275 °C. This is due to the occurrence of the NH_3 decomposition reaction, which is efficiently catalyzed over Rh, as it is shown below.

2.3. NO/NH$_3$-TPR

Figure 5 shows the results obtained during the reduction of NO with NH_3 in the case of Rh-BaO/Al$_2$O$_3$ (Figure 5A) and Pt-BaO/Al$_2$O$_3$ catalysts (Figure 5B). The onset temperature for NO reduction is seen near 200 and 100 °C for the Rh- and Pt-based catalysts, respectively. These temperature thresholds are higher than those measured in the case of H_2, but show the same order of reactivity of the investigated samples, being the Pt-BaO/Al$_2$O$_3$ catalyst more reactive than Rh-BaO/Al$_2$O$_3$. Over both the catalytic systems the initial formation of N_2O and N_2 is observed; then, when complete NO consumption is attained (*i.e.*, near 295 °C and 185 °C over Rh-BaO/Al$_2$O$_3$ and Pt-BaO/Al$_2$O$_3$, respectively) the N_2O concentration drops to zero resulting in complete N_2 selectivity. In the case of the Rh-based sample, above 300 °C an increase in the N_2 concentration is observed, along with H_2 and a correspondent decrease of the NH_3 concentration. This is due to the occurrence of the NH_3 decomposition reaction:

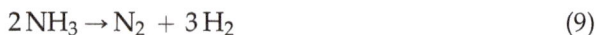

$$2\,NH_3 \rightarrow N_2 + 3\,H_2 \tag{9}$$

which is catalyzed by the Rh component.

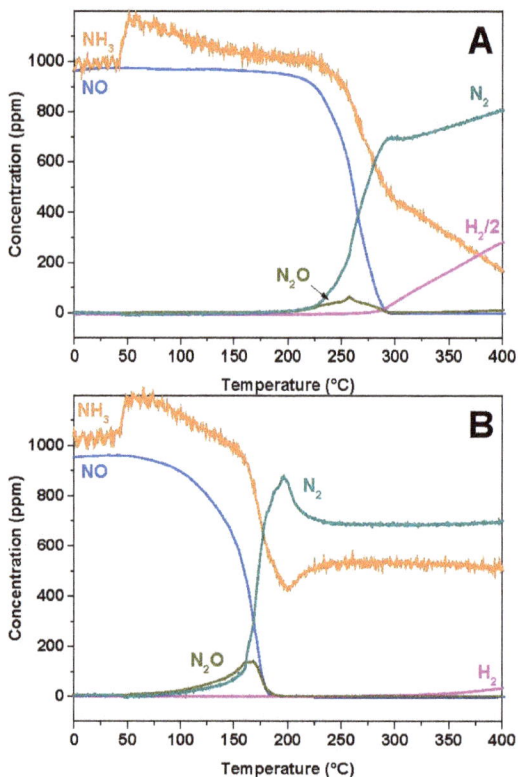

Figure 5. NO/NH_3-TPR experiments with NO (1000 ppm) + NH_3 (1000 ppm) over (**A**) Rh-BaO/Al_2O_3 and (**B**) Pt-BaO/Al_2O_3 from 40–400 °C (4 °C/min) (rig#3).

The results of the FT-IR analysis carried out over the Rh-BaO/Al_2O_3 and the Pt-BaO/Al_2O_3 catalysts are shown in Figure 6A,B respectively. In the case of Rh-BaO/Al_2O_3 (see Figure 6A), upon NH_3 admission, bands at 1095 and 1647 cm^{-1} (low intensity) immediately appear (spectrum a). These bands are related to the symmetric and antisymmetric bending modes of ammonia. Then, upon NO admission, a weak band of nitrites appears at 1212 cm^{-1}. Upon heating, the bands associated with NH_3 decrease and eventually disappear. Besides, the nitrite band slightly increases up to 275 °C and then disappears. In addition to these bands, positive and negative bands are evident in the spectra due to the distortion of carbonate species, as previously discussed.

The Pt-BaO/Al$_2$O$_3$ catalyst shows similar features (see Figure 6B) but a much higher amount of nitrites is formed upon NO admission, with respect to Rh-BaO/Al$_2$O$_3$. Nitrites readily disappear near 200 °C (spectrum g).

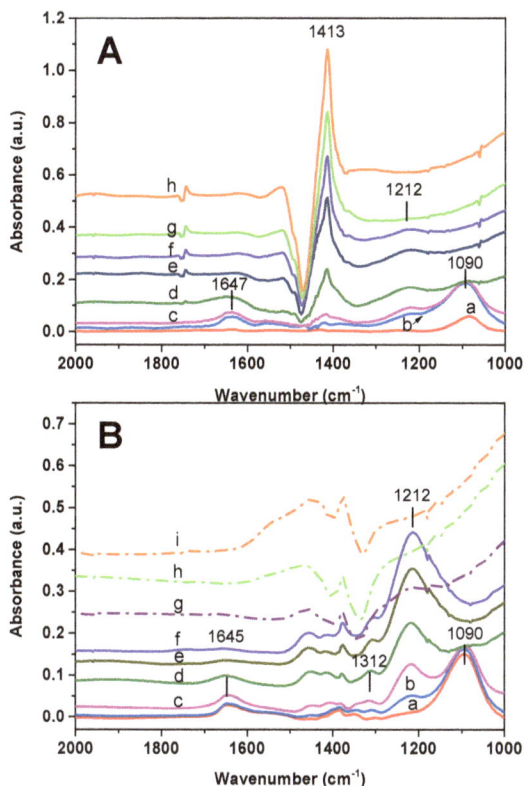

Figure 6. Operando FT-IR spectra during NO/NH$_3$–TPR with NO (1000 ppm) + NH$_3$ (1000 ppm) from 40–400 °C (4 °C/min) over (**A**) Rh-BaO/Al$_2$O$_3$ at: (**a**) 40 °C with NH$_3$; (**b**) 40 °C with NH$_3$+NO; (**c**) 40 °C before temperature ramp; (**d**) 50 °C; (**e**) 104 °C; (**f**) 260 °C; (**g**) 292 °C; (**h**) 400 °C. Panel (**B**) Pt-BaO/Al$_2$O$_3$ at: (**a**) 40 °C with only NH$_3$; (**b**) 40 °C NH$_3$ + NO; (**c**) 40 °C before temperature ramp; (**d**) 104 °C; (**e**) 153 °C; (**f**) 178 °C; (**g**) 203 °C; (**h**) 328 °C; (**i**) 400 °C (rig#3).

The picture obtained from the analysis with NH$_3$ is consistent with that discussed in the case of H$_2$. Upon reduction of the PGM sites by NH$_3$, NO is dissociated into N- and O-adatoms and formation of N$_2$O and nitrites is observed (reactions Equations (2)–(4)). Formation of these species is observed at lower temperatures over the Pt-based sample in view of its higher reactivity, in line with data obtained during the NO/H_2-TPR.

By comparing NO/NH_3-TPR and NO/H_2-TPR data, it is apparent that the onset of N_2O formation is shifted towards higher temperatures (compare Figures 3A and 4A) and also higher amounts of nitrites are formed. This can be explained considering that NH_3 is a much less efficient reductant (*i.e.*, a less efficient O-scavenger) compared to H_2. This results in a higher onset temperature for the reaction and a higher temperature for complete NO dissociation.

In the case of NO/NH_3-TPR experiments, N_2O and N_2 formation involves coupling of N-containing species coming from NO and/or NH_3 molecules. To investigate such aspects, isotopic labeling experiments were carried out with NH_3 and ^{15}NO over both Pt- [13] and Rh-BaO/Al_2O_3 [14]. In both cases the low-T NO consumption (*i.e.*, below the maximum in N_2O concentration) is accompanied by the evolution of double labeled $^{15}N_2O$ and of single labeled N_2 ($^{15}N^{14}N$), with only minor amounts of single labeled N_2O ($^{15}N^{14}NO$). Then, at higher temperatures (near the complete NO consumption) N_2O is no more detected among the reaction products and all the N_2 isotopes are observed.

These results indicate that N_2O formation involves primarily NO molecules only, according to reactions Equations (2) and (3). In this pathway, the role of NH_3 is hence that of keeping the noble metal sites in a reduced state, with formation of Me-H and Me-NH_2 species according to reaction Equation (10):

$$NH_3 + 2\,Me \rightarrow Me\text{-}NH_2 + Me\text{-}H \tag{10}$$

Me-H scavenges O-adatoms, resulting in H_2O formation while Me-NH_2 reacts with (labelled) nitrite species leading to the formation of the single labelled N_2 ($^{15}N^{14}N$), according to reaction Equation (11), where $O^=$ is lattice oxygen:

$$Me\text{-}NH_2 + {}^{15}NO_2 \rightarrow Me + {}^{15}N^{14}N + 3/2\,H_2O + \tfrac{1}{2}O^= \tag{11}$$

This accounts for the formation of single labelled N_2 that accompanies N_2O evolution. At higher temperatures, near complete NO consumption where no N_2O is seen, all N_2 isotopes are detected due to the statistical coupling of N-adatoms coming from NO- and NH_3-decomposition [13].

2.4. H_2-Temperature Programmed Surface Reaction (H_2-TPSR)

Figure 7 shows the results of the H_2-TPSR experiments performed over Rh-BaO/Al_2O_3 and Pt-BaO/Al_2O_3 after NO_x adsorption at 350 °C. The gas phase results are shown in Figure 7A,B for the Rh- and Pt-based catalyst samples, respectively. In the case of the Rh-containing sample, the temperature threshold for hydrogen consumption is seen near 170 °C. Above 200 °C NH_3 evolution is seen, followed by that of N_2 above 300 °C. The initial H_2 consumption is probably related to the reduction of the Rh sites, since it is not accompanied by the release of reaction products.

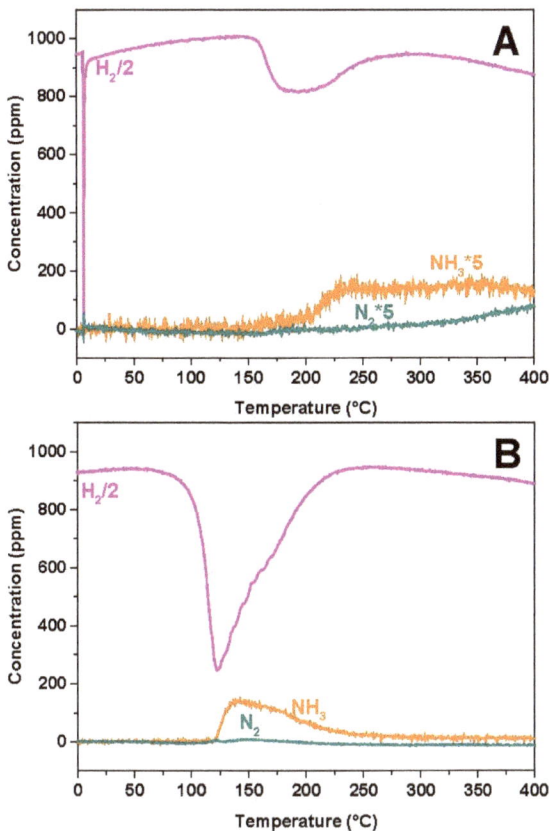

Figure 7. H_2-TPSR experiments with H_2 (2000 ppm) from 40–400 °C (4 °C/min) after NO/O_2 adsorption at 350 °C over (**A**) Rh-BaO/Al$_2$O$_3$ and (**B**) Pt-BaO/Al$_2$O$_3$ (rig#3).

The corresponding FT-IR spectra are shown in Figure 8A. The spectrum recorded prior to H_2 consumption (spectrum a) corresponds to the surface ionic nitrate ($\nu_{asym}(NO_3)$ mode split into two peaks at 1400 and 1330 cm^{-1}; $\nu_{sym}(NO_3)$ mode at 1035 cm^{-1}). The band near 1470 cm^{-1} is an artefact due to the presence of carbonate species, which have been partially displaced upon adsorption of NO_x and therefore lead to the formation of negative peaks in the subtraction spectra. Consequently, FT-IR analysis confirms that NO_x have been stored at 350 °C in the form of nitrate species.

Upon heating, the reduction of nitrates occurs from 200 °C, as it is obvious from the corresponding decrease of the bands. The nitrate bands decrease in intensity with temperature, and no new bands are apparent. At 388 °C (spectrum j) the surface is clean and only a negative band at 1440 cm^{-1} is visible, due to carbonates that have been displaced upon nitrate adsorption from NO/O$_2$ at 350 °C.

These results indicate that, over Rh-BaO/Al$_2$O$_3$, the stored nitrates are reduced by hydrogen selectively into NH$_3$ (and minor amounts of N$_2$), without any formation of NO or N$_2$O.

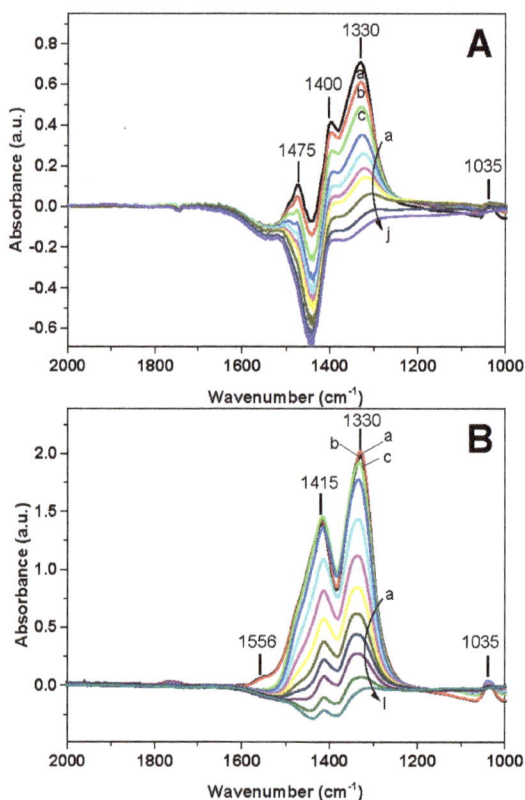

Figure 8. Operando FT-IR spectra obtained during H$_2$-TPSR experiments with H$_2$ (2000 ppm) from 40–400 °C (4 °C/min) after NO/O$_2$ adsorption at 350 °C over (**A**) Rh-BaO/Al$_2$O$_3$ at: (**a**) 157 °C; (**b**) 200 °C; (**c**) 215 °C; (**d**) 232 °C; (**e**) 248 °C; (**f**) 264 °C; (**g**) 281 °C; (**h**) 322 °C; (**i**) 355 °C; (**j**) 388 °C; and (**B**) Pt-BaO/Al$_2$O$_3$ at: (**a**) 40 °C; (**b**) 49 °C; (**c**) 107 °C; (**d**) 115 °C; (**e**) 124 °C; (**f**) 132 °C; (**g**) 140 °C; (**h**) 148 °C; (**i**) 157 °C; (**j**) 165 °C; (**k**) 182 °C; (**l**) 198 °C (rig#3).

Similar results were obtained in the case of the Pt-containing catalyst (Figures 7B and 8B). Gas phase analysis (Figure 7B) shows that H$_2$ consumption is seen starting from 100 °C, *i.e.*, at lower temperatures than Rh-BaO/Al$_2$O$_3$. The H$_2$ consumption is accompanied by the formation of ammonia with traces of nitrogen; no other reaction products (e.g., NO, N$_2$O) are formed. Also in this case the H$_2$ consumption is seen before the evolution of the reaction products, likely indicating the reduction of the catalyst metal sites.

Surface analysis (Figure 8B) showed the presence of the ionic nitrates (ν_{asym} (NO$_3$)) mode split into two peaks at 1417 and 1320 cm^{-1}; ν_{sym}(NO$_3$) mode at 1035 cm^{-1}) and minor amounts of bidentate nitrates (ν(N=O) mode at 1555 cm^{-1}) on the catalyst surface at the beginning of the run (spectrum a). The reduction of nitrates is observed above 115 °C, in line with gas-phase results. Upon increasing the temperature the reduction of nitrates proceeds faster and complete reduction is attained above 200 °C, at a much lower temperature with respect to Rh-BaO/Al$_2$O$_3$. Also in this case, residual carbonates, present on the surface before the NO$_x$ storage, were removed and this results in weak negative bands at 1440 and 1380 cm^{-1}.

H$_2$-TPSR experiments hence indicate that the reduction with H$_2$ of nitrates stored over both Pt- and Rh-containing catalyst samples does not lead to the formation of N$_2$O. The Pt-BaO/Al$_2$O$_3$ sample is more active than the Rh-containing catalyst but in both cases only NH$_3$ was observed as reduction product, along with traces of N$_2$. These results are in line with previous data obtained over the same Pt-containing sample used in this study and published elsewhere [15].

2.5. NH$_3$–TPSR

Figures 9 and 10 show the results obtained during TPSR experiments performed with NH$_3$ after NO$_x$ adsorption at 350 °C. The gas phase results are shown in Figure 9A,B for Rh-BaO/Al$_2$O$_3$ and Pt-BaO/Al$_2$O$_3$, respectively. In the case of the Rh-catalyst (Figure 9A), ammonia desorption is observed at the beginning of the heating ramp. Then the temperature onset for ammonia consumption is seen near 250 °C, *i.e.*, at higher temperatures if compared to H$_2$ (see Figure 7A). The NH$_3$ consumption is accompanied by the formation of nitrogen, whose concentration shows a peak with maximum near 280 °C. At temperatures above 300 °C ammonia is further consumed leading to the evolution of N$_2$ and H$_2$. The initial N$_2$ formation, with maximum near 280 °C, is due to the reduction of the stored nitrates, whereas the nitrogen formation above 300 °C is due to occurrence of the NH$_3$ decomposition reaction. Notably, the reduction of the stored nitrates with ammonia is fully selective to N$_2$, and neither N$_2$O nor NO formation was observed.

FT-IR spectra recorded during the NH$_3$-TPSR are shown in Figure 10A. Ionic nitrates (bands at 1402, 1338, and 1036 cm^{-1}) and adsorbed ammonia (bands at 1095 and 1640 cm^{-1}) are initially present on the catalyst surface at 40 °C (spectrum a). The increase of the temperature leads at first to the disappearance of the 1095 and 1640 cm^{-1} bands, in line with the desorption of NH$_3$ (spectrum b). At further temperature increase, nitrates are reduced and at 400 °C (spectrum i) the surface is free from nitrates.

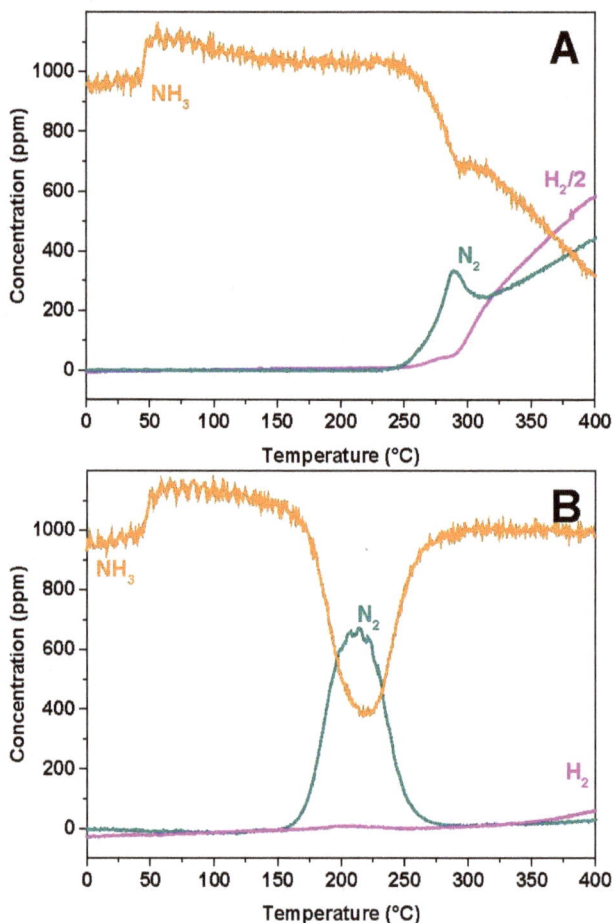

Figure 9. NH$_3$-TPSR experiments with NH$_3$ (1000 ppm) from 40–400 °C (4 °C/min) after NO/O$_2$ adsorption at 350 °C over: (**A**) Rh-BaO/Al$_2$O$_3$ and (**B**) Pt-BaO/Al$_2$O$_3$ (rig#3).

In the case of the Pt-catalyst (Figure 9B), the temperature threshold for ammonia consumption is observed near 150 °C, *i.e.*, at lower temperatures if compared to the Rh-containing sample, and leads to N$_2$ formation with a maximum near 220 °C. At 270 °C nitrogen evolution is completed and the NH$_3$ concentration reaches the inlet concentration value. Also in this case the reaction is fully selective towards nitrogen, and no N$_2$O formation was observed. Above 350 °C the Pt-catalyst shows a limited activity in the dissociation of NH$_3$, as pointed out by the small evolution of H_2 and of nitrogen. Concerning the surface phase (Figure 10B), ionic nitrates (bands at 1415, 1348, and 1035 cm^{-1}), bidentate nitrates (1555 cm^{-1}) and adsorbed ammonia (band at 1090 cm^{-1}) are initially present (40 °C, spectrum a). Upon increasing the temperature, initially bidentate nitrates are transformed into ionic nitrates (spectrum b) and then ionic nitrates decrease. Above 250 °C nitrates are completely removed (spectra j and k).

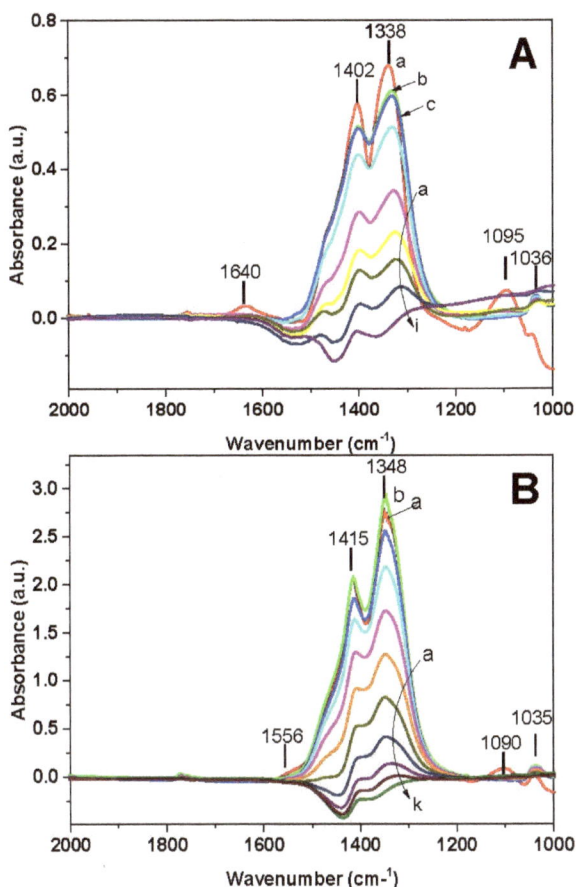

Figure 10. Operando FT-IR spectra obtained during NH$_3$-TPSR experiments with NH$_3$ (1000 ppm) from 40–400 °C (4 °C/min) after NO/O$_2$ adsorption at 350 °C over (**A**) Rh-BaO/Al$_2$O$_3$ at: (**a**) 40 °C; (**b**) 275 °C; (**c**) 283 °C; (**d**) 291 °C; (**e**) 300 °C; (**f**) 308 °C; (**g**) 316 °C; (**h**) 333 °C; (**i**) 400 °C; and (**B**) Pt-BaO/Al$_2$O$_3$ at: (**a**) 40 °C; (**b**) 171 °C; (**c**) 196 °C; (**d**) 204 °C; (**e**) 213 °C; (**f**) 221 °C; (**g**) 229 °C; (**h**) 237 °C; (**i**) 246 °C; (**j**) 254 °C; (**k**) 303 °C (rig#3).

The results hence indicate that over both catalyst samples NH$_3$ is an effective reductant for the stored nitrates, although it is less efficient than H$_2$. The Pt-containing sample is more reactive than the Rh-based catalyst. Notably, no N$_2$O is formed upon the reduction of the stored nitrates, the reaction being very selective towards N$_2$. This apparently rules out the involvement of surface nitrates and/or nitrate containing compounds (e.g., ammonium nitrate) in the formation of N$_2$O, at least under the investigated experimental conditions.

3. Pathways for N$_2$O Formation

The results of the reactivity of gas-phase NO and adsorbed NO$_x$ species with H$_2$ and NH$_3$ clearly illustrated that N$_2$O is formed only during the run with the presence of gas-phase NO. In fact no nitrous oxide formation has been observed both over the Pt- and Rh-BaO/Al$_2$O$_3$ catalysts upon the reduction of stored nitrates, where no gas-phase NO is fed to the catalyst samples.

Notably, N$_2$O is formed even in the absence of reductants (*i.e.*, from NO alone), provided that reduced PGM sites are present over the catalyst surface. Reduced Pt or Rh sites catalyze the NO dissociation reaction into N- and O-adatoms; N-species further interact with undissociated NO molecules (either in the gas-phase or from an adsorbed state) leading to the formation of N$_2$O. This pathway is represented by reactions Equations (2) and (3) previously discussed.

In the absence of any reductant, the reaction is self-inhibited by the formation of oxidized PGM sites (*i.e.*, O-adatoms), formed upon NO dissociation. In this respect, NO itself acts as a weak reductant, being able to scavenge O-adatoms and resulting in the formation of nitrite species adsorbed over the storage component (Ba). The overall stoichiometry of the process is then well described by following NO disproportion reaction Equation (1).

Hence also the storage element participates in the reaction, providing the adsorption sites for nitrites. It may also increase the reactivity of the PGM sites in the NO dissociative chemisorption, as suggested in the literature [10,12]. In fact, experiments carried out over a Ba-free sample showed a much lower N$_2$O formation if compared to the Ba-containing catalysts [14].

Reduced Pt- and Rh-based catalysts show similar reactivity in the N$_2$O formation from NO, N$_2$O being formed at the same temperature, although in the case of the Rh-containing catalyst much lower amounts of N$_2$O were detected. This is likely related to the lower metal dispersion of the Rh-BaO/Al$_2$O$_3$ catalyst if compared to the Pt-containing sample (6.8% *vs.* 48%, see Table 1), and hence to the lower amounts of surface reduced sites.

Table 1. Morphological characterization of catalysts.

Catalyst	Surface Area, m^2/g	Pore Volume, cm^3/g	Pore Radius, Å	Metal Dispersion, %
Pt-BaO/Al_2O_3	133	0.63	93	48
Rh-BaO/Al_2O_3	150	0.76	102	6.8

When a reductant (H_2 or NH_3) is co-fed with NO, N_2O always appears as the first NO reduction product, along with nitrogen. The onset temperature for NO reduction (and N_2O evolution) depends on the nature of the reductant (with H_2 showing a greater reactivity than NH_3) and on the catalyst as well (Pt-based catalyst being more active than the corresponding Rh-containing sample). The pathways involved in N_2O formation are those already depicted for NO alone: the reductant leads to the reduction of the PGM sites, leading to NO dissociation into N- and O-adatoms. However in this case O-adatoms are scavenged by the reductant, and nitrite formation is reduced with respect to the reaction of pure NO. When the reactivity of the reductant is not very high (*i.e.*, near the onset temperature for NO reduction) the PGM sites are not fully reduced and hence there is the chance for N-adspecies to react with undissociated NO molecules, leading to N_2O evolution. At variance, when the reductant maintains the PGM sites in a fully reduced state, NO is readily dissociated and the reduction products are N_2 and NH_3 (when H_2 is used as reductant) or N_2 (reductant NH_3). It is concluded that the formation of N_2O occurs only in the presence of undissociated NO molecules, *i.e.*, when the PGM sites are not fully reduced. Accordingly, the temperature onset for N_2O formation depends on the reactivity of the reductant: it is seen at lower temperature when H_2 is used as a reductant instead of NH_3, due to the higher reactivity of H_2.

Notably, no N_2O formation was observed under our experimental conditions upon reduction with both H_2 and NH_3 of nitrates stored over both Pt- and Rh-BaO/Al_2O_3. Notably, the intermediacy of species like ammonium nitrate in N_2O formation is apparently ruled out since no N_2O was observed in the NH_3/nitrate reaction over both Pt- and Rh-based catalytic materials. It has been suggested that the reduction of the stored NO_x species is involved in the release of NO from the stored NO_x as first step, followed by the reduction of the released NO over the PGM sites [16–19]. The reduction process is initiated by the reduction of the PGM sites by the reductant, which precedes the NO release. The observation that no N_2O is formed in this case may be explained by considering that the rate determining step in the reduction of the stored NO_x is the NO release step [16]; once NO is released, it is readily converted to the reduction products. Accordingly the gas-phase concentration of NO is small and this prevents N_2O formation. However under real operating conditions, where significant temperature effects may take place, significant release of NO may take place and accordingly N_2O formation is likely to occur.

4. Materials and Methods

Model Pt-BaO/Al$_2$O$_3$ (1/20/100 $w/w/w$) and Rh-BaO/Al$_2$O$_3$ (0.5/20/100 $w/w/w$) samples were prepared by incipient wetness impregnation of a commercial γ-alumina (Versal 250 UOP, Des Plaines, IL, USA) support with Pt(NO$_2$)$_2$(NH$_3$)$_2$ (Strem Chemicals, 5% w/w, Newburyport, MA, USA) and Rh(NO$_3$)$_3$ (Sigma Aldrich, 10% w/w, St. Louis, MO, USA) aqueous solutions, respectively, followed by impregnation with Ba(CH$_3$COO)$_2$ (Sigma Aldrich, 99% w/w) aqueous solution. After each impregnation step the samples were dried at 80 °C overnight and calcined at 500 °C for 5 h. The obtained catalysts were characterized by Brunauer-Emmett-Teller (BET) analysis for specific surface area and H$_2$ chemisorption for metal dispersion and results are reported in Table 1.

The reactivity of NO alone and with different reductants (H$_2$, NH$_3$) was investigated *via* temperature programmed reaction (TPR) experiments. In the case of NO alone, the catalysts were heated in the temperature range 40–400 °C (heating rate 10 °C/min) under a flow of NO (1000 ppm in He, NO-TPR). Before performing the experiments, the catalysts were reduced with H$_2$ (2000 ppm in He) at 350 °C for 30 min and cooled to 40 °C in He. Such experiments were carried out in a flow micro-reactor apparatus consisting of an home made quartz tube reactor (7 mm I.D.) connected to a mass spectrometer (Omnistar 200, Pfeiffer Vacuum, Asslar, Germany), a micro-gas chromatograph (Agilent 3000A, Santa Clara, CA, USA) and a ultraviolet (UV) analyzer (Limas 11HW, ABB, Zurich, Switzerland) for the on-line analysis of the outlet gases (rig #1). An amount of 60 mg of catalyst was used with a flow rate of 100 cc/min (at 0 °C and 1 atm).

Information on the surface species formed during the experiments was collected in this case by performing the NO-TPR experiments under the same experimental conditions inside an IR reactor cell (ISRI Infrared Reactor, Granger, IN, USA) by using the catalysts in form of self-supported wafers (15 mg, diameter = 13 mm, thickness = 0.1 mm) (rig #2). The IR measurements were collected with a FT-IR Vertex 70 (Bruker, Billerica, MA, USA) spectrometer with 4 cm^{-1} spectral resolution and accumulation of 64 scans using a mercury-cadmium-telluride (MCT) detector.

The reactivity of NO with H$_2$ and NH$_3$ (NO/H$_2$-TPR and NO/NH$_3$-TPR, respectively) was investigated by heating the catalyst samples in a flow of NO (1000 ppm) with H$_2$ (2000 ppm) or NH$_3$ (1000 ppm) in Ar in the temperature range 40–400 °C (heating rate 4 °C/min). These experiments were carried out in an *operando* FT-IR spectroscopy setup (rig #3) consisting in a "Sandwich" IR cell containing the catalyst (15 mg) in the form of a self-supported wafer; the gases leaving the cell were analyzed by a mass spectrometer (ThermoStar TM GSD 301, Pfeiffer Vacuum) and a chemiluminescence analyzer (42i-HL MEGATEC, Thermo Fisher Scientific, Waltham, MA, USA). Total flow was 25 cm^3 · min^{-1} (at 1 atm and 0 °C). Surface FT-IR spectra were collected with a FT-IR Nicolet Nexus spectrometer (Thermo Fisher Scientific) with 4 cm^{-1} spectral resolution and accumulation of 64 scans using a deuterated-triglycine sulfate (DTGS) detector.

The same apparatus was used to perform temperature programmed surface reaction (H$_2$- and NH$_3$-TPSR) experiments. In this case the catalysts were at first saturated with NO$_x$ species at 350 °C by feeding NO (1000 ppm) in O$_2$ (3% v/v) + Ar. This resulted in the formation of surface nitrate species. Then the catalyst was cooled under He at 40 °C and subsequently heated in flowing H$_2$ (2000 ppm in Ar, H$_2$-TPSR) or NH$_3$ (1000 ppm in Ar, NH$_3$-TPSR) up to 400 °C (heating rate 4 °C/min). Notably, due to the low reductant concentration, experiments were carried out under nearly isothermal conditions, *i.e.*, in the absence of significant temperature effects during the experiments. Further details concerning apparatuses and procedures can be found in [10,15].

Since the fresh catalyst samples were covered by carbonate species, before the experiments few adsorption-reduction cycles under isothermal conditions at 350 °C, (*i.e.*, NO/O$_2$ (1000 ppm NO + O$_2$ 3%, v/v) and H$_2$ (2000 ppm)) were performed. This procedure partially replaces the originally present BaCO$_3$ species with BaO/Ba(OH)$_2$ species [20].

5. Conclusions

In this paper, mechanistic aspects involved in the formation of N$_2$O over Pt-BaO/Al$_2$O$_3$ and Rh-BaO/Al$_2$O$_3$ model NSR catalysts were investigated. In particular, the reactivity of both gas-phase NO and stored nitrates were investigated by using H$_2$ and NH$_3$ as reductants. Transient experiments with simultaneous gas phase and surface analysis using FT-IR spectroscopy were used to correlate the concentration of the gas-phase species with adsorbed surface species.

The data herein obtained provides clear indication of the fact that over both Rh- and Pt-BaO/Al$_2$O$_3$ catalysts, N$_2$O formation involves the presence of gas-phase NO. In fact under our experimental conditions no nitrous oxide formation was observed upon the reduction of the stored nitrate species. The data show that N$_2$O formation involves the coupling of undissociated NO molecules with N-adspecies formed upon NO dissociation onto reduced PGM sites. The reductant keeps the PGM sites in a reduced state by scavenging O-adatoms formed upon NO dissociation, the initial step of the process. However when complete dissociation of gas-phase NO occurs, N$_2$O formation is prevented. In fact N$_2$O formation is observed at temperatures where the reductants start to be active in the reduction of the PGM sites, whereas at higher temperatures, where the reductants effectively keep the PGM sites in a fully reduced state, complete dissociation of NO takes place and N$_2$O formation is not effective. This explains why N_2O formation is observed at low temperatures and then disappears at high temperatures.

Over both Rh- and Pt-BaO/Al$_2$O$_3$ catalysts N_2O formation is observed at lower temperatures with H_2 than with NH_3, in view of the higher reactivity of hydrogen in the reduction of the PGM sites. Also, N_2O formation occurs at lower temperatures onto Pt-containing catalyst if compared to the Rh-based sample, due to the higher reducibility of Pt *vs.* Rh.

In the absence of reducing agents, N_2O can be formed from NO provided that reduced PGM sites are present. In such a case O-adatoms formed upon NO decomposition are scavenged by NO itself leading to the formation of nitrite species so that the formation of N_2O occurs according to the stoichiometry of the NO disproportionation reaction Equation (1).

Finally, the fact that N_2O formation was not observed upon the reduction of stored nitrates with both H_2 and NH_3 is due to the slow rate of the NO release process from the stored species. This limits the NO gas-phase concentration during the reduction of the stored species thus preventing N_2O formation.

Acknowledgments: R. Matarrese gratefully acknowledges "Fondazione Banca del Monte di Lombardia" for financial support for the FT-IR equipment.

Author Contributions: The experimental study was conducted by L.K. and R.M. who were also responsible for preparation of the manuscript. The supervision during catalytic activity was provided by M.D. and L.C. The analysis and interpretation of the data was carried out by all authors in particular M.D., L.L., and P.F.

Conflicts of Interest: The authors declare no conflicts of interest.

References

1. Forzatti, P.; Lietti, L.; Castoldi, L. Storage and Reduction of NO$_x$ Over LNT Catalysts. *Catal. Lett.* **2015**, *145*, 483–504.
2. Castoldi, L.; Bonzi, R.; Lietti, L.; Forzatti, P.; Morandi, S.; Ghiotti, G.; Dzwigaj, S. Catalytic behaviour of hybrid LNT/SCR systems: Reactivity and *in situ* FTIR study. *J. Catal.* **2011**, *282*, 128–144.
3. Bártová, S.; Kočí, P.; Mráček, D.; Marek, M.; Pihl, J.A.; Choi, J.S.; Toops, T.J.; Partridge, W.P. New insights on N$_2$O formation pathways during lean/rich cycling of a commercial lean NO$_x$ trap catalyst. *Catal. Today* **2014**, *231*, 145–154.
4. Choi, J.S.; Partridge, W.P.; Pihl, J.A.; Kim, M.Y.; Kočí, P.; Daw, C.S. Spatiotemporal distribution of NO$_x$ storage and impact on NH$_3$ and N$_2$O selectivities during lean/rich cycling of a Ba-based lean NO$_x$ trap catalyst. *Catal. Today* **2012**, *184*, 20–26.
5. Clayton, R.D.; Harold, M.P.; Balakotaiah, V. Performance Features of Pt/BaO Lean NO$_x$ Trap with Hydrogen as Reductant. *AIChE J.* **2009**, *55*, 687–700.
6. Dasari, P.; Muncrief, R.; Harold, M.P. Cyclic lean reduction of NO by CO in Excess H$_2$O on Pt-Rh/Ba/Al$_2$O$_3$: Elucidating mechanistic features and catalyst performance. *Top. Catal.* **2013**, *56*, 1922–1936.

7. Kočí, P.; Bártová, Š.; Mráček, D.; Marek, M.; Choi, J.S.; Kim, M.Y.; Pihl, J.A.; Partridge, W.P. Effective Model for Prediction of N_2O and NH_3 Formation During the Regeneration of NO_x Storage Catalyst. *Top. Catal.* **2013**, *56*, 118–124.

8. Kumar, A.; Harold, M.P.; Balakotaiah, V. Isotopic studies of NO_x storage and reduction on $Pt/BaO/Al_2O_3$ catalyst using temporal analysis of products. *J. Catal.* **2010**, *270*, 214–223.

9. Mráček, D.; Kočí, P.; Choi, J.S.; Partridge, W.P. New operation strategy for driving the selectivity of NO_x reduction to N_2, NH_3 or N_2O during lean/rich cycling of a lean NO_x trap catalyst. *Appl. Catal. B* **2016**, *182*, 109–114.

10. Kubiak, L.; Righini, L.; Castoldi, L.; Matarrese, R.; Forzatti, P.; Lietti, L.; Daturi, M. Mechanistic aspects of N_2O formation over Pt-based Lean NO_x Trap catalysts. *Top. Catal.* **2015**. submitted.

11. Taylor, K.C. Nitric Oxide Catalysis in Automotive Exhaust Systems. *Catal. Rev.* **1993**, *35*, 457–481.

12. Konsolakis, M.; Yentekakis, I.V. The Reduction of NO by Propene over Ba-Promoted Pt/γ-Al_2O_3 Catalysts. *J. Catal.* **2001**, *198*, 142–150.

13. Lietti, L.; Artioli, N.; Righini, L.; Castoldi, L.; Forzatti, P. Pathways for N_2 and N_2O Formation during the Reduction of NO_x over Pt–Ba/Al_2O_3 LNT Catalysts Investigated by Labelling Isotopic Experiments. *Ind. Eng. Chem. Res.* **2012**, *51*, 7597–7605.

14. Kubiak, L.; Righini, L.; Castoldi, L.; Matarrese, R.; Forzatti, P.; Lietti, L.; Daturi, M. In-depth insights into N_2O formation over Rh-Ba/Al_2O_3 catalyst under operando FT-IR conditions and micro reactor study. Unpublished. 2016.

15. Lietti, L.; Nova, I.; Forzatti, P. Role of ammonia in the reduction by hydrogen of NO_x stored over Pt-Ba/Al_2O_3 lean NO_x trap catalysts. *J. Catal.* **2008**, *257*, 270–282.

16. Castoldi, L.; Righini, L.; Matarrese, R.; Lietti, L.; Forzatti, P. Mechanistic aspects of the release and the reduction of NO_x stored on Pt-Ba/Al_2O_3. *J. Catal.* **2015**, *328*, 270–279.

17. Epling, W.S.; Campbell, L.E.; Yezerets, A.; Currier, N.W.; Parks, J.E. Overview of the fundamental reactions and degradation mechanisms of NO_x storage/reduction catalysts. *Catal. Rev. Eng.* **2004**, *46*, 163–245.

18. Kabin, K.S.; Muncrief, R.L.; Harold, M.P. NO_x storage and reduction on a Pt/BaO/alumina monolithic storage catalyst. *Catal. Today* **2004**, *96*, 79–89.

19. Muncrief, R.L.; Kabin, K.S.; Harold, M.P. NO_x storage and reduction with propylene on Pt/BaO/alumina. *AIChE J.* **2004**, *50*, 2526–2540.

20. Lietti, L.; Forzatti, P.; Nova, I.; Tronconi, E. NO_x Storage Reduction over PtBa/γ-Al_2O_3 Catalyst. *J. Catal.* **2001**, *204*, 175–191.

Methanol Reforming over Cobalt Catalysts Prepared from Fumarate Precursors: TPD Investigation

Eftichia Papadopoulou and Theophilos Ioannides

Abstract: Temperature-programmed desorption (TPD) was employed to investigate adsorption characteristics of CH_3OH, H_2O, H_2, CO_2 and CO on cobalt-manganese oxide catalysts prepared through mixed Co-Mn fumarate precursors either by pyrolysis or oxidation and oxidation/reduction pretreatment. Pyrolysis temperature and Co/Mn ratio were the variable synthesis parameters. Adsorption of methanol, water and CO_2 was carried out at room temperature. Adsorption of H_2 and H_2O was carried out at 25 and 300 °C. Adsorption of CO was carried out at 25 and 150 °C. The goal of the work was to gain insight on the observed differences in the performance of the aforementioned catalysts in methanol steam reforming. TPD results indicated that activity differences are mostly related to variation in the number density of active sites, which are able to adsorb and decompose methanol.

Reprinted from *Catalysts*. Cite as: Papadopoulou, E.; Ioannides, T. Methanol Reforming over Cobalt Catalysts Prepared from Fumarate Precursors: TPD Investigation. *Catalysts* **2016**, *6*, 33.

1. Introduction

Mixed cobalt-manganese fumarate salts are useful precursors leading to catalysts with different structure depending on the type of surrounding atmosphere during activation [1]. Thus, activation in air leads to burn-off of the fumarate group and concomitant formation of mixed $Co_xMn_{1-x}O_y$ spinel oxides, while activation in inert gas leads to pyrolysis of the fumarate group and formation of species with lower oxidation state, such as metallic cobalt, mixed oxides of Co^{2+} and Mn^{2+} and residual carbon. Combination of *in-situ* XRD, H_2-TPR and methanol-TPR has shown that catalysts produced by pyrolysis are almost fully reduced [1]. Thus, catalysts derived from pyrolysis do not need prereduction and are more active than those with an initial spinel structure in the reaction of steam reforming of ethanol or methanol [2]. State-of-the-art catalysts for steam reforming of methanol are copper-based and operate at 250–300 °C, while ethanol reforming requires higher temperatures of the order of 600 °C [3–9]. Cobalt is a less efficient catalyst than copper in the steam reforming of methanol operating at temperatures around 400 °C [2].

From a mechanistic point of view, adsorption is a key step in catalytic reactions. Hence, study of adsorption and desorption of relevant molecules on catalytic surfaces can provide insight on the population and intrinsic properties of active sites. TPD, in particular, is a standard technique via which one can obtain information concerning: (i) adsorption site homogeneity, as reflected in the presence of one or more desorption peaks; (ii) strength of the adsorbate-surface bond, as reflected in the peak temperature of desorbed species; and (iii) number density of adsorption sites, as reflected in the amount of desorbed species. It is especially useful in comparative parametric studies of a catalyst family.

Adsorption of methanol on a Co(0001) surface takes place as methoxide via OH bond scission. During heating, a small amount of methanol desorbs molecularly, while the majority of methoxide decomposes to CO and hydrogen [10]. Infrared spectra produced by adsorbed species formed during the exposure of silica-supported Co to methanol have been obtained by Bliholder *et al.* [11]. Methanol was found to adsorb to a small extent on the silica support -probably as a methoxide, while varying mixtures of methoxide, acyl, and chemisorbed CO species were produced on Co. The interaction between Co_3O_4 or CoO with methanol under either atmospheric or high vacuum conditions was examined by Natile *et al.* [12]. Methanol was found to chemisorb mainly molecularly on cobalt oxide surfaces, while its dissociation became evident at higher temperatures. In the case of Co_3O_4, the presence of formate and formaldehyde species was evident in the temperature range 200–350 °C, whereas under high vacuum conditions, formaldehyde and several decomposition and fragmentation products were observed along with carbon oxides.

Adsorption of water on a hexagonal Co(1120) surface was studied by means of photoelectron spectroscopies (XPS, UPS) by Grellner *et al.* [13]. Molecular adsorption of water at 100 K was accompanied by the formation of small amounts of OH in the submonolayer range. When the temperature is increased, desorption of the multilayer occurs first at 150 K and OH remains on the surface. Disproportionation of OH takes place at 270 K leaving oxygen on the surface. A systematic study of the adsorption and dissociation of water on transition and noble metal dimers was presented by Heras *et al.* [14].

Activated hydrogen chemisorption on unsupported and supported (silica, alumina, titania, magnesia, and carbon supports) cobalt catalysts prepared by a variety of techniques has been reported by Bartholomew *et al.* [15–17]. The surface interaction of CO, CO_2 and H_2 with the perovskite-type oxide $LaCoO_x$ has been studied as a function of reduction temperature using XPS and TPD by Tejuca *et al.* [18]. Hydrogen was found to adsorb on cobalt both weakly and strongly (desorption peaks at 70 °C and above 200 °C). Dissociative adsorption yielding hydroxyl groups was also detected. Narayanan *et al.* [19] studied H_2 adsorption at 100 °C on Co/Al_2O_3 catalysts with varying cobalt content (10%–50%). The amount of adsorbed H_2 was

15 $\mu mol \cdot g^{-1}$ at 20% cobalt loading and increased to 67 $\mu mol \cdot g^{-1}$ for 50% cobalt loading. Hydrogen adsorption/desorption characteristics on Co-TiN nanocomposite particles have been studied using TPD by Sakka et al. [20]. Hydrogen desorption was observed in the temperature range 100–320 °C. Activated chemisorption of hydrogen on prereduced $MnFe_2O_4$ spinel oxides was reported by Soong et al. [21]. Hydrogen desorption at 570–630 °C was attributed to chemisorbed species on MnO.

Gauthier et al. [22] investigated CO adsorption on PtCo(111) surfaces by scanning tunneling microscopy. It was found that CO molecules reside exclusively on top of Pt sites and never on Co. High-pressure, in-situ diffuse reflectance, Fourier Transform Infrared Spectroscopy was employed by Jiang et al. [23] to study CO adsorption on samples derived from precipitated cobalt-manganese oxides of different Co/Mn ratios reduced by H_2. According to their results the adsorption features of the samples vary significantly with manganese loading. CO linearly adsorbed, bridged and multiple-bridged on $Co°$ sites was identified. Mohana et al. [24] found that CO adsorption at room temperature on cobalt particles supported on MgO leads mainly to the formation of linearly adsorbed species, while the disproportionation reaction accompanied by carbon deposition already takes place at room temperature. Carbon deposition on cobalt catalysts in Fischer-Tropsch and steam reforming reactions is well-documented [25–31]. Regarding methanol reforming, the pathway of carbon deposition on cobalt catalysts appears to be the Boudouard reaction since CO is the main reaction product. The structure of carbon deposits originating from CO depends on reaction temperature with amorphous and filamentous carbon prevailing at low temperatures (350–600 °C), which are relevant to methanol steam reforming [31]. Formation of filamentous carbon does not lead to catalyst deactivation but rather to reactor plugging leading to excessive pressure drop.

In the present study, TPD experiments of pre-adsorbed CH_3OH, H_2O, H_2, CO_2 and CO were employed in order to examine the adsorptive properties of cobalt catalysts prepared through mixed cobalt-manganese fumarate precursors by activation under oxidative or reducing conditions. The objective of this study was to investigate the effect of catalyst synthesis parameters on the corresponding adsorptive properties. More specifically, the examined catalyst synthesis parameters were: (i) the activation type, pyrolysis or calcination; (ii) the Co/Mn ratio; and (iii) the pyrolysis temperature. These are the main parameters influencing the structure [1] and activity [2] of the specific catalysts in methanol steam reforming. To our knowledge, an extensive investigation of the adsorption of a variety of catalysis relevant molecules on Co-Mn based catalysts has not been reported previously.

2. Results and Discussion

2.1. TPD of Adsorbed CO or CO_2

No significant CO adsorption was found on the catalysts (both pyrolyzed and preoxidized ones) following exposure to CO at room temperature. As an example, TPD results following CO adsorption on the CoMn11AFp600 catalyst at 25 and 150 °C are presented in Figure 1a,b. The curve called "blank" corresponds to the amounts of CO and CO_2 observed during TPD without any prior CO adsorption. CO and CO_2 production in the blank experiment is due to residual decomposition of organic species originating from the fumarate precursor. It can be observed that after adsorption of CO at room temperature, there is minimal desorption of CO during TPD (less than 20 ppm of CO in the gaseous stream) and no CO_2 desorption. Following CO adsorption at 150 °C, both CO and CO_2 were found in the TPD profile. CO desorbs in the form of two peaks at 50 °C and 140 °C, while CO_2 appears with a main peak at 200 °C followed by a shoulder at 350 °C and a smaller peak at 500 °C. The origin of CO_2 could be either the Boudouard reaction or oxidation of adsorbed CO by surface oxygen. Based on literature findings, the occurrence of the Boudouard reaction is highly probable. For example, formation of carbon on a Co/Al_2O_3 catalyst by CO disproportionation at 230 °C has been reported by Nakamura *et al.* [32]. This temperature is in the same range as the one shown in Figure 1b regarding CO_2 formation. CO disproportionation on Co/MgO catalysts has been found to take place already at room temperature [24]. Since the catalyst samples of this work, however, contain carbon in their composition (residual carbon from fumarate pyrolysis), it is not possible to measure any carbon formed on the catalysts from CO disproportionation via subsequent temperature-programmed oxidation.

The corresponding results of TPD following CO_2 adsorption at 25 °C on catalysts prepared via oxidation and oxidation/reduction or pyrolysis are given in Figure 2a,b, respectively. CO_2 profiles from oxidized/reduced catalysts (Figure 2a) show a main peak at ~100 °C and a smaller high temperature peak above 500 °C. CO_2 profiles from pyrolyzed catalysts are quite complicated and broad from 30 to 600 °C with multiple desorption peaks. The population of sites that adsorb CO_2 strongly appears to decrease with increase of pyrolysis temperature and this leads to concomitant decrease of the amount of desorbed CO_2 with increase of pyrolysis temperature. More specifically, the amount of desorbed CO_2 is 240 $\mu mol \cdot g^{-1}$ for CoMn11AFp500 and decreases to 100 and 57 $\mu mol \cdot g^{-1}$ for catalysts prepared by pyrolysis at 600 and 700 °C, respectively. At the same time, the specific surface area of pyrolyzed samples is more or less independent of pyrolysis temperature in the range of 200–220 $m^2 \cdot g^{-1}$, as measured by the BET method. This indicates that pyrolysis temperature affects mostly active sites for adsorption of CO_2 and not the exposed surface area in general. The amount of adsorbed CO_2 on catalysts prepared by oxidation/reduction is

17–19 $\mu mol \cdot g^{-1}$, *i.e.*, considerably smaller than the one found over the pyrolyzed catalysts. This is in line with the smaller specific surface area of these samples by one order of magnitude.

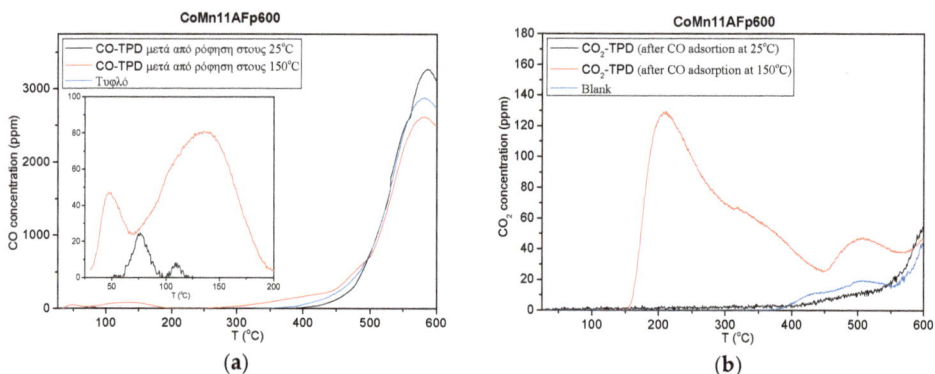

Figure 1. TPD following CO adsorption at 25 °C or 150 °C on CoMn11AFp600 catalyst: (**a**) CO desorption; and (**b**) CO_2 desorption.

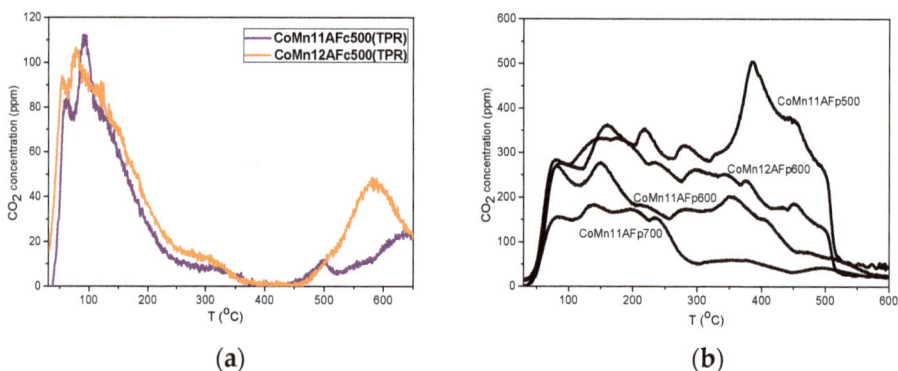

Figure 2. TPD of CO_2 following its adsorption at 25 °C on (**a**) CoMn1*x*AFc500 (TPR), *x* = 1, 2; and (**b**) CoMn11AFp*i*, *i* = 500, 600, 700 °C and CoMn12AFp600.

2.2. TPD of Adsorbed H_2

H_2-TPD studies were carried out following H_2 adsorption at 25 or 300 °C under a flow of pure hydrogen. Representative TPD profiles of hydrogen are presented in Figure 3. The oxidized/reduced sample (CoMn11AFc500(TPR)) adsorbed no measurable amount of hydrogen at room temperature and trace amounts at 300 °C. The pyrolyzed cobalt-only sample indicates the presence of rather weakly bound hydrogen desorbing with peak at ~70 °C. Increase of adsorption temperature to 300 °C leads to population of more strongly-bound adsorbed hydrogen, as evidenced by the appearance of a shoulder at 100–200 °C. TPD profiles from the manganese-only sample show the presence of strongly-bound hydrogen desorbing in the range of 400–650 °C after adsorption at room temperature. Increase of adsorption temperature to 300 °C leads to the appearance of an intermediate-strength state with desorption at ~300 °C. The TPD profile from the CoMn11AFp600 sample, which contains both cobalt and manganese, incorporates features that are attributable to the presence of both cobalt crystallites and MnO. Strongly-bound hydrogen desorbs at 460 °C and is larger in quantity compared to the manganese-only sample. Therefore, the observed profile is not just the sum of isolated contributions of cobalt and MnO species, but is influenced by mutual interactions. Table 1 presents the amounts of desorbed hydrogen during TPD after adsorption at 25 and 300 °C. The oxidized catalyst which had been reduced prior to H2 adsorption adsorbs no H_2 at room temperature and 10 μmol·g^{-1} at 300 °C, which is one to two orders of magnitude smaller than the corresponding values of catalysts from the same precursor prepared by pyrolysis. Concerning the effect of adsorption temperature, it is observed that the amount of desorbed hydrogen increases by up to 350% with increase of adsorption temperature from 25 to 300 °C. The smallest increase is found for the sample adsorbing the largest amount of hydrogen at room temperature, *i.e.*, CoMn11AFp600. The amounts of desorbed hydrogen in samples containing both cobalt and manganese are considerably larger than the ones found over single-component samples, indicating the presence of synergy and creation of additional adsorption sites. Contrary to what was found in the case of CO_2 adsorption, the amount of adsorbed hydrogen does not decrease with increase of pyrolysis temperature, indicating that at least some of the adsorption sites for hydrogen and CO_2 are not identical. Using the data in Table 1, estimates of the maximum dispersion of cobalt in the various catalysts can be provided assuming that no adsorption takes place on MnO sites in the case of Co-Mn catalysts. Taking that the stoichiometry of hydrogen adsorption is one hydrogen atom per one surface cobalt atom, the dispersion (H/Co) ratio is ~2% for the cobalt-only catalyst and becomes even higher than 20% for Co-Mn catalysts pyrolyzed at 600 °C.

H2-TPD results for the cobalt-only catalysts are in agreement to those reported by Popova and Babenkova for thermal desorption of preadsorbed hydrogen on a-Co and b-Co prepared by formate decomposition in a hydrogen flow at 300, 350 and 600 °C, whereas hydrogen desorption was completed at 300 °C [24].

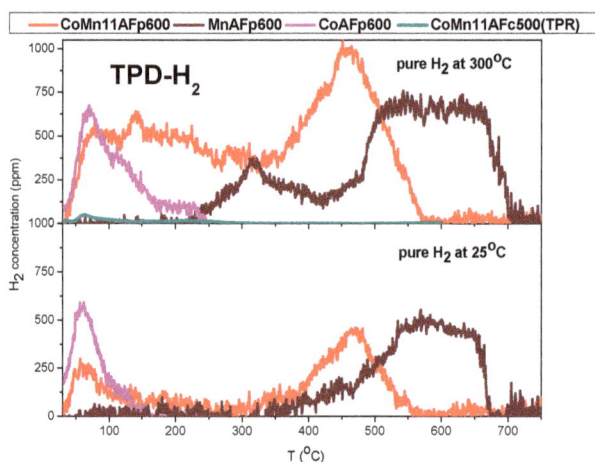

Figure 3. H$_2$-TPD profiles following its adsorption at 25 °C and 300 °C on CoMn11AFp600, CoAFp600, MnAFp600 and CoMn11AFc500 (TPR) samples.

Table 1. Amounts of hydrogen desorbed during TPD after adsorption at 25 and 300 °C.

Catalyst	H$_2$-TPD Adsorption at 25 °C		H$_2$-TPD Adsorption at 300 °C	
	μmol H$_2$ g^{-1}	H/Co	μmol H$_2$ g^{-1}	H/Co
CoMn11AFp500	174	0.058	492	0.164
CoMn11AFp600	189	0.060	661	0.210
CoMn11AFp700	328	0.104	371	0.118
CoMn12AFp600	195	0.096	478	0.236
CoMn11AFc500(TPR)	0	0	10	0.0034
CoAFp600	34	0.00116	61	0.0212
MnAFp600	100	0.034 *	195	0.066 *

* H/Mn.

2.3. TPD of Adsorbed H$_2$O

Water adsorption was carried out at 25 and 300 °C. A characteristic feature of water adsorption at 300 °C over catalysts prepared by pyrolysis is the accompanying appearance of hydrogen in the gas phase indicative of reactive adsorption according to:

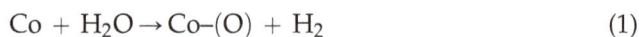

$$Co + H_2O \rightarrow Co-(O) + H_2 \qquad (1)$$

TPD profiles of H$_2$O and H$_2$ after water adsorption at room temperature over pyrolyzed catalysts are shown in Figure 4. H$_2$O profiles are characterized by a main peak with maximum at 90–100 °C followed by a broad descending feature up to 450 °C. Increase of pyrolysis temperature leads to decrease of the amount of adsorbed water. This trend is analogous to what was found for CO_2 adsorption (Section 2.1). The profiles of H_2 during TPD after water adsorption at room temperature over pyrolyzed catalysts are also presented in Figure 4. Hydrogen desorption is observed mostly above 400 °C with CoMn11AFp600 catalyst showing also minor hydrogen desorption at 60 °C. Since TPD profiles following H_2 adsorption are characterized by hydrogen peaks in the 400–600 °C range, it is not clear whether appearance of hydrogen in the gas phase is desorption or reaction limited. However, taking into account that water decomposition does take place at 300 °C, it is most probable that the profile of hydrogen is desorption limited. In addition, the absence of hydrogen in the low-temperature range implies that adsorbed water decomposition takes place at temperatures higher than ~150 °C. The amount of produced hydrogen during H_2O-TPD was 0.2–0.3 mmol·g^{-1} while the amount of desorbed hydrogen after its adsorption at room temperature was 0.17–0.33 mmol·g^{-1}. This implies that the extent of adsorbed water decomposition is related to the number of surface centers available for adsorption of produced H_2.

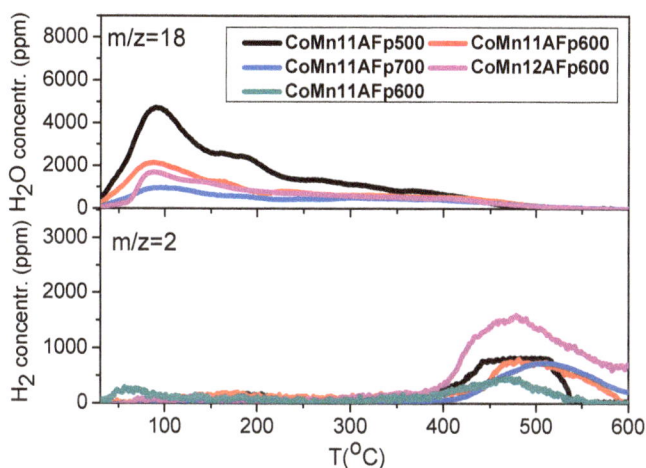

Figure 4. TPD profiles of H_2O and H_2 after water adsorption at 25 °C over pyrolyzed catalysts.

Water adsorption at 300 °C was accompanied by gaseous H_2 production for all pyrolyzed catalysts with the exception of MnAFp600. TPD profiles of H_2O and H_2 after water adsorption at 300 °C over pyrolyzed catalysts are shown in Figure 5. TPD profiles of water consist of a main peak at 60–100 °C followed by a tail up to 500 °C. The profiles of hydrogen during TPD are quite broad and they show measurable desorption of hydrogen at the whole temperature range from 30 to 600 °C. These are in general agreement to TPD profiles obtained following adsorption of hydrogen at 300 °C.

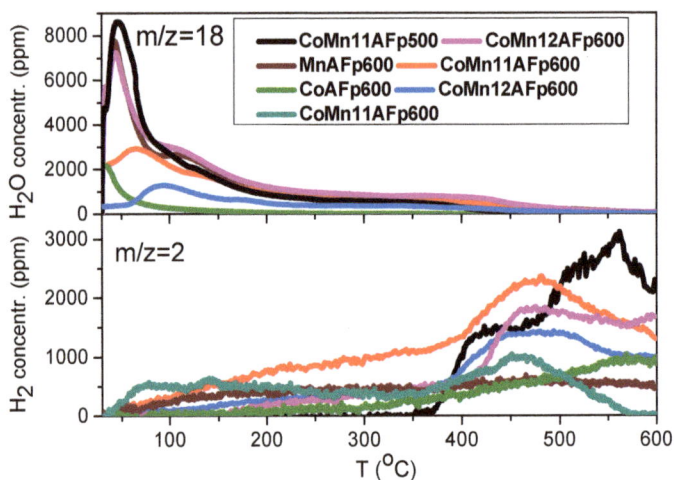

Figure 5. TPD profiles of H_2O and H_2 after water adsorption at 300 °C over pyrolyzed catalysts.

Quantitative results of the total adsorbed water (mmol· g^{-1}) at 300 °C (sum of water absorbed at 300 °C and during cooling to RT), hydrogen produced during H_2O adsorption, and subsequent TPD as a function of catalyst content in Mn are given in Figure 6a,b. The total adsorbed water is four times higher on MnAFp600 than on CoAF600, although this difference is mostly due to water adsorbed during cooling. The highest amount of water adsorption is, however, observed over the Co-Mn catalysts. In an analogous manner, the amount of formed H_2, *i.e.*, the sum of hydrogen produced during adsorption and desorbed during TPD, is considerably higher over the Co-Mn catalysts compared to the cobalt-only catalyst and especially to the Mn-only sample.

The CoAFp600 catalyst contains 11.6 mmol Co· g^{-1} and produced 0.4 mmol H_2· g^{-1} during water adsorption and TPD. Assuming one adsorbed hydrogen atom per surface cobalt atom, the resulting H/Co ratio is 0.069. The corresponding H/Co ratio for hydrogen adsorption at 300 °C is 0.021 (Table 1). In the case of the CoMn11AFp600 catalyst, the total amount of hydrogen was 0.8 mmol· g^{-1}, which compares well with the amount of hydrogen adsorbed during hydrogen adsorption (0.66 mmol· g^{-1}).

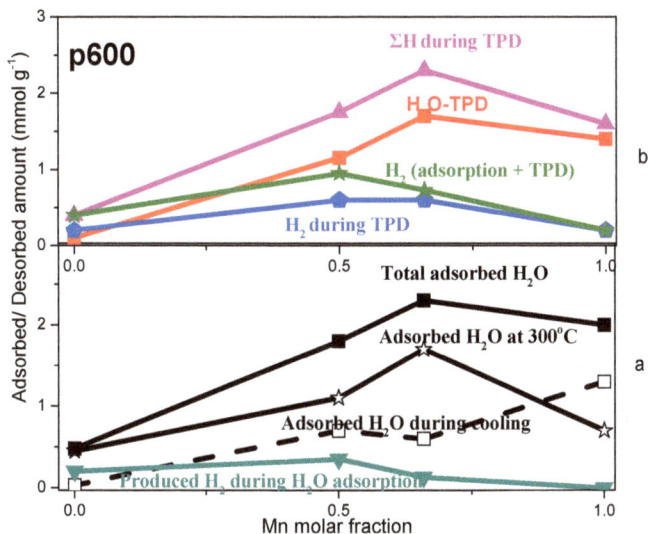

Figure 6. (a) Total absorbed H$_2$O at 300 °C and H$_2$ produced during water adsorption; and (b) H$_2$O and H$_2$ desorbed during TPD from CoAF, MnAF, CoMn11AF and CoMn12AF pyrolyzed at 600 °C.

The effect of pyrolysis temperature of the CoMn11AF catalyst on adsorbed/desorbed water and H$_2$ amounts is presented in Figure 7a,b, respectively. It can be observed that although the amount of total adsorbed water does not change appreciably with variation of pyrolysis temperature, there are changes in the "strength" of water adsorption. Thus, the amount of water desorbed during TPD decreases, because there is an increase in the amount of weakly adsorbed water, which desorbs already at room temperature before initiation of the TPD run. Similarly, the amount of water adsorbing during cooling decreases and the amount of water adsorbing at 300 °C increases with increase of pyrolysis temperature. This also relates to the fact that the extent of cobalt reduction increases with increase of pyrolysis temperature.

The effect of catalyst activation procedure on quantitative behavior of water adsorption is presented in Figure 8. Three catalysts are compared, all activated at 500 °C: CoMn11AFp500 prepared by pyrolysis, CoMn11AFc500(AIR) prepared by oxidation of the precursor and CoMn11AFc500(TPR) prepared from CoMn11AFc500(AIR) by subsequent reduction by H$_2$ up to 600 °C. The two latter catalysts adsorb significantly less water than CoMn11AFp500 and, in addition, they produce almost no hydrogen during exposure to water at 300 °C.

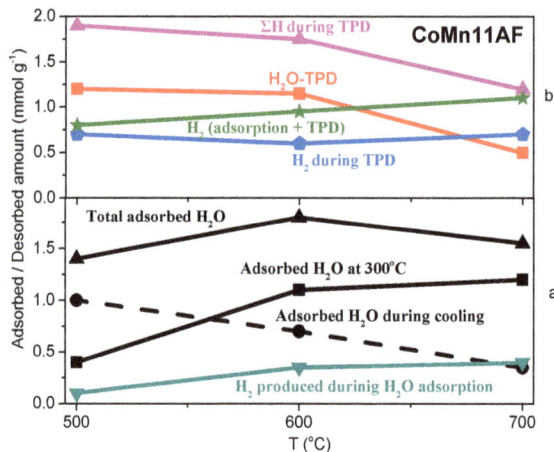

Figure 7. (a) Total absorbed H_2O at 300 °C and H_2 produced during water adsorption; and (b) H_2O desorbed and H_2 producing during TPD. CoMn11AFpi with i = 500, 600, 700 °C.

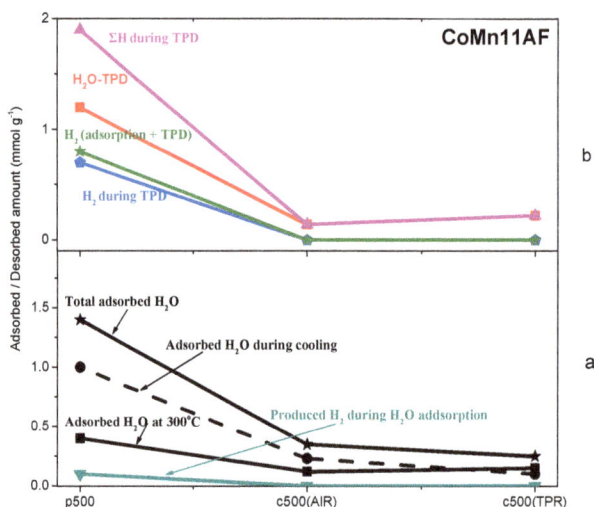

Figure 8. (a) Total absorbed H_2O at 300 °C and H_2 produced during water adsorption; and (b) H_2O desorbed and H_2 producing during TPD from CoMn11AFc500, CoMn11AFc500(AIR) and CoMn11AFc500(TPR).

2.4. TPD of Adsorbed CH_3OH

CH_3OH, CO, H_2, CO_2 and H_2O profiles during CH_3OH-TPD for CoMn11F samples pyrolyzed at 500 or 700 °C are shown in Figure 9a,b, respectively. Dashed lines in Figure 9a represent CO_2, H_2, CO and H_2O production during the blank experiment (without prior methanol adsorption). The appearance of these molecules is due to residual pyrolysis of the fumarate precursor. TPD profiles following adsorption of methanol at room temperature indicated the presence of CO, CO_2, H_2 and H_2O in addition to methanol. Methanol desorption profiles are composed of a main peak at 90 °C and a less intense second peak in the form of a shoulder at 160 °C. Methanol desorption is completed at 300 °C for CoMn11AFp500 and at 210 °C for CoMn11AFp700. At the same time, methanol decomposition to CO and H_2 takes place above 150 °C for both catalysts. The hydrogen peak is shifted by ~10 °C to the right compared to CO peak, probably due to readsorption effects. At the maximum CO production temperature (230 °C), CO_2 starts also appearing in the gas phase. One possible explanation for the production of CO_2 is reaction of CO with catalyst surface oxygen. This hypothesis is supported by the fact that less CO_2 is produced over the CoMn11AFp700 catalyst, which is in a more reduced state due to its activation at higher temperature. In addition, CO_2 production is accompanied by water production (at least for CoMn11p500), which also indicates oxidation of hydrogen by surface oxygen. Another possibility is that (part of) CO_2 is produced via the Boudouard reaction. This hypothesis cannot be checked by oxidation of surface carbon (also produced during Boudouard), since pyrolyzed catalysts already contain residual carbon in their structure. Judging from the fact that water production is minimal over the CoMn11AFp700, the occurrence of the Boudouard reaction cannot be disregarded.

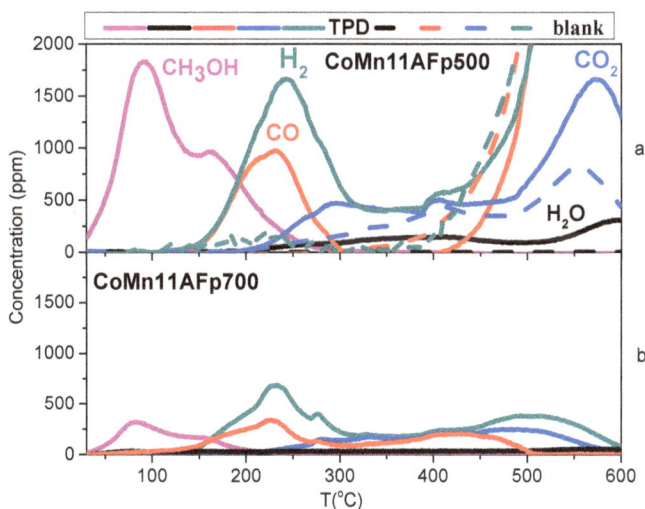

Figure 9. CH_3OH-TPD on CoMn11AFp500 and CoMn11AFp700.

CH$_3$OH, CO, H$_2$, CO$_2$ and H$_2$O profiles during CH$_3$OH-TPD from CoMn11AF and CoMn12AF catalysts prepared from pyrolysis at 600 °C are shown in Figure 10a,b, respectively. For both catalysts methanol is desorbed with a main peak at 90 °C and a tail extending up to 250 °C. The CO profile of CoMn11AFp600 includes two main peaks at 210 and 390 °C and is completed at 500 °C. Decrease of cobalt content leads to CO production with a similar profile, but in this case its production extends even above 600 °C. The hydrogen profile follows generally the CO profile. For both catalysts, CO$_2$ production is also detected and is more intense on CoMn11AFp600. Overall, the profiles indicate the presence of more than one adsorbed species of methanol and have interference by CO$_2$ readsorption effects (Figure 2b).

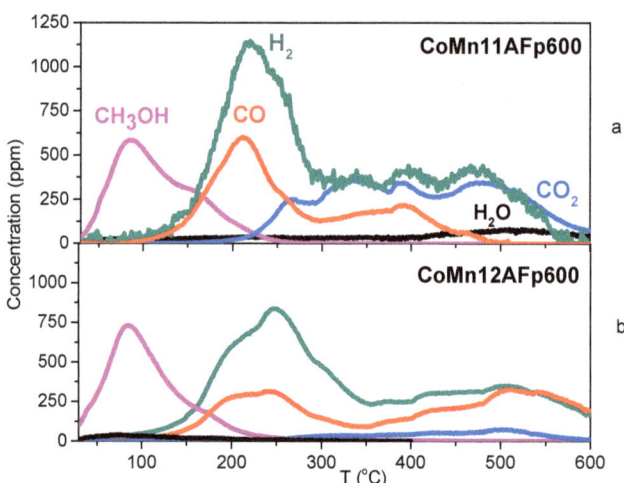

Figure 10. CH$_3$OH-TPD on CoMn11AFp600 and CoMn12AFp600.

CH$_3$OH, CO, H$_2$, CO$_2$ and H$_2$O profiles during CH$_3$OH-TPD for the catalyst prepared by oxidative pretreatment (CoMn11AFc500(AIR)) or oxidation/reduction (CoMn11c500(TPR)) are shown in Figure 11. Methanol adsorbed on the oxidized catalyst mainly acts as a reducing agent during TPD. Hence, CO$_2$ and H$_2$O are produced from the oxidation of adsorbed methanol by surface oxygen. Oxidation of methanol takes place in the form of two peaks, indicative either of the presence of two different modes of adsorbed methanol or of stepwise reduction of the surface. The peaks of oxidation products, CO$_2$ and H$_2$O, do not coincide evidently due to readsorption effects. The high temperature CO$_2$ peak is accompanied by production of small quantities of CO and H$_2$, which implies that active sites for methanol decomposition have been created only at that point (and not after the first CO$_2$ peak). On the contrary, the TPD profile of CoMn11AFc500 (TPR) corresponds to decomposition of adsorbed methanol towards CO and H$_2$ in the range of 150–250 °C.

The production of small amounts of CO_2 can be attributed to additional surface reduction of the catalyst. Comparison of Figures 9–11 shows that the onset and the main peak of methanol decomposition to CO and H_2 take place in the same temperature range for both the oxidized/reduced catalyst and all pyrolyzed catalysts. One important difference is that pyrolyzed catalysts contain additional states of adsorbed methanol which decompose at higher temperatures up to 500–600 °C.

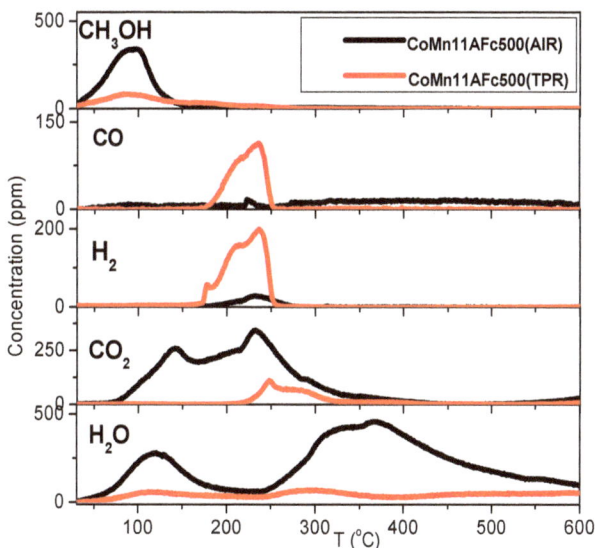

Figure 11. CH_3OH-TPD on CoMn11AFc500(AIR) and CoMn11AFc500(TPR).

The quantitative analysis of methanol adsorption and TPD experiments is presented in Figure 12. The amounts of adsorbed methanol are indicated with asterisks, while the amounts of desorbed CH_3OH, CO, CO_2, H_2 and H_2O during TPD are given in column form. The following observations can be made concerning Figure 12:

- With the exception of catalyst CoMn11AFp500, which adsorbs 1 $mmol \cdot g^{-1}$, all catalysts adsorb methanol in the range of 0.2–0.5 $mmol \cdot g^{-1}$. The lowest quantity is found over the oxidized/reduced sample.
- Less than half of adsorbed methanol desorbs molecularly.
- Increase of pyrolysis temperature and decrease of cobalt content lead to decrease of adsorbed methanol.
- The amounts of CO and CO_2 produced during methanol TPD are 2–4 times higher over the pyrolyzed catalysts compared to those prepared via oxidation or oxidation/reduction.

- The oxygen mass balance between output and input shows a surplus indicating that adsorbed methanol acts as a reducing agent scavenging lattice oxygen from the catalysts. The amount of adsorbed methanol that gets oxidized towards CO_2 and H_2O depends on the oxidation state of the catalyst surface.
- The carbon mass balance is overall satisfied (error <10%) with the exception of the CoMn11AFp500 catalyst, because its reported values correspond to temperatures below 400 °C (at higher temperatures interference from residual pyrolysis does not allow for reliable measurement).

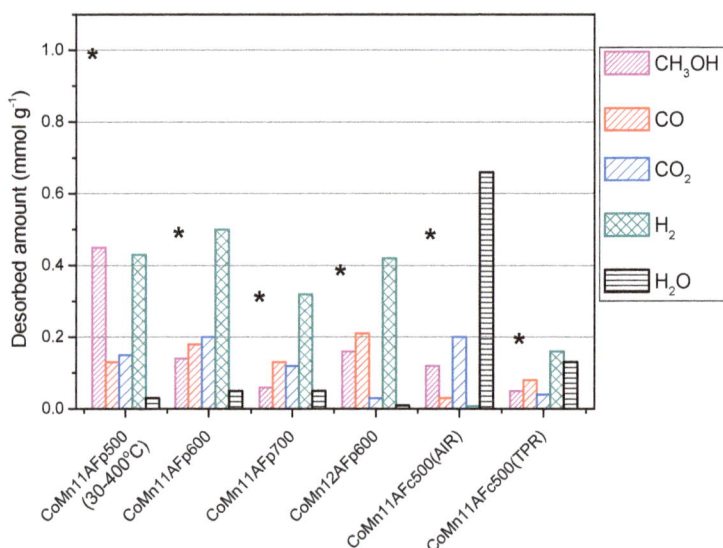

Figure 12. Desorbed amounts of CH_3OH, CO, CO_2, H_2 and H_2O during TPD after CH_3OH adsorption. Adsorbed amount of methanol is shown with asterisks (*).

2.5. Discussion

Application of the catalysts examined in the present work in methanol steam reforming has shown that the catalysts obtained via pyrolysis are more active than those prepared via oxidative pretreatment [1,2]. The former are already in reduced state exemplified by the presence of metallic cobalt and a mixed $Co_xMn_{1-x}O$ phase, while the latter consist initially of a mixed Co-Mn spinel oxide phase. When exposed to the reforming reaction mixture, however, the spinel oxide phase gets gradually reduced and the catalyst becomes activated. It is understandable that surface reduction of the spinel phase is a prerequisite for switching on the catalytic function. The findings of the present work during TPD of methanol indicate that adsorbed methanol acts as the reduction agent and the preoxidized catalyst is indeed initially inactive in methanol decomposition (Figure 11). On the other hand, all pyrolyzed catalysts, as well as the oxidized/reduced catalyst, are active in the decomposition of preadsorbed methanol towards CO as well as CO_2. The appearance of CO_2 takes place at higher temperatures than CO and its profile is matched well by the profile of coproduced H_2 (Figures 9 and 10). This implies that both CO_2 and H_2 originate from a common surface species, which is most probably formate. At this point, it would be interesting to compare the observed behavior of cobalt catalysts in methanol-TPD with the one of copper catalysts and, more specifically, CuO/Al_2O_3 [33]. The TPD profile from CuO/Al_2O_3 contains minor amounts of methanol (170 °C) and production of CO_2 and H_2 with peak at 200–250 °C. Small amounts of HCOOH are also produced in the same temperature range. CO is observed at higher temperatures with peak at 350 °C. Therefore, the main difference in the function of cobalt and copper catalysts is that CuO forms surface formate from adsorbed methanol, which decomposes at relatively low temperature to CO_2 and hydrogen. The observed CO at higher temperatures may be attributed to the reverse water-gas shift reaction. The cobalt catalysts of the present work, on the other hand, produce CO and H_2 at 200–250 °C, while CO_2 and H_2 production from formate decomposition is observed at much higher temperatures. These results exemplify nicely the superiority of copper *versus* cobalt in steam reforming of methanol: copper creates formate species which decompose at low temperatures to CO_2 and H_2, while cobalt decomposes methanol to CO and H_2 and the created formate species are stable so that they act rather as spectators occupying a fraction of the active sites [34].

The higher activity of pyrolyzed catalysts can be attributed mainly to their higher number density of active sites since peak temperatures of CO during TPD after methanol adsorption are similar in pyrolyzed and oxidized/reduced catalysts. During pyrolysis of the fumarate precursors, residual carbon is left behind and helps stabilize cobalt and MnO crystallites against extensive sintering. The resulting specific surface area of pyrolyzed catalysts is thus approximately one order of magnitude higher than the one of spinel oxide samples [1]. The higher available area for adsorption has been confirmed for water (Figure 8) and CO_2 (Figure 2) adsorption, whereas pyrolyzed samples adsorb up to one order of magnitude more water or CO_2 per unit weight. This applies to a lesser extent for methanol adsorption, whereas pyrolyzed samples adsorb up to 5 times more methanol. It is understandable that a certain fraction of these adsorbed molecules on pyrolyzed catalysts may reside on the carbonaceous support, whose surface composition in terms of oxygen content is strongly dependent on pyrolysis temperature. This may explain the observed variations in the adsorbed amounts of water, methanol and CO_2 as a function of pyrolysis temperature, which causes no change in the BET surface area otherwise. Comparison of relative amounts of adsorbed hydrogen and methanol on pyrolyzed and oxidized/reduced catalysts shows considerable discrepancies. Using hydrogen adsorption as a measure of cobalt dispersion and hence of active sites would lead to a wrong conclusion regarding differences among the various samples as related to activity in methanol reforming. In this respect, methanol is a more suitable probe, since it is the molecule of interest in the target reaction. Following this train of thought, one should further consider not the total amount of adsorbed methanol, but rather the amount of irreversibly adsorbed methanol that decomposes during heating. In this respect, the concentration of active sites in pyrolyzed catalysts is smaller than the one expected if one takes BET surface area or the amount of adsorbed hydrogen as indicators.

Concerning the role of water, it has been found in the present work that the catalysts are effective in dissociating the water molecule at temperatures of interest for the reforming reaction getting themselves oxidized in the process. The extent of water adsorption and dissociation is directly related to available sites responding to hydrogen adsorption (H_2-TPD experiments). Although one could envisage, based on this finding, methanol reforming taking place through a surface reduction-oxidation mechanism, $i.e.$, surface reduction by methanol leading to CO_2 and H_2O production followed by surface reoxidation by water leading to hydrogen production, this does not appear to be the case. Indeed, the product distribution during methanol reforming over all cobalt catalysts examined indicates a mechanism comprising methanol decomposition and the water–gas shift reaction and CO selectivity is higher or at best equal to the one predicted by thermodynamic equilibrium.

3. Experimental Section

3.1. Catalyst Preparation

The precursor compounds for catalyst synthesis were mixed fumarate salts of cobalt and manganese, which were prepared by mixing aqueous solutions of cobalt and manganese acetate with a solution of fumaric acid in ethanol followed by drying at 120 °C. The corresponding catalysts were prepared by pyrolysis of the salts under inert gas with a linear heating rate of 5 °C/min to the target temperature, 500, 600 or 700 °C, soak for 5 min and cooling down to adsorption temperature. The pyrolyzed catalysts are named CoMn1xAFpTTT, with TTT being the pyrolysis temperature and 1x (x = 1 or 2) being the Co/Mn atomic ratio. For example, the CoMn11AFp600 catalyst has a Co/Mn atomic ratio of 1/1 and has been prepared by pyrolysis at 600 °C. Samples were also prepared via oxidative treatment of the salts at 500 °C for 2 h for comparison purposes (named CoMn11AFc500). Calcined catalysts that had been reduced before adsorption are named CoMn11AFc500 (TPR).

3.2. Catalyst Characterization

Catalysts were characterized by nitrogen physisorption (BET), H_2-TPR, CH_3OH-TPR and in-$situ$ XRD. Characterization results have been reported in [1,2].

3.3. TPD Experiments

Temperature-programmed desorption (TPD) of pre-adsorbed CH_3OH, H_2O, H_2, CO_2 and CO at room temperature were performed. TPD of pre-adsorbed H_2O and H_2 at 300 °C were performed also. Temperature Programmed Desorption (TPD) experiments were carried out at atmospheric pressure in a fixed-bed reactor system with two independent gas lines equipped with mass flow controllers (Aera GmbH, Kirchheim, Germany). A mass spectrometer (Omnistar/Pfeiffer Vacuum, Asslar, Germany) was used for on-line monitoring of effluent gases. Prior to each CO or CO_2 adsorption experiment, the calcined catalysts were reduced by 3% H_2/He mixture at 400 °C for 20 min with a linear heating rate of 5 °C·min^{-1}. For all the other molecules examined reduction of calcined catalysts was performed by 3% H_2/He mixture at 600 °C for 30 min. For all molecules whose adsorption was studied at room temperature following completion of the adsorption, indicated by stable signals in the mass spectrometer, the reactor was purged with pure He until all signals met their baselines. Then, the TPD run was started under a helium flow of 50 cm^3·min^{-1} with a heating rate of 10 °C·min^{-1}. For all molecules whose adsorption was studied at 300 °C, cooling from adsorption temperature to 30 °C was performed in the presence of the same gaseous flow. After cooling, the reactor was purged with 50 cm^3·min^{-1} He and TPD was started as soon as the signals had been stabilized. For all catalysts, blank experiments were also performed. Adsorption of CO and CO_2 was carried out under a flow of 1.1% CO/He (50 cm^3·min^{-1}) or 1.1% CO_2/He (50 cm^3·min^{-1}) mixture, respectively. Hydrogen adsorption was performed under pure hydrogen flow of 20 cm^3·min^{-1} for 10 min. Water adsorption was carried out using a 5400 ppm H_2O/He mixture (50 cm^3·min^{-1}). Methanol adsorption was carried out using a 17,000 ppm CH_3OH/He mixture. The following masses were recorded in the mass spectrometer during all TPD experiments: 18 (H_2O), 28 (CO), 44 (CO_2), 31 and 32 (CH_3OH), 15 (CH_4), 29 and 30 (HCHO), 45 (CH_3OCH_3) and 49 (HCOOH).

4. Conclusions

The employment of TPD in order to investigate the interaction of CO, CO_2, H_2, H_2O and CH_3OH with cobalt catalysts prepared from cobalt-manganese fumarate precursors via pyrolysis or oxidation has led to the following findings:

- Adsorption of CO and H_2 is activated. Although activated hydrogen adsorption on cobalt is well established, activated adsorption of CO has not been reported previously.
- Hydrogen appears to adsorb both on cobalt and MnO components. Taking into account literature results concerning cobalt and MnO and results of the present work concerning cobalt, MnO and Co-MnO samples, it is inferred that

hydrogen desorbing below 250 °C originates from cobalt crystallites, hydrogen desorbing above 500 °C originates from MnO, while hydrogen desorbing in the intermediate temperature range (250–500 °C) probably originates from sites created at the interface of Co and MnO or from a mixed reduced oxide phase.

- Water adsorption is dissociative at an adsorption temperature of 300 °C, but not at 25 °C, leading to surface oxidation of the catalyst. Hydrogen produced from water dissociation remains partially adsorbed on the catalyst surface confirming that part of hydrogen is quite strongly bound on the catalysts.

- Reaction paths of adsorbed methanol during TPD include decomposition to CO and H_2, as well as creation of rather stable surface formates, which decompose at higher temperatures to CO_2 and H_2. Adsorbed methanol acts as a reducing agent during TPD leading to catalyst reduction.

- Differences of the pyrolyzed and oxidized/reduced catalysts appear to be mainly in the number density of active sites, which, however, is not directly analogous to differences in specific surface area.

Acknowledgments: This work was carried out in the frame of the ACENET project "Hydrogen from bio-alcohols: An efficient route for hydrogen production via novel reforming catalysts" (ACE.07.009). The authors acknowledge funding from the General Secretariat for Research and Technology of the Ministry of Education, Lifelong Learning and Religious Affairs (Greece).

Author Contributions: E.P. and T.I. conceived and designed the experiments; E.P. performed the experiments; E.P. and T.I. analyzed the data; T.I. wrote the paper.

Conflicts of Interest: The authors declare no conflict of interest.

References

1. Papadopoulou, E.; Ioannides, T. Steam reforming of methanol over cobalt catalysts: effect of cobalt oxidation state. *Int. J. Hydrogen Energy* **2015**, *40*, 5251–5255.
2. Papadopoulou, E.; Delimaris, D.; Denis, A.; Machocki, A.; Ioannides, T. Alcohol reforming on cobalt-based catalysts prepared from organic salt precursors. *Int. J. Hydrogen Energy* **2012**, *37*, 16375–16381.
3. Yong, S.T.; Ooi, C.W.; Chai, S.P.; Wu, X.S. Review of methanol reforming-Cu-based catalysts, surface reaction mechanisms, and reaction schemes. *Int. J. Hydrogen Energy* **2013**, *38*, 9541–9552.
4. Sá, S.; Silva, H.; Brandão, L.; Sousa, J.M.; Mendes, A. Catalysts for methanol steam reforming-A review. *Appl. Catal. B* **2010**, *99*, 43–57.
5. Palo, D.R.; Dagle, R.A.; Holladay, J.D. Methanol Steam Reforming for Hydrogen Production. *Chem. Rev.* **2007**, *107*, 3992–4021.
6. Agrell, J.; Birgersson, H.; Boutonnet, M.; Melián-Cabrera, I.; Navarro, R.M.; Fierro, J.L.G. Production of hydrogen from methanol over Cu/ZnO catalysts promoted by ZrO_2 and Al_2O_3. *J. Catal.* **2003**, *219*, 389–403.
7. De Wild, P.J.; Verhaak, M.J.F.M. Catalytic production of hydrogen from methanol. *Catal. Today* **2000**, *60*, 3–10.

8. Matter, P.H.; Braden, D.J.; Ozkan, U.S. Steam reforming of methanol to H_2 over nonreduced Zr-containing CuO/ZnO catalysts. *J. Catal.* **2004**, *223*, 340–351.

9. Tartakovsky, L.; Baibikov, V.; Veinblat, M. Comparative Performance Analysis of SI Engine Fed by Ethanol and Methanol Reforming Products. *SAE Tech. Pap.* **2013**.

10. Habermehl-Cwirzen, K. *An Insight: Studies of Atomic and Molecular Adsorption on Co(0001)*; Helsinki University of Technology: Helsinki, Finland, 2006.

11. Blyholder, G.; Wyatt, W.V. Infrared spectra and structures of some C,H,O compounds adsorbed on silica-supported iron, cobalt, and nickel. *J. Phys. Chem.* **1966**, *70*, 1745–1750.

12. Natile, M.M.; Glisenti, A. Study of surface reactivity of cobalt oxides: interaction with methanol. *Chem. Mater.* **2002**, *14*, 3090–3099.

13. Grellner, F.; Klingenberg, B.; Borgmann, D.; Wedler, G. Electron spectroscopic study of the interaction of oxygen with Co(1120) and of coadsorption with water. *J. Electron Spectrosc.* **1995**, *71*, 107–115.

14. Heras, J.M.; Papp, H.; Spiess, W. Face specificity of the H_2O adsorption and decomposition on Co surfaces: A LEED, UPS, sp and TPD study. *Surf. Sci.* **1982**, *117*, 590–604.

15. Reuel, R.C.; Bartholomew, C.H. The stoichiometries of H_2 and CO adsorptions on cobalt: Effects of support and preparation. *J. Catal.* **1984**, *85*, 63–77.

16. Bartholomew, C.H.; Reuel, R.C. Cobalt-support interactions: Their effects on adsorption and CO hydrogenation activity and selectivity properties. *Ind. Eng. Chem. Prod. Res. Dev.* **1985**, *24*, 56–61.

17. Zowtiak, J.M.; Bartholomew, C.H. The kinetics of H_2 adsorption on and desorption from cobalt and the effects of support thereon. *J. Catal.* **1983**, *83*, 107–120.

18. Tejuca, L.G.; Bell, A.T.; Fierro, J.L.G.; Pena, M.A. Surface behavior of reduced LaCoO₃ as studied by TPD of CO, CO_2 and H_2 probes and by XPS. *Appl. Surf. Sci.* **1988**, *31*, 301–316.

19. Narayanan, S.; Unnikrishnan, R.P. Comparison of hydrogen adsorption and aniline hydrogenation over co-precipitated Co/Al_2O_3 and Ni/Al_2O_3 catalysts. *J. Chem. Soc. Faraday Trans.* **1997**, *93*, 2009–2013.

20. Sakka, Y.; Ohno, S. Hydrogen desorption characteristics of composite Co-TiN nanoparticles. *Appl. Surf. Sci.* **1996**, *100/101*, 232–237.

21. Soonq, Y.; Rao, V.U.S.; Zarochak, M.F.; Gormley, R.J.; Zhang, B. Temperature-programmed desorption study on manganese-iron catalysts. *Appl. Catal.* **1991**, *78*, 97–108.

22. Gauthier, Y.; Schmidt, M.; Padovani, S.; Lundgren, E.; Bus, V.; Kresse, G.; Redinger, J.; Vagra, P. Adsorption sites and ligand effect for CO on an alloy surface: A direct view. *Phys. Rev. Lett.* **2001**, *87*, 036103.

23. Jiang, M.; Koizumi, N.; Ozaki, T.; Yamada, M. Adsorption properties of cobalt and cobalt-manganese catalysts studied by *in situ* diffuse reflectance FTIR using CO and CO+H_2 as probes. *Appl. Catal. A* **2001**, *209*, 59–70.

24. Mohana Rao, K.; Scarano, D.; Spoto, G.; Zecchina, A. CO adsorption on cobalt particles supported on MgO: An IR investigation. *Surf. Sci.* **1988**, *204*, 319–330.

25. Rostrup-Nielsen, J.R. Catalytic Steam Reforming. *Catalysis* **1984**, *5*, 1–117.

26. Baker, R.T.K.; Kim, M.S.; Chambers, A.; Park, C.; Rodriguez, N.M. The relationship between metal particle morphology and the structural characteristics of carbon deposits. *Stud. Surf. Sci. Catal.* **1997**, *111*, 99–109.

27. Ōya, A.; Ōtani, S. Catalytic graphitization of carbons by various metals. *Carbon* **1979**, *17*, 131–137.

28. Budiman, A.W.; Song, S.A.; Chang, T.S.; Shin, C.H.; Choi, M.J. Dry Reforming of Methane Over Cobalt Catalysts: A Literature Review of Catalyst Development. *Catal. Surv. Asia* **2012**, *16*, 183–197.

29. Tsakoumis, M.; Ronning, M.; Qyvind, B. Deactivation of cobalt based Fischer-Tropsch catalysts: A review. *Catal. Today* **2010**, *154*, 162–182.

30. Bartholomew, C.H. Mechanisms of catalyst deactivation. *Appl. Catal. A* **2001**, *21*, 17–60.

31. Bartholomew, C.H. Carbon Deposition in Steam Reforming and Methanation. *Catal. Rev. Sci. Eng.* **1982**, *24*, 67–112.

32. Nakamura, J.; Tanaka, K.; Toyoshima, I. Reactivity of deposited carbon on Co-Al_2O_3 catalyst. *J. Catal.* **1987**, *108*, 55–62.

33. Tagawa, T.; Pleizier, G.; Amenomiya, Y. Methanol synthesis from CO_2+H_2: Characterization of catalysts by TPD. *Appl. Catal.* **1985**, *18*, 285–293.

34. Meunier, F.C.; Reid, D.; Goguet, A.; Shekhtman, S.; Hardacre, C.; Burch, R.; Deng, W.; Flytzani-Stephanopoulos, M. Quantitative analysis of the reactivity of formate species seen by DRIFTS over a Au/Ce(La)O_2 water-gas shift catalyst: First unambiguous evidence of the minority role of formates as reaction intermediates. *J. Catal.* **2007**, *247*, 269–279.

Hydrogen Production by Ethanol Steam Reforming (ESR) over CeO$_2$ Supported Transition Metal (Fe, Co, Ni, Cu) Catalysts: Insight into the Structure-Activity Relationship

Michalis Konsolakis, Zisis Ioakimidis, Tzouliana Kraia and George E. Marnellos

Abstract: The aim of the present work was to investigate steam reforming of ethanol with regard to H$_2$ production over transition metal catalysts supported on CeO$_2$. Various parameters concerning the effect of temperature (400–800 °C), steam-to-carbon (S/C) feed ratio (0.5, 1.5, 3, 6), metal entity (Fe, Co, Ni, Cu) and metal loading (15–30 wt.%) on the catalytic performance, were thoroughly studied. The optimal performance was obtained for the 20 wt.% Co/CeO$_2$ catalyst, achieving a H$_2$ yield of up to 66% at 400 °C. In addition, the Co/CeO$_2$ catalyst demonstrated excellent stability performance in the whole examined temperature range of 400–800 °C. In contrast, a notable stability degradation, especially at low temperatures, was observed for Ni-, Cu-, and Fe-based catalysts, ascribed mainly to carbon deposition. An extensive characterization study, involving N$_2$ adsorption-desorption (BET), X-ray diffraction (XRD), Scanning Electron Microscopy (SEM/EDS), X-ray Photoelectron Spectroscopy (XPS), and Temperature Programmed Reduction (H$_2$-TPR) was undertaken to gain insight into the structure-activity correlation. The excellent reforming performance of Co/CeO$_2$ catalysts could be attributed to their intrinsic reactivity towards ethanol reforming in combination to their high surface oxygen concentration, which hinders the deposition of carbonaceous species.

Reprinted from *Catalysts*. Cite as: Konsolakis, M.; Ioakimidis, Z.; Kraia, T.; Marnellos, G.E. Hydrogen Production by Ethanol Steam Reforming (ESR) over CeO$_2$ Supported Transition Metal (Fe, Co, Ni, Cu) Catalysts: Insight into the Structure-Activity Relationship. *Catalysts* **2016**, *6*, 39.

1. Introduction

Energy is a vital element in our everyday lives, mostly generated, however, from fossil fuels. Furthermore, our continuous dependence on fossil fuels is strongly coupled with natural resource depletion and serious environmental implications, such as greenhouse effect, stratospheric ozone depletion, photochemical smog, *etc.* Therefore, the need for renewable energy is becoming ever more urgent. In view of this fact, solar, wind, and biomass have become promising renewable energy sources (RES). Although significant technological progress has been accomplished in their efficient energy conversion, their share in the world energy mix is limited due to their intermittent character and site-dependence. On the other hand, hydrogen has been acknowledged as an ideal energy currency for sustainable energy development. Hydrogen can be employed as feedstock in a fuel cell to directly generate electricity at high efficiencies and low environmental footprint. In this regard, hydrogen production in a clean and renewable manner is essential for a sustainable energy future [1–4].

Nowadays, the steam reforming of natural gas is the most frequently employed and economically feasible method for hydrogen production [1–3]. This process, however, is based on a fossil resource and is linked with CO_2 emissions, unless carbon capture and storage (CCS) techniques are employed, which inevitably increase the complexity and overall costs. The rising concerns, related to the reduction of atmospheric pollutants in conjunction with the continuously increasing energy demands, have stimulated the research toward hydrogen production from renewable sources. In this direction, ethanol is among the most promising feedstocks, due to its relatively high hydrogen content, wide availability, non-toxicity, as well as storage and handling safety. Moreover, ethanol can be generated in a renewable manner through biomass (energy crops, agricultural and industrial wastes, forestry residues, and the organic fraction of municipal solid waste) fermentation [5–7]. The employment of bio-fuels in hydrogen production incorporates important environmental advantages, since the generated CO_2 can be recycled through photosynthesis during plant growth, resulting in a carbon neutral process.

Hydrogen production by reforming processes is usually achieved either by steam or auto-thermal reforming. Compared to auto-thermal reforming, ethanol steam reforming (ESR) has received more focus due to the higher H_2 yields obtained through this route [8–12]:

$$CH_3CH_2OH_{(g)} + 3H_2O_{(g)} \rightarrow 6H_{2(g)} + 2CO_{2(g)}, \Delta H^0{}_{298} = +173.1 \, kJ/mol \quad (1)$$

Furthermore, the water gas shift (WGS) reaction, taking place at excess H_2O conditions, can further increase the H_2 yield:

$$CO_{(g)} + H_2O_{(g)} \rightarrow CO_{2(g)} + H_{2(g)}, \Delta H^0{}_{298} = -41 \, kJ/mol \quad (2)$$

Nevertheless, the overall reforming process consists of a complex network of reactions, such as ethanol dehydrogenation, dehydration, or decomposition, which can lead to the formation of several byproducts (methane, ethylene, acetaldehyde, acetone, coke, etc.) [10,13].

Furthermore, technological advances in the direct utilization of bio-fuels, like bio-ethanol, in Solid Oxide Fuel cells (SOFCs) have stimulated the research toward the development of high efficiency fuel processing electro-catalysts with adequate activity and durability. Under these perspectives, several catalytic systems, involving mainly base (e.g., Ni, Co, Cu) and precious (e.g., Pt, Rh, Ru, Pd, and Ir) metal supported catalysts have been reported for ESR [10,12]. Although the noble metal-based catalysts demonstrate sufficient reforming activity in a wide temperature range [10,12,14–19], their large-scale implementation in practice is limited by their significant high cost [20]. Hence, the development of active and stable non-precious metal catalysts for ESR is highly desirable. Ni-based catalysts are the most commonly employed catalytic systems, due to their inherent activity on C-C and C-H bond cleavage [21]. However, their insufficient tolerance to carbon deposition in conjunction with the sintering of Ni particles at elevated temperatures, have been considered the main barriers for industrial application [22]. Recent studies, however, have shown that highly active and stable Ni-based catalysts can be obtained by appropriately adjusting their physicochemical properties by means of advanced synthesis procedures and/or structural/surface promotion (e.g., [23,24]). Strong Ni-ceria interactions can be accounted for the enhanced ESR activity [25–27].

Regarding the catalyst support, it should generally possess adequate chemical and mechanical properties to provide the required activity and stability under reaction conditions. Most importantly, supporting carriers should provide strong metal-support interactions and high oxygen mobility (redox properties), preserving the active phase from sintering and coking deactivation. Ceria has been widely employed in several catalytic reactions, involving the WGS reaction [28], NOx reduction [29,30], oxidation or partial oxidation of hydrocarbons [31], steam reforming [32], *etc.*, owing to its excellent redox properties. The mobile oxygen related to the ceria lattice, is considered to be responsible for the oxidation of deposited carbonaceous fragments, thus protecting the catalyst surface from poisoning [33–36]. In addition, mobile oxygen can activate water, with regard to the formation of hydroxyl groups, resulting in higher ESR efficiency [25,33]. In a comprehensive study by Xu *et al.* [25], utilizing both *in situ* and *ex situ* characterization techniques it was revealed that metallic Ni and Ce(III) entities were the active components under ESR conditions. Ce(III) facilitates the decomposition of H_2O to –OH groups, which are essential for the transformation of $C_xH_yO_z$ intermediates to CO_2 and H_2O, whereas Ni promotes the adsorption of ethanol and dissociation of C–C bonds. Moreover, strong Ni-ceria interactions perturb the electronic and chemical properties of Ni ad-atoms, resulting in an inferior ability of Ni with regard to CO methanation [26,27].

Besides Ni-based catalysts, various transition metals have been also tested for ESR, with Co-based catalysts being among the most efficient [34–46]. In particular, Co/Mg/Al hydrotalcites are very active and stable for ESR, due to the formation of highly reactive Co^{2+} species instead of less active metallic cobalt [42–44]. Moreover, Co/CeO_2 catalysts demonstrated excellent ESR activity, ascribed mainly to the high oxygen storage capacity (OSC) of ceria, which suppresses the carbon deposition (e.g., [35,36,41]). In this regard, very active Co-based catalysts can be designed by properly adjusting the local surface structure of Co species by means of support and/or active phase modification [35–46]. For instance, it was found that the addition of Rh as a promoter at small amounts (0.1 wt.%) on 2.0 wt.% Co/CeO_2 catalysts can notably increase the ESR performance [45,46]. Rhodium, through hydrogen spillover phenomena, facilitates the reduction of both cobalt oxides and ceria, which then is reflected on steam reforming performance. The presence of Rh in Co/Ceria catalysts hinders the formation of acetone as well as the build-up of strongly bonded carbide species, leading to improved activity and stability [45,46].

In the light of the above aspects, the present manuscript aims to comparatively explore ESR performance over different CeO_2-supported transition metal (Ni, Fe, Cu, Co) catalysts. The effect of several operation parameters, related to reaction temperature (400–800 °C), time on stream (24 h), S/C feed ratio (0.5, 1.5, 3, 6), and metal loading (15–30 wt.%) on the activity and stability performance is systematically examined. Furthermore, an extensive characterization study—N_2 adsorption-desorption (BET), X-ray diffraction (XRD), Scanning Electron Microscopy (SEM/EDX), X-ray Photoelectron Spectroscopy (XPS), and Temperature Programmed Reduction (H_2-TPR)—was carried out to attain a possible structure-activity relationship. Although recent efforts on ESR are focusing to develop efficient catalysts with low metal contents (e.g., [45,46]), relatively high metal loadings (15–30 wt.%) are employed in the present study, motivated by the potential application of these materials as anodic electrodes in internal reforming ethanol fed SOFCs.

2. Results

2.1. Characterization Studies

2.1.1. Textural Characterization (BET Analysis)

Table 1 presents the textural characteristics (surface area, pore volume, pore diameter) of as prepared catalysts. It is obvious that the addition of transition metals to the ceria carrier has a detrimental effect on the resulted BET area, which, in general, is exacerbated upon increasing the metal content. However, the decrease of BET area upon increasing the cobalt content is not monotonic. The latter can be attributed to the different crystallite size of Co_3O_4 and CeO_2 phases in Co/CeO_2 catalysts, as determined by XRD analysis presented in the sequence. The size of Co_3O_4 is about 3-fold higher than CeO_2 indicating the segregation of cobalt species over the catalyst surface (Table 2). Thus, a significant decrease of bare ceria area (71.5 m^2/g) is expected upon increasing Co loading, due to the blockage of ceria pores. However, this decrease could not be monotonic, since above a certain Co content the formation of large Co_3O_4 particles hampers the dispersion of Co_3O_4 entities into CeO_2 pores and consequently any further decrease of BET area. The same trend was observed upon increasing Cu content in Cu/CeO_2 catalysts [47].

Table 1. Textural characteristics of M/CeO_2 (M: Fe, Co, Ni, Cu) catalysts.

Sample	SBET (m^2/g)	Total Pore Volume (cm^3/g)	Average Pore Diameter (nm)
CeO_2	71.5	0.27	15.4
15 wt. % Co/CeO_2	64.2	0.28	17.8
20 wt. % Co/CeO_2	33.4	0.13	16.0
25 wt. % Co/CeO_2	42.2	0.19	18.2
30 wt. % Co/CeO_2	44.9	0.21	18.7
20 wt. % Ni/CeO_2	57.6	0.54	37.7
20 wt. % Cu/CeO_2	44.6	0.15	13.1
20 wt. % Fe/CeO_2	57.0	0.26	18.3

Table 2. Structural characteristics of M/CeO_2 (M: Fe, Ni, Co, Cu) catalysts.

Samples	Phase Detected	Crystallite Size (nm)	Lattice
15 wt.% Co/CeO_2	CeO_2	11.5	Cubic
	Co_3O_4	30.8	Cubic
20 wt.% Co/CeO_2	CeO_2	10.2	Cubic
	Co_3O_4	37.7	Cubic
25 wt.% Co/CeO_2	CeO_2	13.6	Cubic
	Co_3O_4	32.3	Cubic
30 wt.% Co/CeO_2	CeO_2	10.4	Cubic
	Co_3O_4	37.7	Cubic
20 wt.% Fe/CeO_2	CeO_2	10.6	Cubic
	Fe_2O_3	34.1	Rhombohedral
20 wt.% Ni/CeO_2	CeO_2	11.2	Cubic
	NiO	23.2	Cubic
20 wt.% Cu/CeO_2	CeO_2	9.3	Cubic
	CuO	43.5	Monoclinic

Regarding the M/CeO_2 catalysts with a constant metal loading (20 wt.%), Ni- and Fe-based catalysts possess the higher BET area ($\sim$57 m^2/g), while the Cu- and Co-based samples exhibited lower values of surface area, $i.e.$, 44.6 and 33.4 m^2/g, respectively. These variations in the textural characteristics could be related with the different crystallite size of the metal oxides formed in each case, as is discussed later.

2.1.2. Structural Characterization (XRD Analysis)

Figure 1A depicts the XRD spectra of M/CeO_2 catalysts, whereas Figure 1B presents the corresponding spectra of Co-based catalysts of different metal loading. Table 2 lists the phases that were detected for each sample, the corresponding structure, and approximate crystallite size as determined by Scherrer analysis. All metals crystallized in the form of oxides, *i.e.*, iron as Fe_2O_3, cobalt as Co_3O_4, nickel as NiO, and copper as CuO, with their average particle size following the order: $CuO > Co_3O_4 > Fe_2O_3 > NiO$ (Table 2). Regarding CeO_2 crystallites their size increases with the decrease of the corresponding size of the metal oxide phase. Concerning the Co-based catalysts of different metal content, cobalt is again presented in the form of oxide (Co_3O_4), regardless of the Co loading. The crystallite size of CeO_2 remains almost constant upon increasing the Co content. Additionally, the size of Co_3O_4 phase is increased from 30.8 to 37.7 nm with the increment of Co content from 15 to 20 wt.%. However, no further increase was observed for higher Co loadings, confirming the stabilization of BET area values above a certain Co loading.

2.1.3. Reducibility Studies (H_2-TPR)

Temperature-programmed reduction (TPR) experiments were performed to gain insight into the role of metal entity on the reducibility of M/CeO_2 catalysts. Figure 2 depicts the TPR profiles of M/CeO_2 catalysts, in terms of hydrogen consumption as a function of temperature.

In the Cu/CeO_2 sample the overlapping peaks at 100–300 °C temperature interval are assigned to Cu oxide species reduction along with the surface oxygen reduction [48–52]. Specifically, the peak at 133 °C is attributed to the reduction of Cu ions in close proximity to CeO_2, while, the features at 215 °C and 267 °C are assigned to Cu entities not closely associated with ceria [52].

In the Co/CeO_2 sample no obvious peaks were detected at temperatures lower than 300 °C, denoting that Co species, in contrast to Cu, are more difficult to reduce. The Co/CeO_2 reduction profile was comprised of several overlapping features in the *ca.* 300–400 °C temperature range. These peaks are related to the reduction of surface oxygen groups along with the stepwise reduction of Co oxides to metallic cobalt [53].

Ni/CeO_2 catalysts mainly exhibit two reduction peaks at 231 and 330 °C. The peak at 231 °C is attributed to adsorbed surface oxygen reduction, while the peak at 330 °C to the reduction of the bulk NiO and surface CeO_2 [54–57].

Concerning the reduction profile of Fe/CeO_2 sample, it has been widely recognized that hematite (Fe_2O_3) reduction to metallic iron (Fe^0) takes place through the formation of magnetite (Fe_3O_4) and wustite (FeO). Wustite is metastable and disproportionates into magnetite and metallic iron (4FeO → Fe$_3$O$_4$ + Fe) below 620 °C [58]. Commonly, bare iron oxide shows overlapping features at *ca.* 300–800 °C, corresponding to the stepwise reduction of hematite to metallic iron [58–60]. The Fe/CeO$_2$ sample exhibits in addition a main peak at 296 °C, which can be attributed to surface oxygen reduction as well as to the facilitation of the stepwise reduction of Fe oxides to metallic iron. Based on the main hydrogen TPR peaks, the following order of reducibility was recorded: Cu/CeO$_2$ > Fe/CeO$_2$ > Ni/CeO$_2$ > Co/CeO$_2$.

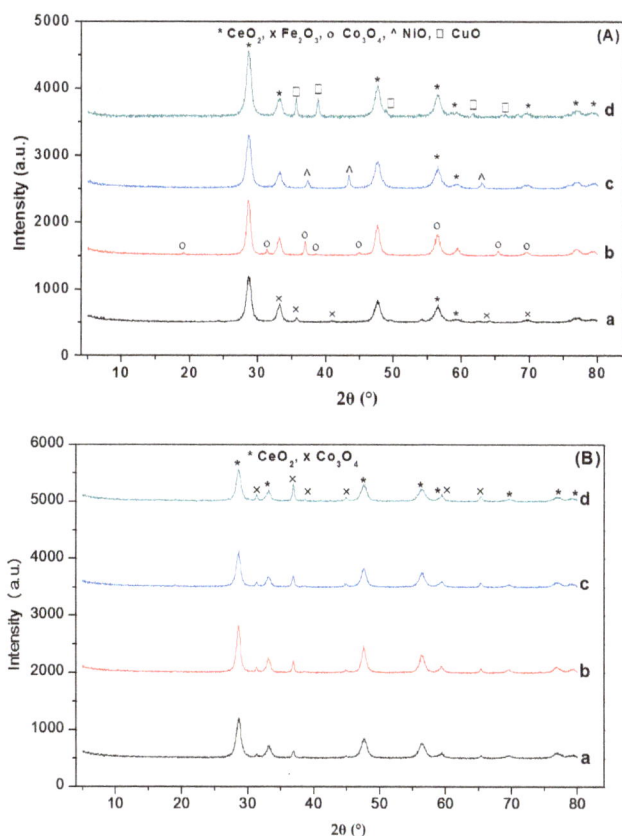

Figure 1. X-ray powder diffraction patterns of 20 wt.% M/CeO$_2$ catalysts (**A**) where M stands for Fe (a), Co (b), Ni (c) and Cu (d), and of Co/CeO$_2$ catalysts (**B**) with 15 (a), 20 (b), 25 (c) and 30 wt.% (d) Co loading.

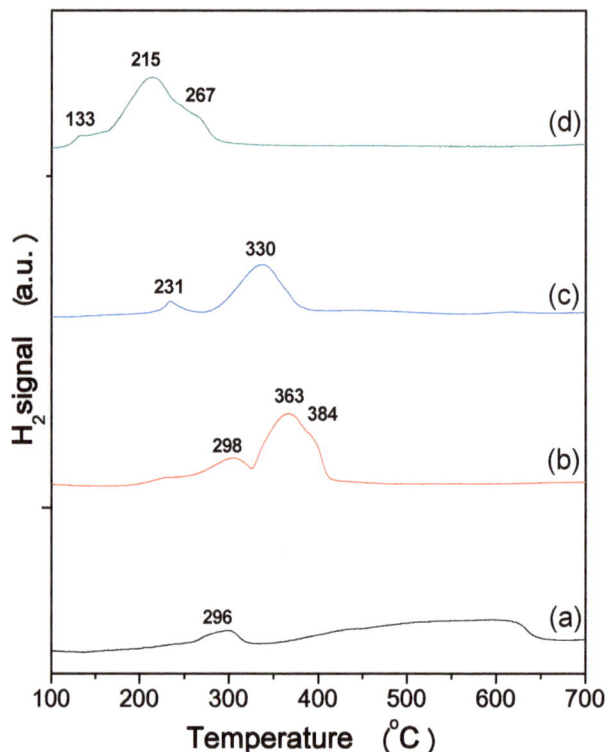

Figure 2. Temperature programmed reduction profiles of 20 wt.% Fe (a), Co (b), Ni (c) and Cu (d) catalysts supported on CeO_2.

2.1.4. Surface Characterization (XPS Analysis)

XPS measurements were carried out next to obtain insights into the elemental oxidation states and surface composition. Figure 3 depicts the XPS spectra in the 2p region of Fe, Co, Ni, and Cu catalysts supported on ceria. In the case of the Fe/CeO_2 sample, peaks in the region of 709–713 eV with satellites at *ca.* 725 eV appear. Curve-fitting revealed the contribution of Fe^{2+} and Fe^{3+} ions at 709–711 eV and 711–713 eV, respectively [61,62]. The presence of a certain amount of Fe^{2+} ions probably implies the co-existence of Fe_2O_3 (as indicated by XRD) with lower valence iron oxides (FeO/Fe_3O_4). The Co 2p spectrum of Co/CeO_2 catalyst is characterized by a main peak at 780.6 eV accompanied by a low intensity satellite and a spin-orbit doublet $Co2p_{1/2}$-$Co2p_{3/2}$ of 15.2 eV. These characteristics point to the formation of Co^{3+} species in Co_3O_4-like phase [54]. The $Ni2p_{3/2}$ can be analyzed into two components corresponding to different Ni chemical states. The peak at *ca.* 854 eV is usually assigned to NiO, whereas the higher BE peak at 855–856 eV to Ni bonded with OH groups, *i.e.*, $Ni(OH)_2$ and/or NiOOH [63,64]. It should be noted, however, the difficulty of unambiguously assigning the different chemical states of Ni oxides due to the complexity of Ni 2p spectra arising from multiplet splitting, shake-up satellites, and plasmon loss structures [65]. In any case the majority of Ni species seems to be related with NiO in accordance with the XRD findings. The spectrum of the Cu/CeO_2 sample is characterized by a Cu $2p_{3/2}$ band at 933.8 eV and shake-up satellites at *ca.* 944.0 eV, typical characteristics of Cu^{2+} species in CuO-like phase [20,48,66–70]. The above assignments are in line with the XRD findings, which revealed the formation of the corresponding metal oxides in M/CeO_2 samples (Table 2).

Figure 4 depicts the O1s spectra of M/CeO_2 samples. Curve-fitting, based on a mixture of Lorentzian and Gaussian curves, revealed three components. The low binding energy peak (O_I) at ~529 eV can be ascribed to lattice oxygen, the intermediate peak (O_{II}) at ~531 eV to surface O, OH groups and oxygen vacancies and the high energy band (O_{III}) at ~534 eV to adsorbed water [68,71,72]. In the case of Ni/CeO_2 the low BE peak at *ca.* 527 eV could be due to the differential charging of the oxide. Based on the area of O_I, O_{II} and O_{III} envelopes, a relative comparison between the M/CeO_2 samples could be obtained (Table 3). It is evident that the ratio between the lattice oxygen to the sum of surface oxygen species notably changes with the metal type. Co/CeO_2 catalysts possess the highest amount of O_I species (70%), which are related to lattice oxygen in ceria and cobalt oxides. These differences in the population of surface oxygen species are expected to affect the redox reactions between the gaseous reactants and solid carbonaceous deposits with lattice oxygen ions, as is discussed in the sequence.

Table 3. Percentage of surface oxygen species on M/CeO_2 catalysts.

Sample	%		
	O_I	O_{II}	O_{III}
Fe/CeO_2	48	45	7
Co/CeO_2	70	24	6
Ni/CeO_2	46	54	-
Cu/CeO_2	38	55	7

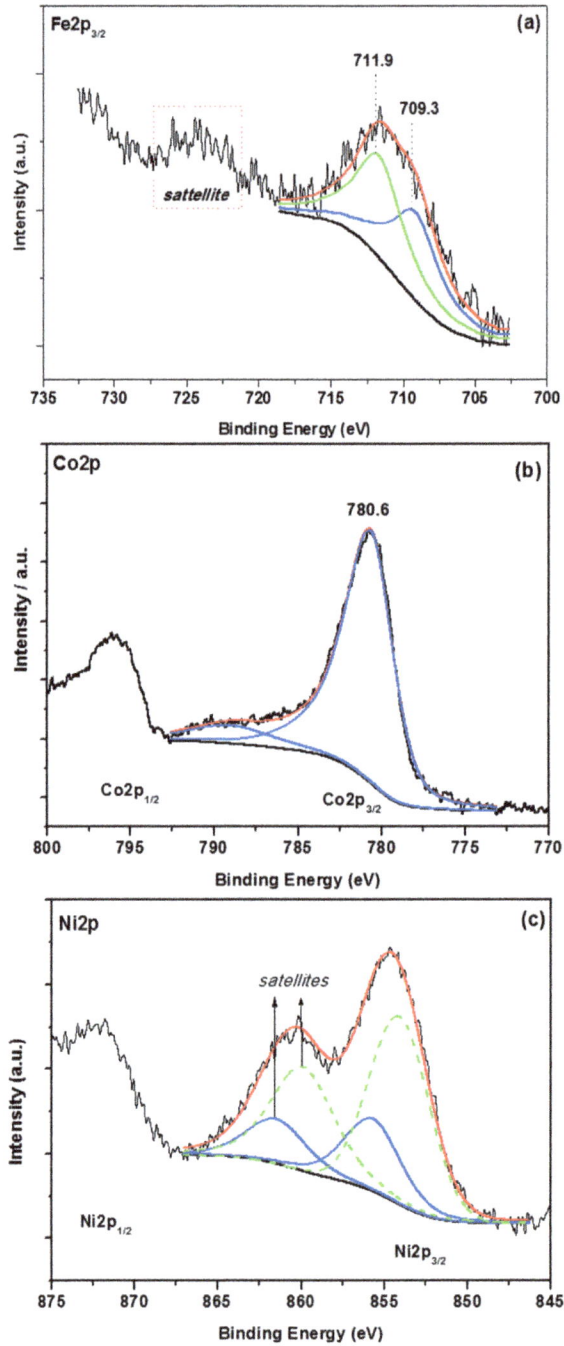

Figure 3. *Cont.*

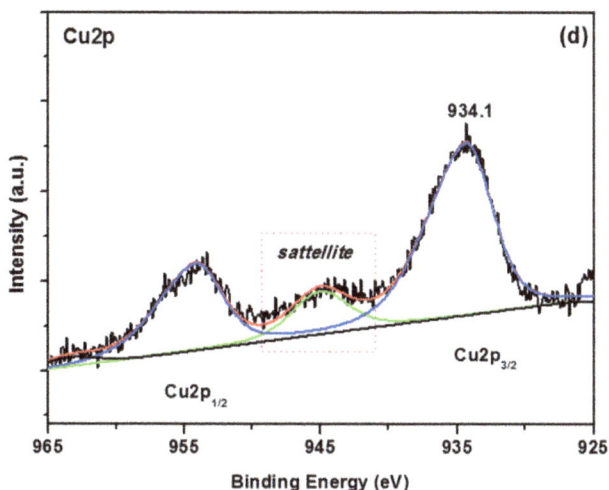

Figure 3. Core level 2p spectra of 20 wt.% Fe (**a**); Co (**b**); Ni (**c**) and Cu (**d**) catalysts supported on CeO$_2$.

The surface composition of M/CeO$_2$ samples obtained by XPS analysis, along with the bulk atomic concentrations (at. %), is presented in Table 4. Comparison of the XPS and nominal values of transition metal (Fe, Co, Ni, Cu) atomic concentrations reveals a decrease in metal surface species for all samples. These findings can be understood to a good extent considering surface characterization studies over Co-Ce binary oxides [53], where it was demonstrated the impoverishment of catalyst surface in Co species during the calcination procedure. The latter has been mainly attributed to the low surface energy of cobalt compared to ceria, resulted in a localization of ceria on the outer surface. In a similar manner, it has been revealed that the surface concentration of metal ions (such as Ni and Co) supported on ceria-based materials can be notably decreased due to the incorporation of metal ions into the support lattice [73–75]. This particular configuration, involving the coexistence of metal ions in the outer surface and inside the support structure, could largely favor the oxygen mobility, with large consequences in the catalytic activity (e.g., [75]).

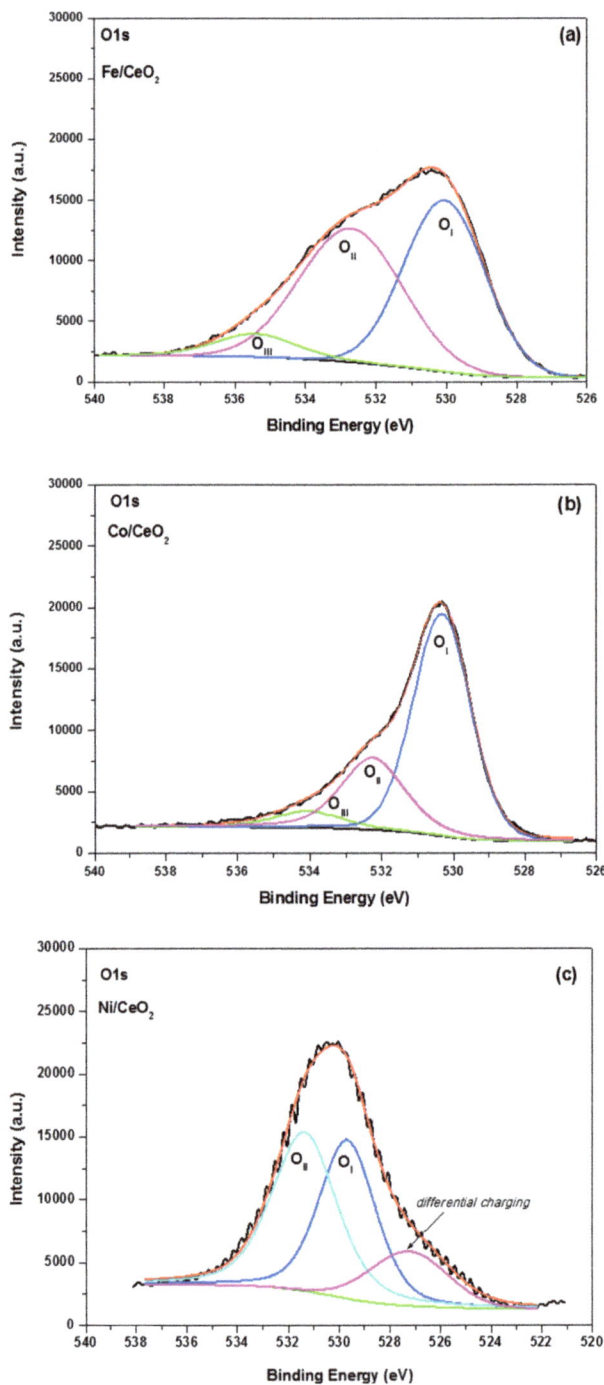

Figure 4. *Cont.*

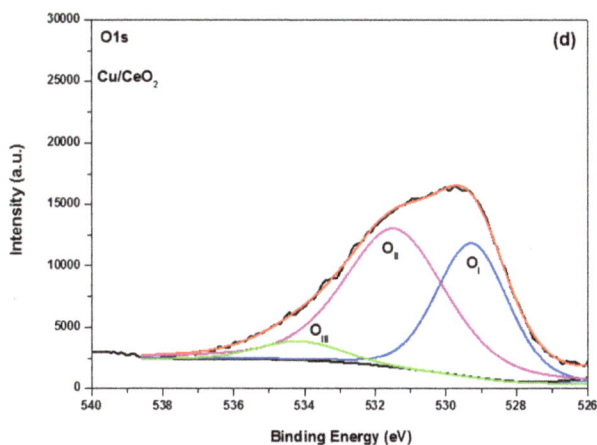

Figure 4. O1s spectra of Fe (**a**); Co (**b**); Ni (**c**) and Cu (**d**) catalysts supported on CeO$_2$.

Table 4. Bulk and surface—X-ray Photoelectron Spectroscopy (XPS)—atomic concentrations (at%) for 20 wt.% M/CeO$_2$ catalysts.

Samples	XPS [a]				Nominal [b]			
	M	**Ce**	**O**	**M/Ce**	**M**	**Ce**	**O**	**M/Ce**
Fe/CeO$_2$	7.7	29.6	62.7	0.26	15.6	20.3	64.1	0.77
Co/CeO$_2$	11.0	25.5	63.5	0.43	15.5	21.3	63.2	0.73
Ni/CeO$_2$	15.4	23.5	61.1	0.65	16.4	22.4	61.2	0.73
Cu/CeO$_2$	8.2	34.0	57.8	0.24	15.5	23.0	61.5	0.68

[a] estimated from XPS spectra; [b] calculated from the nominal catalyst composition, assuming that M is oxidized to M$_x$O$_y$ during the preparation procedure, as indicated by XRD analysis.

2.2. Catalytic Activity Studies

2.2.1. Effect of Metal Entity on Steam Reforming Performance

Ethanol Steam Reforming (ESR) is a multi-molecular reaction involving several pathways producing hydrogen, CO, CO$_2$ and numerous by-products, such as methane, ethylene, acetaldehyde, acetone, coke, *etc.* [76]. The simplified reaction network, presented in Figure 5, can be considered for the different product distribution during ESR [77,78]. This mainly involves apart from the ESR (Equation (1)) and water gas shift (Equation (2)) reactions, the ethanol decomposition (Equation (3)), ethanol dehydrogenation (Equation (4)), acetaldehyde decarbonylation (Equation (5)), acetone formation via acetaldehyde condensation, followed by decarboxylation (Equation (6)), ethanol dehydration (Equation (7)), methanation (Equation (8)), and carbon deposition (Equation (9)) reactions:

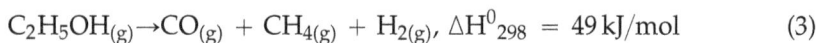

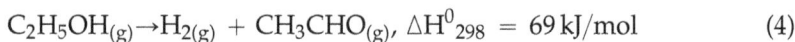

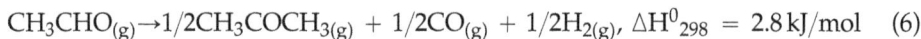

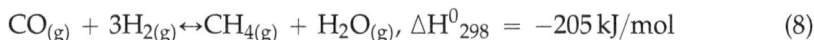

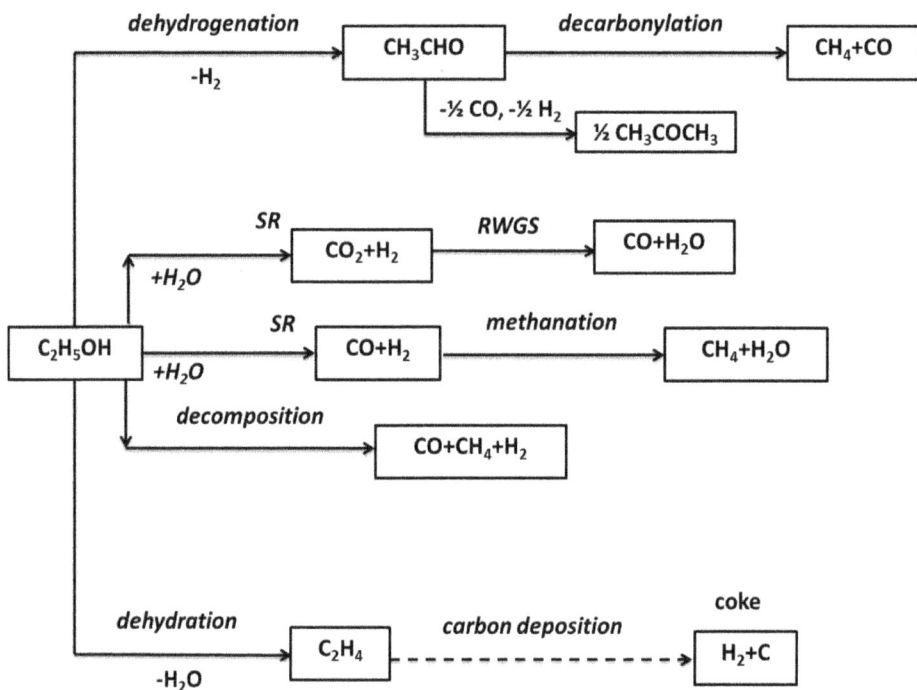

$$C_2H_5OH_{(g)} \rightarrow CO_{(g)} + CH_{4(g)} + H_{2(g)}, \ \Delta H^0_{298} = 49 \, kJ/mol \tag{3}$$

$$C_2H_5OH_{(g)} \rightarrow H_{2(g)} + CH_3CHO_{(g)}, \ \Delta H^0_{298} = 69 \, kJ/mol \tag{4}$$

$$CH_3CHO_{(g)} \rightarrow CH_{4(g)} + CO_{(g)}, \ \Delta H^0_{298} = -19 \, kJ/mol \tag{5}$$

$$CH_3CHO_{(g)} \rightarrow 1/2CH_3COCH_{3(g)} + 1/2CO_{(g)} + 1/2H_{2(g)}, \ \Delta H^0_{298} = 2.8 \, kJ/mol \tag{6}$$

$$C_2H_5OH_{(g)} \rightarrow C_2H_{4(g)} + H_2O_{(g)}, \ \Delta H^0_{298} = 45 \, kJ/mol \tag{7}$$

$$CO_{(g)} + 3H_{2(g)} \leftrightarrow CH_{4(g)} + H_2O_{(g)}, \ \Delta H^0_{298} = -205 \, kJ/mol \tag{8}$$

$$2CO_{(g)} \rightarrow C_{(s)} + CO_{2(g)}, \ \Delta H^0_{298} = -171.5 \, kJ/mol \tag{9}$$

Figure 5. Schematic representation of the reaction network in the ethanol steam reforming (ESR) process.

Figure 6 depicts the effect of temperature (400–800 °C) on ethanol conversion and product selectivity during the ESR over M/CeO$_2$ catalysts, at a constant S/C feed ratio of 3. The reforming process resulted mainly to H$_2$, CO, CO$_2$, and CH$_4$.

Minor quantities of C_2H_4, C_2H_6, and traces of CH_3CHO and CH_3COCH_3 are also observed. The H_2 selectivity follows, in general, a downward shift with increasing temperature, in contrast to CO selectivity. The particular high hydrogen selectivities at lower temperatures (400–500 °C) in the case of Fe-, Co- and Cu-based samples, which exceed the expected theoretical values from reaction stoichiometry, maybe due to the thermal decomposition and/or parallel reactions taking place at low temperatures. The latter keeps in step with the extended carbon formation at low temperatures, as will be discussed in the sequence. The selectivity towards CO_2 and CH_4 decreases up to 600 °C and then gradually level-off until 800 °C. The variation of H_2, CO, CO_2, and CH_4 selectivity upon increasing temperature can be realized by taking into account the different reactions involved in ESR (Figure 5). At low temperatures the WGS reaction (Equation (2)) is favored, increasing the selectivity towards H_2 and CO_2. In contrast, at high temperatures the reverse WGS reaction enhances the production of CO and H_2O at the expense of H_2 and CO_2. The decrease of CH_4 formation at higher temperatures may be ascribed to its reforming to CO and H_2 [32].

The impact of catalyst type (Fe, Co, Ni, Cu) and temperature on product distribution is more evident in Figure 7, which comparatively depicts the major product yield at 400, 600 and 800 °C. For comparison purposes the corresponding results over bare CeO_2 are also depicted. The superior performance of M/CeO_2 catalysts compared to bare CeO_2 is clearly shown. The latter implies the pronounced effect of transition metal and/or metal-support interfacial sites on ethanol steam reforming, as discussed in the sequence. At 400 °C the best performance, in terms of H_2 yield, is achieved by Co- and Cu-based catalysts both exhibiting a H_2 yield of ~66%. Fe- and Ni-based catalysts demonstrated lower H_2 yields of 59% and 48%, respectively. Ni/CeO_2 catalysts display very high CH_4 yield (24%), compared to all other M/CeO_2 catalysts (~0.4%). This can be attributed to the acetaldehyde decarbonylation to CH_4 (Equation (5)) over Ni sites at low temperatures [77,79–83]. At 600 °C, the Co-, Ni-, and Cu-based catalysts demonstrated a similar H_2 yield of ~62%, with Fe/CeO_2 catalyst exhibiting a slightly better yield (~67%). It is worth mentioning that at 600 °C CO formation is in general favored at the expense of CO_2, probably implying participation of the reverse WGS reaction on the ESR reaction network (Figure 5). At 800 °C the Co- and Ni-based catalysts showed H_2 yields of 65% and 62%, respectively, followed by a CO yield of 35% and 38%. Therefore, in these catalysts, the ESR reaction leads almost exclusively to H_2 and CO. In contrast, the Fe- and Cu-based catalysts as well as the bare CeO_2 exhibited significantly lower syngas yields and higher amounts of CH_4, implying an inferior reforming performance at high reaction temperatures.

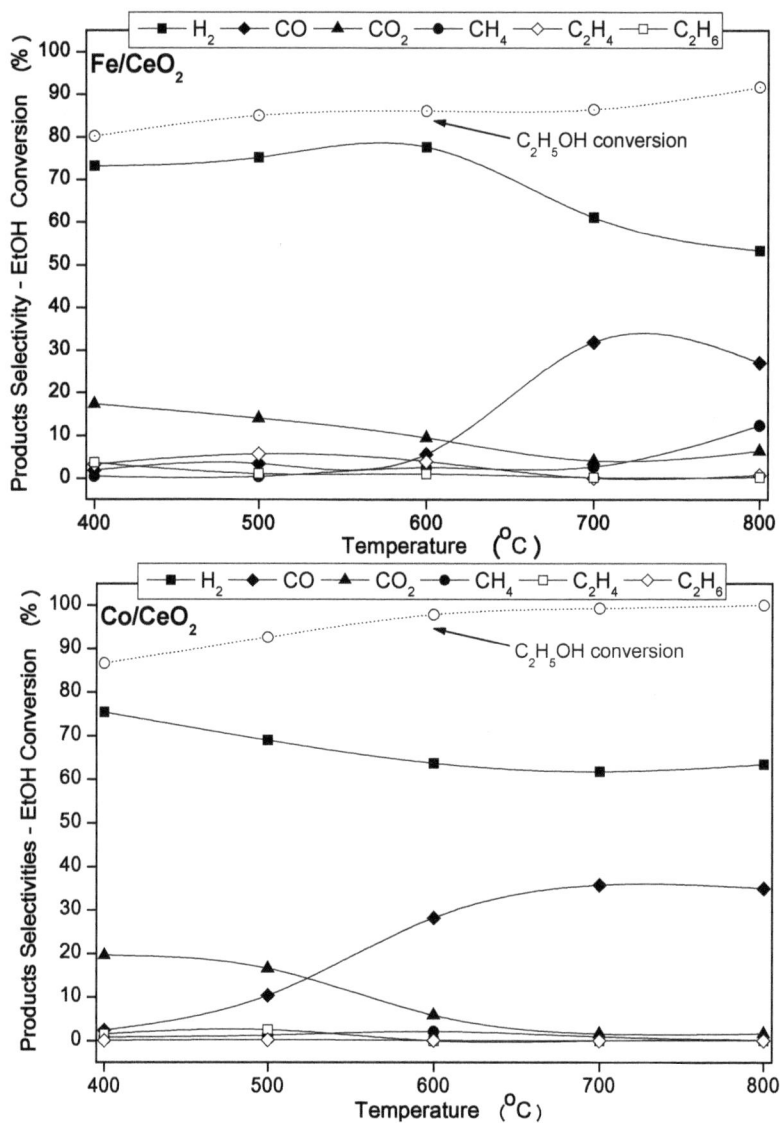

Figure 6. *Cont.*

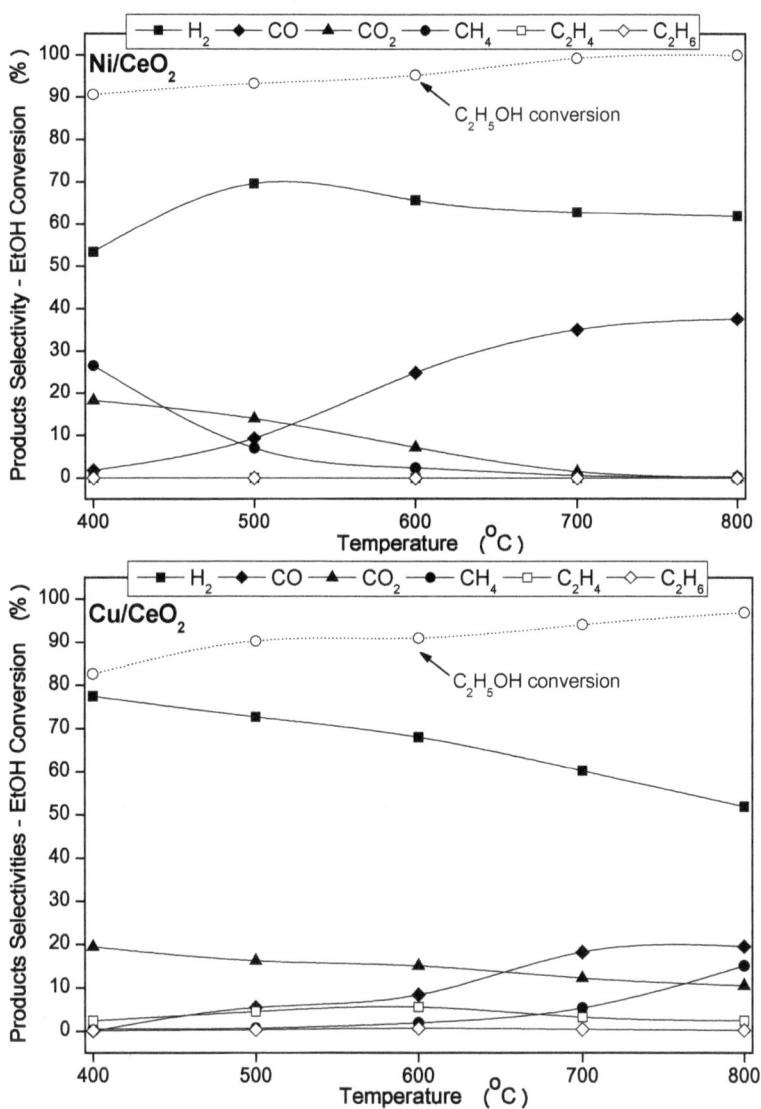

Figure 6. Effect of temperature on product selectivity and C_2H_5OH conversion over M/CeO_2 catalysts. Reaction conditions: S/C = 3, m_{cat} = 250 mg, F_T = 150 cm^3/min.

2.2.2. Effect of Metal Loading

Figure 8 depicts the effect of Co loading on the main products yield at 400, 600, and 800 °C. It is evident that in the absence of metal (bare CeO_2), the yields toward the reformate products (CO and H_2) are much lower compared to the corresponding ones over Co/CeO_2 catalysts. At 400 °C the optimum H_2 yield

(~66%) is obtained for 20 wt.% metal loading, whereas at higher temperatures both the absolute products yields and product distribution are almost independent of the metal loading (0–30 wt.%).

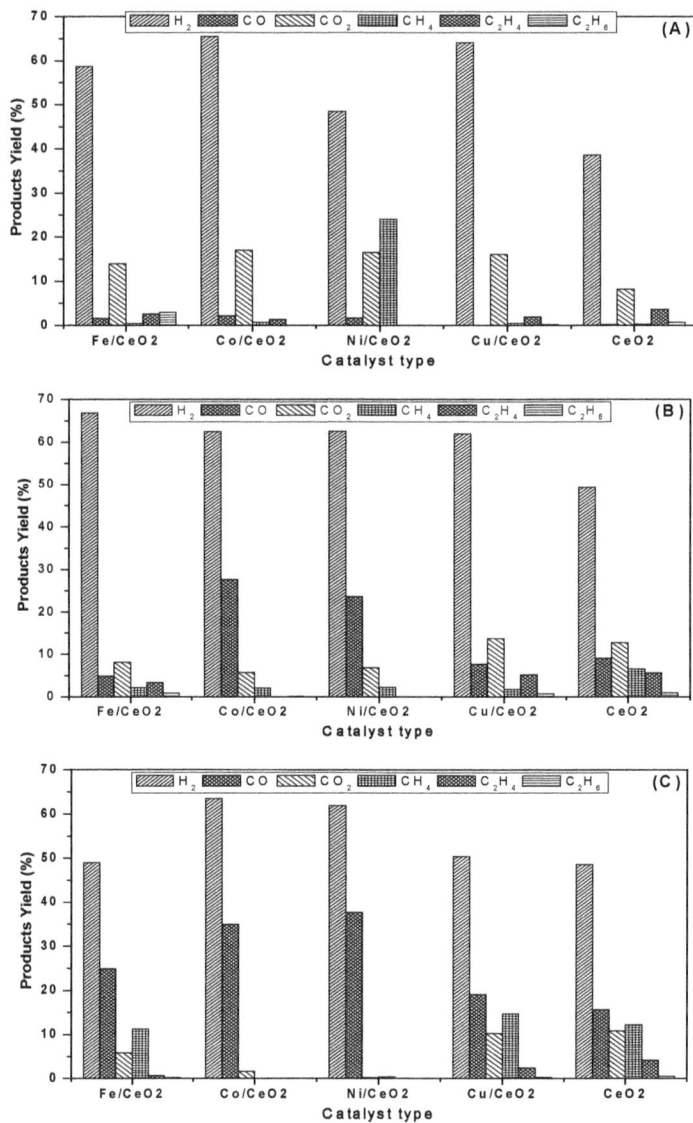

Figure 7. Effect of metal type on major product yield at 400 °C (**A**); 600 °C (**B**) and 800 °C (**C**). Reaction conditions: S/C = 3, m_{cat} =250 mg, F_T = 150 cm^3/min.

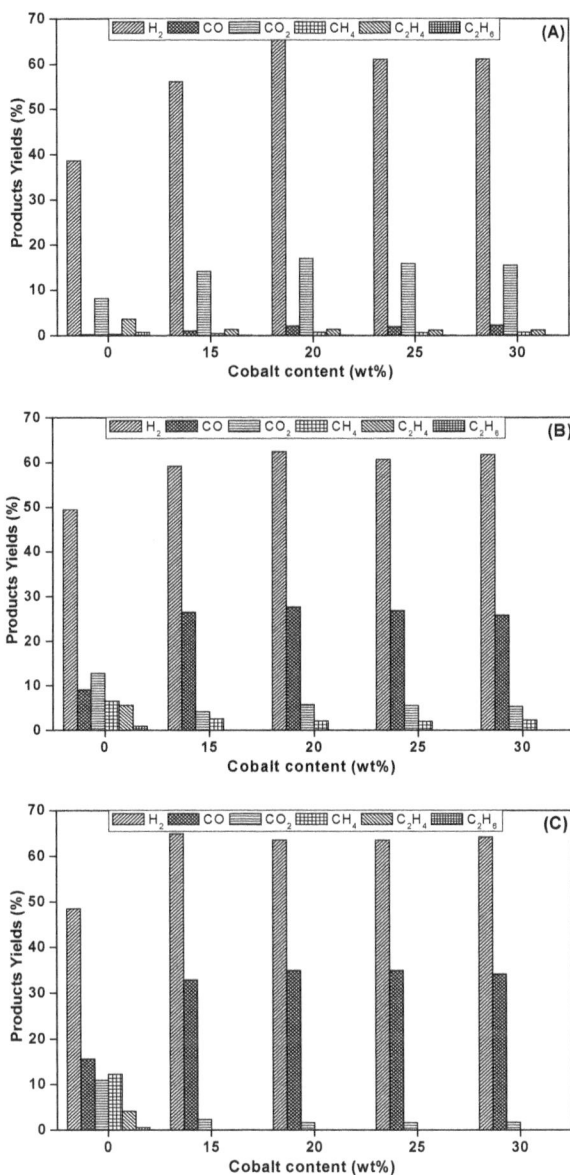

Figure 8. Effect of Co loading in Co/CeO_2 catalysts on products yield at 400 °C (**A**); 600 °C (**B**) and 800 °C (**C**). Reaction conditions: S/C = 3, m_{cat} = 250 mg, F_T = 150 cm^3/min.

2.2.3. Effect of S/C Ratio

Figure 9 presents the effect of S/C (0.5, 1.5, 3, 6) feed ratio on the major product yield achieved over the 20 wt.% Co/CeO_2 catalyst. At low temperatures, *i.e.*, at 400 °C, the optimum performance is obtained for S/C reactants ratio of 3 (~66%), whereas for temperatures higher than 400 °C the H_2 yield is maximized for a S/C feed ratio of 6. At 800 °C the yield toward H_2 is equal to 77% for a S/C ratio of 6, as compared to approximately 63% for lower S/C ratios. However, it is worth noticing, that for a S/C ratio of 3 the largest fuel production was experienced, *i.e.*, sum of H_2 and CO, as the yield with regard to CO_2 and hydrocarbons is essentially negligible. Therefore, the S/C ratio equal to 3 can be regarded as optimal, in terms of ethanol energy exploitation.

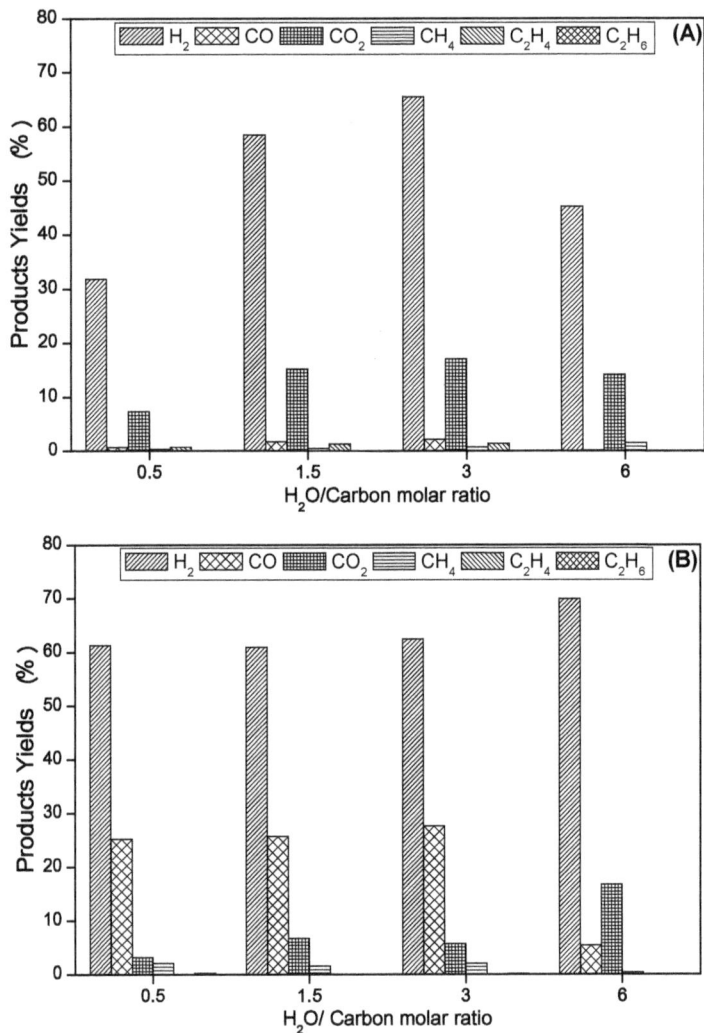

Figure 9. *Cont.*

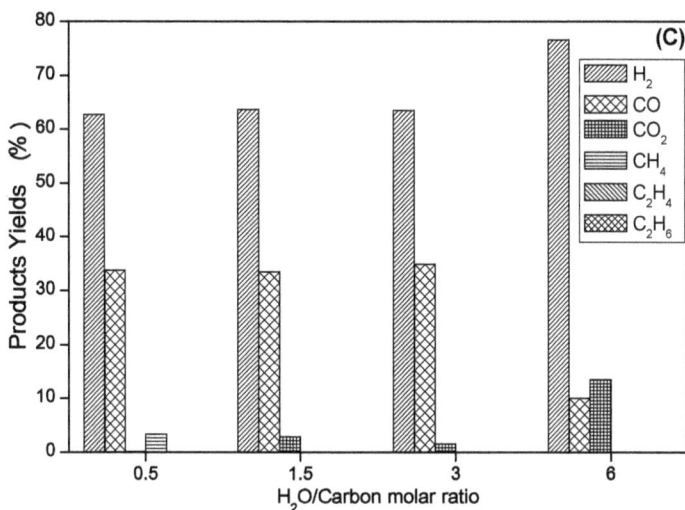

Figure 9. Effect of H_2O/Carbon (S/C) feed molar ratio on product yield at 400 °C (**A**); 600 °C (**B**) and 800 °C (**C**) over 20 wt.% Co/CeO$_2$ catalyst. Reaction conditions: m_{cat} = 250 mg, F_T = 150 cm^3/min.

2.2.4. Stability Experiments

Short-term (24 h) stability experiments were then carried out over Co-, Ni-, Fe- and Cu-CeO$_2$ catalysts to assess the catalyst life time characteristics. Stability experiments were performed at low (400 °C), intermediate (600 °C) and high (800 °C) temperatures to reveal the effect of temperature on stability performance. Figure 10 depicts the H$_2$ production rate at 400 °C (A), 600 °C (B) and 800 °C (C) *versus* time on stream (TOS) at a steam-to-carbon (S/C) ratio equal to 3. In these experiments the catalyst was first exposed to pure He up to the desired temperature, and then the standard feed mixture was introduced into the reactor.

The superiority of Co/CeO$_2$ catalysts in terms of H$_2$ production and stability is obvious. Co-based catalysts demonstrated an excellent stability performance at 600 and 800 °C whereas a slight degradation was observed at 400 °C. At low temperatures (400 °C) the formation rate of hydrogen was gradually decreased over Ni/CeO$_2$ catalysts during the first 13 h, remaining then stable. A more rapid deactivation was obtained with Cu- and Fe-based catalysts, which were totally deactivated after about 12 and 4 h, respectively. A similar picture was recorded at intermediate temperatures (Figure 10B), although in this case the deactivation was milder. Ni/CeO$_2$ catalysts were stable in the first 14 h, declined then steadily. Cu- and Fe-based catalysts followed a downgrade trend in the first 10 h, stabilized then at very low levels. In contrast, at high temperatures, *i.e.*, 800 °C (Figure 10C), all catalysts demonstrated an adequate stability performance throughout the period of 24 h.

113

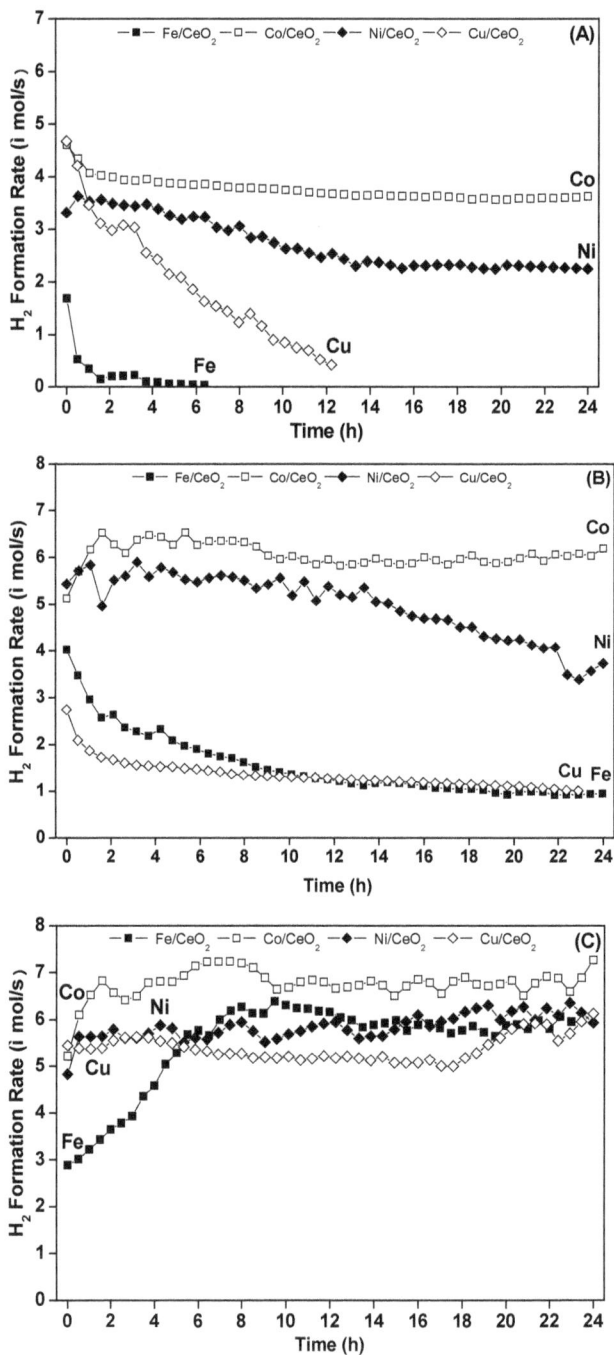

Figure 10. Stability performance of M/CeO_2 catalysts (M: Fe, Co, Ni, Cu) at 400 °C (**A**); 600 °C (**B**) and 800 °C (**C**) during ESR. Reaction conditions: S/C=3, m_{cat} = 250 mg, F_T = 150 cm^3/min.

3. Discussion

The present findings clearly demonstrated the superior steam reforming performance of Co/CeO$_2$ catalysts, in terms of H$_2$ production and stability, in the examined temperature range of 400–800 °C. In contrast all other catalysts (Fe-, Ni-, Cu-CeO$_2$) exhibited satisfactory stability only at high temperatures (800 °C), whereas at low and intermediate temperatures a gradual deactivation was observed, being more intense for Cu- and Fe-based catalysts. The optimum performance of high metal loading Co-based catalysts, compared to conventional Ni-based catalyst, should be also emphasized. The latter is of major importance with regard to the direct utilization of ethanol in SOFCs, where highly active, conductive and robust electro-catalysts are required [84].

To better understand the influence of temperature and catalyst nature on stability performance indicative morphological studies (SEM/EDS) were carried out over the fast deactivated Cu/CeO$_2$ catalysts and very stable Co/CeO$_2$ catalysts (not shown for brevity's sake). In general, a significant amount of carbon was detected over Cu/CeO$_2$ exposed for 24 h to reaction conditions at 400 °C, which, however, decreased at 600 °C and was almost eliminated at 800 °C. To be specific, a small amount of deposited carbon, mostly in the form of carbon fibers, was identified at elevated temperatures. Moreover, no significant changes in the particle size were recorded upon temperature increase.

Given that the Cu/CeO$_2$ catalyst maintains its activity at 800 °C (Figure 10C) it could be argued that neither catalyst particle sintering nor the formation of carbon fibers could affect the catalysts' stability at high temperatures. In this regard, it has been shown that filamentous carbon does not directly lead to the poisoning of CeO$_2$-based catalysts in ESR, although it delaminates metal particles from the support; amorphous carbon tends to block active sites leading to deactivation [85]. Therefore, the deactivation of Cu/CeO$_2$ catalysts at low (400 °C), and to a lesser extent at intermediate (600 °C) temperatures, can be mainly ascribed to amorphous carbon deposition. The improved stability performance at 600 °C, compared to 400 °C, can be attributed to the facile gasification of carbonaceous deposits at elevated temperatures [85–87].

Similar conclusions, in relation to the impact of reaction temperature on the carbon deposition and particle size of Co/CeO$_2$ catalysts were obtained by means of SEM/EDS analysis. The particle size remained practically unaffected by temperature increase to 600 °C, whereas it was slightly increased after reaction at 800 °C. Amorphous carbon was revealed by EDS analysis and elemental mapping for spent Co/CeO$_2$ at 400 °C, although at a relatively lesser amount compared to Cu/CeO$_2$. Carbon was also identified at 600 °C, but at a significantly lower amount to that observed at 400 °C. At higher temperature of 800 °C the carbon was almost absent.

In the light of the above results, it could be stated that Co/CeO_2 catalyst displayed the optimum reforming performance in terms of H_2 yield and life-time stability. In contrast, an inferior performance was obtained over Ni-, Cu- and Fe-based catalysts. In a similar manner, it was found that neither Cu nor Ni supported on ZnO catalysts were stable in ESR conditions [24]. On Cu/ZnO the dehydrogenation of ethanol into acetaldehyde was favored, whereas on Ni/ZnO the decomposition of ethanol to CH_4 and CO_x took place.

The present findings can be mainly interpreted by taking into account the surface and redox properties of transition metal-based catalysts. TPR studies revealed the lower reducibility of Co/CeO_2 catalyst amongst the investigated samples. The latter, could be possibly related to the strong metal-support interactions, maintaining the Co species in an oxidized state ($Co^{\delta+}$ species). This factor could be considered responsible for the adequate ESR performance of Co/CeO_2 catalysts, in agreement with relevant literature studies [42–44,88–91].

On the other hand, although Cu-based materials are highly reducible under mild conditions, they are almost inactive toward reforming reactions, leading mainly to ethanol dehydrogenation. In this regard, it has been reported that the over-reduction of CeO_2-based materials under H_2-rich environments, can lead to the blockage of the redox mechanism involved in ESR and the carbon gasification reaction network [92–94].

In addition, the high concentration of lattice oxygen in Co/CeO_2, as compared to Fe-, Cu- and Ni-CeO_2 catalysts, can be further considered for their superior ESR performance. To this end, the key role of oxygen availability in both H_2 yield and stability of Co-based catalysts has been revealed [36]; the gas-solid interactions between the adsorbed hydrocarbon species (C_nH_m) and the surface oxygen groups (O_x) can facilitate hydrocarbon species gasification instead of their decomposition to carbon.

The inhibition of carbon deposition by gas-solid interactions between hydrocarbons and lattice oxygen has been already demonstrated in several catalytic systems [95–98]. In this regard, isotopic tracer and nuclear reaction analysis (NRA) clearly suggested that methane is selectively oxidized by lattice oxygen ions to produce carbon monoxide during the catalytic partial oxidation (CPOX) of methane to synthesis gas on $Rh/Ce_{0.56}Zr_{0.44}O_{2-x}$ catalysts [97].

Based on the above aspects it can be stated that the intrinsic reactivity of Co entities, along with the strong metal-support interactions in Co/CeO_2 composites, maybe accounted for the high oxidation degree of cobalt entities, as well as for their lower reducibility under reaction conditions. Furthermore, the high concentration of lattice oxygen in the Co/CeO_2 catalyst can be further regarded for the inhibition of carbon deposition through the gas-solid interactions between the adsorbed hydrocarbon species and the lattice oxygen. These factors could be considered equally crucial for the transformation of the deposited carbonaceous species into gaseous products during the stream reforming process.

The present results are in agreement with those obtained over Cu- and Co-based catalysts during ethanol [90] or iso-octane [48,93] steam reforming. Highly reducible Cu-based catalysts are almost inactive for ESR, leading mainly to ethanol dehydrogenation [90]. In contrast, Co/SiO_2 catalysts demonstrated superior reforming performance, despite their limited reducibility at low temperatures due to the strong metal-support interactions [90]. Along the same lines, it has been demonstrated that the oxygen storage capacity of the support can greatly affect the reforming performance of Co-based catalysts by promoting the gasification of carbon deposits [88,91]. To this end, the superior ESR performance of Co/CeO_2 catalysts as compared to ZrO_2- or Al_2O_3-based samples has been ascribed to the abundance of redox sites on the catalyst surface [36,40,41]. More specifically, it has been clearly revealed by means of steady-state reaction experiments coupled with post-reaction characterization techniques that the high oxygen mobility in Co/CeO_2 samples—linked with the high oxygen storage capacity (OSC) of ceria—not only suppresses the carbon deposition, but also allows delivery of oxygen to ethoxy species, thus promoting the complete oxidation of carbon to CO_2 [36].

In the light of the above analysis, it is evident that in order to fully understand the complex and dynamic ESR mechanism of ceria-based transition metal catalysts, it is essential to know the local surface structure and chemistry of active entities under real "working" conditions. This requires the combination of advanced *ex situ* characterization studies with *in situ* operando measurements. In this regard, the key role of metal-ceria interactions on the ESR performance, has been verified by combining *in situ* and *ex situ* characterization techniques [25,41].

4. Materials and Methods

4.1. Materials Synthesis

A series of CeO_2 supported transition metal catalysts (Fe, Co, Ni, Cu) were synthesized using the incipient wet impregnation method. Initially, the bare CeO_2 was prepared by diluting CeO_2 precursor salt ($Ce(NO_3)_3 \cdot 6H_2O$, Sigma Aldrich, Taufkirchen, Germany) in double distilled water. The whole contents were heated and stirred until the evaporation of water. The resulted sample was dried at 100 °C for 16 h and then calcined at 600 °C for 2 h (heating rate: 5 °C/min). The ceria-based catalysts were then prepared via the impregnation of calcined ceria in an aqueous metal (Fe, Co, Ni, Cu) solution, employing the corresponding nitrate salts (Sigma Aldrich, Taufkirchen, Germany) as precursors, at the appropriate concentration so as to yield 20 wt.% metal loading. To explore the impact of metal loading a series of cobalt-based catalysts with metal contents of 15, 20, 25 and 30 wt.% were also prepared following the same synthesis route. The resulted solutions were heated at first under stirring, then dried overnight at 120 °C and calcined at 600 °C for 2 h.

4.2. Characterization Studies

4.2.1. Textural Characterization (BET)

The textural characteristics of as prepared composites were determined by N_2 adsorption/desorption isotherms at 77 K, employing the multipoint BET analysis in a Tristar Micromeritics 3000 flow apparatus. BET area was measured according to the Brunauer-Emmett-Teller (BET) method at the relative pressure in the range of 0.005–0.99. The total pore volume was estimated by the nitrogen volume at the highest relative pressure. The Barret-Joyner-Halenda (BJH) method was employed to obtain the average pore diameter. Prior to measurements the catalysts were degassed at 250 °C for 16 h.

4.2.2. Structural Characterization

The crystalline structure was determined using the X-ray powder diffraction (XRD) method. A Siemens D 500 diffractometer was employed for the XRD measurements with a Cu Kα radiation (λ = 0.154 nm) operated at 40 kV and 30 mA. Diffractograms were collected in the 2θ = 10°–80° range at a scanning rate of 0.04° over 2 s. The DIFFRAC plus Basic data evaluation software was employed to identify the diffraction peaks and through the Scherrer equation to calculate the crystallite sizes of the detected phases within an experimental error of 10%.

4.2.3. Morphological Characterization

The morphology of fresh and spent catalysts was investigated by scanning electron microscopy (SEM) using a JEOL 6300 microscope, coupled with EDX (Oxford Link ISIS-2000) for the determination of local elemental composition.

4.2.4. Redox Characterization (H_2-TPR)

TPR studies were elaborated in a flow system assembled with a quadrupole mass spectrometer (Omnistar, Balzers). The TPR tests were carried out in a quartz micro-reactor loaded with *ca.* 100 mg of catalyst. Typically, ~100 mg of the material is placed in the reactor and heated to 1000 °C with a heating rate of 10 °C/min under 5% H_2/He flow (50 cm^3/min). All samples were heated at 600 °C for 1 h under 5% O_2/He flow and then cooled down to room temperature, prior to the TPR measurements.

4.2.5. Surface Characterization (XPS)

The photoemission experiments were performed in an ultrahigh vacuum system (UHV), comprised of a fast entry specimen assembly, a sample preparation chamber and an analysis chamber. The base pressure in both chambers was kept at 1×10^{-9} mbar. Non-monochromatized Al K_α line at 1486.6 eV and an analyzer pass energy of 97 eV, giving a full width at half maximum (FWHM) of 1.7 eV for the Au $4f_{7/2}$ peak, were employed in all XPS studies. The XPS core level spectra were analyzed using a fitting routine, which can decompose each spectrum into individual mixed Gaussian-Lorentzian peaks after a Shirley background subtraction. The calibration of the analyzer's kinetic energy scale was accomplished according to ASTM-E 902–88. The binding energy (BE) scale was calibrated by attributing the main C1s peak at 284.8 (adventitious carbon) eV. The powdered samples were pressed into pellets with a thickness of 1 mm and 1 cm diameter before being introduced into the ultrahigh vacuum system.

4.3. Catalytic Activity Measurements

The apparatus used to perform the ESR experiments has been already described in our previous relevant studies regarding the iso-octane steam reforming over Cu supported on rare earth oxides catalysts [48,93]. In brief, it is comprised of: (i) a liquid reactants feed unit equipped with He mass flow controllers and the gas-liquid saturator units; (ii) a fixed bed U-shaped quartz tubular reactor (9.6 mm, i.d.) loaded with 250 mg catalyst admixed with an equal amount of quartz particles and (iii) a gas analysis system.

A mixture of ethanol and water vapors of different steam-to-carbon (S/C) molar ratios was introduced into the reactor by bubbling He (99.999% purity, Air Liquid) through two separate vessels containing liquid C_2H_5OH (99.5% purity, Riedel-de Haen) and twice distilled water. The S/C feed ratio was adjusted by controlling the temperature and He-bubbled flow in the H_2O and ethanol saturators (based on their corresponding vapor pressures). The inlet volumetric flow rate was held constant at 150 cm^3/min, corresponding to a weight/flow (W/F) and Gas Hourly Space Velocity (GHSV) of 0.1 g.s/cm^3 and 21,000 h^{-1} (based on a catalyst bed volume of about 0.45 cm^3), respectively. The feed concentration used during the activity and stability experiments is: P_{EtOH} = 2 kPa, P_{H2O} = 2, 6, 12, 24 kPa, corresponding to a S/C feed ratio of 0.5, 1.5, 3, 6, respectively. Short-term durability experiments were performed by continuously monitoring the product rate as a function of time-on-stream (TOS, 24 h) at three different temperatures, *i.e.*, 400, 600, and 800 °C.

An on-line gas chromatograph (HP 6890), equipped with thermal conductivity and flame ionization detectors, was employed to qualitatively and quantitatively analyze reactants and products. The separation of H_2, CH_4, and CO was achieved with a Molecular Sieve 5A column (10 ft × 1/8 in), whereas CO_2, H_2O, C_2H_4,

C_2H_6, C_2H_5OH, CH_3CHO and CH_3COCH_3 were separated in a Poraplot Q column (10 ft × 1/8 in) in a series bypass configuration.

The equations used to calculate the ethanol conversion (X_{EtOH}), product selectivity (S_i), and yield (Y_i), have as follows:

$$X_{EtOH} \, (\%) = \frac{[EtOH]_{inlet} - [EtOH]_{outlet}}{[EtOH]_{inlet}} \times 100 \tag{10}$$

$$S_i = \frac{r_i}{\sum r_i} \tag{11}$$

$$Y_i = X_{EtOH} * S_i \tag{12}$$

where [EtOH] denotes the ethanol concentration and r_i the formation rate of product i (mol/s) considering the following produced chemical species: H_2, CO, CO_2, CH_4, C_2H_4, C_2H_6, CH_3CHO, CH_3COCH_3.

The formation rate of product i, r_i, is calculated on the basis of its % v/v concentration in the reactor outlet, according to the equation:

$$r_i (mol/s) = \frac{\text{Concentration} \, (\%v/v) \cdot F_T \, (cm^3/min)}{100 \cdot 60 \, (s/min) \cdot v_m \, (cm^3/mol)} \tag{13}$$

where F_T, is the total flow rate and v_m, is the gas molar volume at STP conditions (298 K and 1 bar).

5. Conclusions

In the present work, the steam reforming of ethanol over transition metal catalysts (Fe, Co, Ni, Cu) supported on CeO_2 was examined. Various parameters, concerning the effect of operation temperature, steam-to-carbon ratio, and metal loading on the catalytic activity and stability were explored. The optimal catalytic performance was observed for the 20 wt.% Co/CeO_2 catalyst, which offers H_2 yields up to 66% at a steam-to-carbon feed ratio of 3. Stability experiments demonstrated the excellent stability of Co/CeO_2 catalysts in the temperature range of 400–800 °C. An inferior stability performance, particularly at low temperatures, was observed for all the other tested catalysts, following the order: Co >Ni >Cu >Fe. The excellent stability of Co/CeO_2 catalysts can be mainly assigned to the strong metal-support interactions which in turn leads to an increased population of surface oxygen species, related to lattice oxygen in ceria and cobalt oxides. This factor is thought to be responsible for the facile gasification of the adsorbed species, thus preventing catalyst deactivation. The enhanced performance of Co-based catalysts, compared to a conventional Ni-based catalyst, is of major importance towards the direct utilization of ethanol in SOFCs, where highly active and robust electro-catalysts need to be developed.

Acknowledgments: This research was supported by the European Union (European Social Fund—ESF) and Greek national funds through the Operational Program "Education and Lifelong Learning" of the National Strategic Reference Framework (NSRF)—Research Funding Program: Heracleitus II. Investing in knowledge society through the European Social Fund. The Laboratory of Surface Characterization at FORTH/ICE-HT in Patras, Greece, is acknowledged for performing the XPS measurements.

Author Contributions: M.K. and G.M. conceived and designed the project. M.K. participated in the analysis and interpretation of characterization results. Tz.K. performed catalysts synthesis, whereas Z.I. carried out catalyst characterization and evaluation. The manuscript was written through the contribution of all authors. All authors approved the final version of the manuscript.

Conflicts of Interest: The authors declare no conflict of interest.

References

1. Goltsov, V.A.; Veziroglu, T.N.; Goltsova, L.F. Hydrogen civilization of the future—A new conception of the IAHE. *Int. J. Hydrogen Energy* **2006**, *31*, 153–159.

2. Ni, M.; Leung, D.Y.C.; Leung, M.K.H.; Sumathy, K. An overview of hydrogen production from biomass. *Fuel Process. Technol.* **2006**, *87*, 461–472.

3. Ni, M.; Leung, M.K.H.; Sumathy, K.; Leung, D.Y.C. Potential of renewable hydrogen production for energy supply in Hong Kong. *Int. J. Hydrogen Energy* **2006**, *31*, 1401–1412.

4. Ni, M.; Leung, M.K.H.; Leung, D.Y.C.; Sumathy, K. A review and recent developments in photocatalytic water-splitting using TiO$_2$ for hydrogen production. *Renew. Sustain. Energy Rev.* **2007**, *11*, 401–425.

5. Galbe, M.; Zacchi, G.A. A review of the production of ethanol from softwood. *Appl. Microbiol. Biotechnol.* **2002**, *59*, 618–628.

6. Dien, B.S.; Cotta, M.A.; Jeffries, T.W. Bacteria engineered for fuel ethanol production: Current Status. *Appl. Microbiol. Biotechnol.* **2003**, *63*, 258–266.

7. Sun, Y.; Cheng, J.Y. Hydrolysis of lignocellulosic materials for ethanol production: A Review. *Bioresour. Technol.* **2002**, *83*, 1–11.

8. Cai, W.; Wang, F.; van Veen, A.C.; Provendier, H.; Mirodatos, C.; Shen, W. Autothermal reforming of ethanol for hydrogen production over an Rh/CeO$_2$ catalyst. *Catal. Today* **2008**, *138*, 152–156.

9. Zhang, B.; Tang, X.; Li, Y.; Xu, Y.; Shen, W. Hydrogen production from steam reforming of ethanol and glycerol over ceria-supported metal catalysts. *Int. J. Hydrogen Energy* **2006**, *32*, 2367–2373.

10. Ni, M.; Leung, D.Y.C.; Leung, M.K.H. A review on reforming bio-ethanol for hydrogen production. *Int. J. Hydrogen Energy* **2007**, *32*, 3238–3247.

11. Nishiguchi, T.; Matsumoto, T.; Kanai, H.; Utani, K.; Matsumura, Y.; Shen, W.J.; Imamura, S. Catalytic steam reforming of ethanol to produce hydrogen and acetone. *Appl. Catal. A* **2005**, *279*, 273–277.

12. Contreras, J.L.; Salmones, J.; Colin-Luna, J.A.; Nuno, L.; Quintana, B.; Cordova, I.; Zeifert, B.; Tapia, C.; Fuentes, G.A. Catalysts for H₂ production using the ethanol steam reforming (a review). *Int. J. Hydrogen Energy* **2014**, *39*, 18835–18853.

13. Haryanto, H.; Fernando, S.; Murali, N.; Adhikari, S. Current Status of Hydrogen Production Techniques by Steam Reforming of Ethanol: A Review. *Energy Fuels* **2005**, *19*, 2098–2106.

14. Frusteri, F.; Freni, S.; Spadaro, L.; Chiodo, V.; Bonura, G.; Donato, S.; Cavallaro, S. H₂ production for MC fuel cell by steam reforming of ethanol over MgO supported Pd, Rh, Ni and Co catalysts. *Catal. Commun.* **2004**, *5*, 611–615.

15. Erdohelyi, A.; Rasko, J.; Kecskes, T.; Toth, M.; Domok, M.; Baan, K. Hydrogen formation in ethanol reforming on supported noble metal catalysts. *Catal. Today* **2006**, *116*, 367–376.

16. Liguras, D.K.; Kondarides, D.I.; Verykios, X.E. Production of hydrogen for fuel cells by steam reforming of ethanol over supported noble metal catalysts. *Appl. Catal. B* **2003**, *43*, 345–354.

17. Cavallaro, S.; Chiodo, V.; Freni, S.; Mondello, N.; Frusteri, F. Performance of Rh/Al₂O₃ catalyst in the steam reforming of ethanol: H₂ Production for MCFC. *Appl. Catal. A* **2003**, *249*, 119–128.

18. Diagne, C.; Idriss, H.; Pearson, K.; Gomez-Garcia, M.A.; Kiennemann, A.R. Efficient hydrogen production by ethanol reforming over Rh catalysts effect of addition of Zr on CeO₂ for the oxidation of CO to CO₂. *C.R. Chim.* **2004**, *7*, 617–622.

19. Mathure, P.V.; Ganguly, S.; Patwardhan, A.V.; Saha, R.K. Steam reforming of ethanol using a commercial nickel-based catalyst. *Ind. Eng. Chem. Res.* **2007**, *46*, 8471–8479.

20. Konsolakis, M.; Ioakeimidis, Z. Surface/structure functionalization of copper-based catalysts by metal-support and/or metal–metal interactions. *Appl. Surf. Sci.* **2014**, *320*, 244–255.

21. Gates, S.M.; Russell, J.N.; Yates, J.T. Bond activation sequence observed in the chemisorption and surface reaction of ethanol on Ni(lll). *Surf. Sci.* **1986**, *171*, 111–134.

22. Sehested, J. Four challenges for nickel steam-reforming catalysts. *Catal. Today* **2006**, *111*, 103–110.

23. Patel, M.; Jindal, T.K.; Pant, K.K. Kinetic Study of steam reforming of ethanol on Ni-Based ceria—Zirconia catalyst. *Ind. Eng. Chem. Res.* **2013**, *52*, 15763–15771.

24. Homs, N.; Llorca, J.; Ramírez de la Piscina, P. Low-temperature steam-reforming of ethanol over ZnO-supported Ni and Cu catalysts: The Effect of Nickel and Copper Addition to ZnO-Supported Cobalt-Based Catalysts. *Catal. Today* **2006**, *116*, 361–336.

25. Xu, W.; Liu, Z.; Johnston-Peck, A.C.; Senanayake, S.D.; Zhou, G.; Stacchiola, D.J.; Stach, E.A.; Rodriguez, J.A. Steam reforming of ethanol on Ni/CeO₂: Reaction Pathway and Interaction between Ni and the CeO₂ Support. *ACS Catal.* **2013**, *3*, 975–984.

26. Senanayake, S.D.; Evans, J.; Agnoli, S.; Barrio, L.; Chen, T.-L.; Hrbek, J.; Rodriguez, J.A. Water–gas shift and CO methanation reactions over Ni–CeO₂(111) catalysts. *Top. Catal.* **2011**, *54*, 34–41.

27. Carrasco, J.; Lopez-Duran, D.; Liu, Z.; Duchon, T.; Evans, J.; Senanayake, S.D.; Crumlin, E.J.; Matolin, V.; Rodriguez, J.A.; Ganduglia-Pirovano, M.V. *In situ* and theoretical studies for the dissociation of water on an active Ni/CeO_2 catalyst: Importance of Strong Metal–Support Interactions for the Cleavage of O–H Bonds. *Angew. Chem. Int. Ed.* **2015**, *54*, 1–6.

28. Polychronopoulou, K.; Kalamaras, C.M.; Efstathiou, A.M. Ceria-Based Materials for Hydrogen Production Via Hydrocarbon Steam Reforming and Water-Gas Shift Reactions. *Rec. Pat. Mat. Sci.* **2011**, *4*, 122–145.

29. Ilieva, L.; Pantaleo, G.; Sobczak, J.W.; Ivanov, I.; Venezia, A.M.; Andreev, D. NO reduction by CO in the presence of water over gold supported catalysts on CeO_2_Al_2O_3 mixed support, prepared by mechanochemical activation. *Appl. Catal. B* **2007**, *76*, 107–114.

30. Baudin, F.; da Costa, P.; Thomas, C.; Calvo, S.; Lendresse, Y.; Schneider, S.; Delacroix, F.; Plassat, G.; Mariadassou, G.D. NO_x reduction over CeO_2-ZrO_2 supported iridium catalyst in the presence of propanol. *Top. Catal.* **2004**, *30/31*, 97–101.

31. De Lima, S.M.; da Cruz, I.O.; Jacobs, G.; Davis, B.H.; Mattos, L.V.; Noronha, F.B. Steam reforming, partial oxidation, and oxidative steam reforming of ethanol over $Pt/CeZrO_2$ catalyst. *J. Catal.* **2008**, *257*, 356–368.

32. Laosiripojana, N.; Assabumrungrat, S. Catalytic steam reforming of ethanol over high surface area CeO_2: The Role of CeO_2 as an Internal Pre-Reforming Catalyst. *Appl. Catal. B* **2006**, *66*, 29–39.

33. Zhang, C.; Li, S.; Li, M.; Wang, S.; Ma, X.; Gong, J. Enhanced Oxygen Mobility and Reactivity for Ethanol Steam Reforming. *AIChE J.* **2012**, *58*, 516–525.

34. Wang, H.; Liu, Y.; Wang, L.; Qin, Y. Study on the Carbon Deposition in Steam Reforming of Ethanol over Co/CeO_2 Catalyst. *Chem. Eng. J.* **2008**, *145*, 25–31.

35. Song, H.; Ozkan, U.S. Changing the Oxygen Mobility in Co/Ceria Catalysts by Ca Incorporation: Implications for Ethanol Steam Reforming. *J. Phys. Chem. A* **2010**, *114*, 3796–3801.

36. Song, H.; Ozkan, U.S. Ethanol steam reforming over Co-based catalysts: Role of Oxygen Mobility. *J. Catal.* **2009**, *261*, 66–74.

37. Llorca, J.; Homs, N.; Sales, J.; Ramirez de la Piscina, P. Efficient production of hydrogen over supported cobalt catalysts from Ethanol Steam Reforming. *J. Catal.* **2002**, *209*, 306–317.

38. Llorca, J.; Ramirez de la Piscina, P.; Dalmonb, J.-A.; Sales, J.; Homs, N. CO-free hydrogen from steam-reforming of bioethanol over ZnO-supported cobalt catalysts. Effect of the metallic precursor. *Appl. Catal. B* **2003**, *43*, 355–369.

39. Casanovas, A.; Roig, M.; de Leitenburg, C.; Trovarelli, A.; Llorca, J. Ethanol steam reforming and water gas shift over Co/ZnO catalytic honeycombs doped with Fe, Ni, Cu, Cr and Na. *Int. J. Hydrogen Energy* **2010**, *35*, 7690–7698.

40. Song, H.; Zhang, L.; Watson, R.B.; Braden, D.; Ozkan, U.S. Investigation of bio-ethanol steam reforming over cobalt-based catalysts. *Catal. Today* **2007**, *129*, 346–354.

41. Soykal, I.I.; Sohn, H.; Ozkan, U.S. Effect of Support Particle Size in Steam Reforming of Ethanol over Co/CeO_2 Catalysts. *ACS Catal.* **2012**, *2*, 2335–2348.

42. Espinal, R.; Taboada, E.; Molins, E.; Chimentao, R.J.; Medina, F.; Llorca, J. Cobalt hydrotalcites as catalysts for bioethanol steam reforming. The promoting effect of potassium on catalyst activity and long-term stability. *Appl. Catal. B* **2012**, *127*, 59–67.

43. Espinal, R.; Anzola, A.; Adrover, E.; Roig, M.; Chimentao, R.; Medina, F.; Lopez, E.; Borio, D.; Llorca, J. Durable ethanol steam reforming in a catalytic membrane reactor at moderate temperature over cobalt hydrotalcite. *Int. J. Hydrogen Energy* **2014**, *39*, 10902–10910.

44. Espinal, R.; Taboada, E.; Molins, E.; Chimentao, R.J.; Medina, F.; Llorca, J. Ethanol Steam reforming over hydrotalcite-derived Co Catalysts doped with Pt and Rh. *Top. Catal.* **2013**, *56*, 1660–1671.

45. Ferencz, Z.; Erdőhelyi, A.; Baán, K.; Oszkó, A.; Óvári, L.; Kónya, Z.; Papp, C.; Steinrück, H.-P.; Kiss, J. Effects of support and Rh additive on Co-based catalysts in the ethanol steam reforming reaction. *ACS Catal.* **2014**, *4*, 1205–1218.

46. Varga, E.; Ferencz, Z.; Oszkó, A.; Erdőhelyi, A.; Kiss, J. Oxidation states of active catalytic centers in ethanol steam reforming reaction on ceria based Rh promoted Co catalysts: An XPS Study. *J. Mol. Catal. A* **2015**, *397*, 127–133.

47. Gamarra, D.; Munuera, G.; Hungría, A.B.; Fernández-García, M.; Conesa, J.C.; Midgley, P.A.; Wang, X.Q.; Hanson, J.C.; Rodríguez, J.A.; Martínez-Arias, A. Structure-activity relationship in nanostructured copper-ceria-based preferential CO oxidation catalysts. *J. Phys. Chem. C* **2007**, *111*, 11026–11038.

48. Al-Musa, A.; Al-Saleh, M.; Ioakimidis, Z.; Ouzounidou, M.; Yentekakis, I.V.; Konsolakis, M.; Marnellos, G.E. Hydrogen production by iso-octane steam reforming over Cu catalysts supported on Rare Earth Oxides (REOs). *Int. J. Hydrogen Energy* **2014**, *39*, 1350–1363.

49. Breen, B.; Ross, J.R.H. Methanol reforming for fuel-cell applications: Development of Zirconia-Containing Cu-Zn-Al Catalysts. *Catal. Today* **1999**, *511*, 521–533.

50. Turco, M.; Bagnasco, G.; Cammarano, C.; Senese, P.; Costantino, U.; Sisani, M. Cu/ZnO/Al$_2$O$_3$ catalysts for oxidative steam reforming of methanol: The Role of Cu and the Dispersing Oxide Matrix. *Appl. Catal. B* **2007**, *77*, 46–57.

51. Tang, X.; Zhang, B.; Li, Y.; Xu, Y.; Xin, Q.; Shen, W. CuO/CeO$_2$ catalysts: Redox Features and Catalytic Behaviors. *Appl. Catal. A* **2005**, *288*, 116–125.

52. Papavasiliou, J.; Avgouropoulos, G.; Ioannides, T. Effect of dopants on the performance of CuO-CeO$_2$ catalysts in methanol steam reforming. *Appl. Catal. B* **2007**, *69*, 226–234.

53. Konsolakis, M.; Sgourakis, M.; Carabineiro, S.A.C. Surface and redox properties of cobalt–ceria binary oxides: On the Effect of Co Content and Pretreatment Conditions. *Appl. Surf. Sci.* **2015**, *341*, 48–54.

54. Zhang, C.; Li, S.; Wu, G.; Gong, J. Synthesis of stable Ni-CeO$_2$ catalysts via ball-milling for ethanol steam reforming. *Catal. Today* **2014**, *233*, 53–60.

55. Shan, W.; Luo, M.; Ying, P.; Shen, W.; Can, L. Reduction property and catalytic activity of Ce$_{1-x}$Ni$_x$O$_2$ mixed oxide catalysts for CH$_4$ oxidation. *Appl. Catal. A* **2003**, *246*, 1–9.

56. Zhang, B.; Tang, X.; Li, Y.; Cai, W.; Xu, Y.; Shen, W. Steam reforming of bio-ethanol for the production of hydrogen over ceria-supported Co, Ir and Ni catalysts. *Catal. Commun.* **2006**, *7*, 367–372.

57. Frusteri, F.; Freni, S.; Chiodo, V.; Donato, S.; Bonura, G.; Cavallaro, S. Steam and auto-thermal reforming of bio-ethanol over MgO and CeO_2 Ni supported catalysts. *Int. J. Hydrogen Energy* **2006**, *31*, 2193–2199.

58. Perez-Alonso, F.J.; Melián-Cabrera, I.; López Granados, M.; Kapteijn, F.; Fierro, J.L.G. Synergy of $Fe_xCe_{1-x}O_2$ mixed oxides for N_2O decomposition. *J. Catal.* **2006**, *239*, 340–346.

59. Jin, Y.; Datye, A.K. Phase Transformations in Iron Fischer-Tropsch catalysts during temperature-programmed reduction. *J. Catal.* **2000**, *196*, 8–17.

60. Voskoboinikov, T.V.; Chen, H.Y.; Sachtler, W.M.H. On the nature of active sites in Fe/ZSM-5 catalysts for NO_x abatement. *Appl. Catal. B* **1998**, *19*, 279–287.

61. Gálvez, M.E.; Ascaso, S.; Stelmachowsk, P.; Legutko, P.; Kotarba, A.; Moliner, R.; Lázaro, M.J. Influence of the surface potassium species in $Fe-K/Al_2O_3$ catalysts on the soot oxidation activity in the presence of NO_x. *Appl. Catal. B* **2014**, *152–153*, 88–98.

62. Reddy, G.K.; Boolchand, P.; Smirniotis, P.G. Sulfur tolerant metal doped Fe/Ce catalysts for high temperature WGS reaction at low steam to CO ratios—XPS and Mössbauer spectroscopic study. *J. Catal.* **2011**, *282*, 258–269.

63. Zyryanova, M.M.; Snytnikov, P.V.; Gulyaev, R.V.; Amosov Yu, I.; Boronin, A.I.; Sobyanin, V.A. Performance of Ni/CeO_2 catalysts for selective CO methanation in hydrogen-rich gas. *Chem. Eng. J.* **2014**, *238*, 189–197.

64. Nesbitt, H.W.; Legrand, D.; Bancroft, G.M. Interpretation of Ni2p XPS spectra of Ni conductors and Ni insulators. *Phys. Chem. Miner.* **2000**, *27*, 357–366.

65. Biesinger, M.C.; Payne, B.P.; Lau, L.W.M.; Gerson, A.; Smart, R.S.C. X-ray photoelectron spectroscopic chemical state quantification of mixed nickel metal, oxide and hydroxide systems. *Surf. Interface Anal.* **2009**, *41*, 324–332.

66. Kundakovic, L.; Flytzani-Stephanopoulos, M. Reduction characteristics of copper oxide in cerium and zirconium oxide systems. *Appl. Catal. A* **1998**, *171*, 13–29.

67. Li, W.; Flytzani-Stephanopoulos, M. Total Oxidation of Carbon-Monoxide and Methane over Transition Metal Fluorite Oxide Composite Catalysts: II. Catalyst Characterization and Reaction-Kinetics. *J. Catal.* **1995**, *153*, 317–332.

68. Ayastuy, J.L.; Gurbani, A.; González-Marcos, M.P.; Gutiérrez-Ortiz, M.A. Selective CO oxidation in H_2 streams on $CuO/Ce_xZr_{1-x}O_2$ catalysts: Correlation between Activity and Low Temperature Reducibility. *Int. J. Hydrogen Energy* **2012**, *37*, 1993–2006.

69. Lamonier, C.; Bennani, A.; D'Huysser, A.; Aboukaïs, A.; Wrobel, G. Evidence for different copper species in precursors of copper-cerium oxide catalysts for hydrogenation reactions: An X-Ray Diffraction, EPR and X-Ray Photoelectron Spectroscopy Study. *Faraday Trans.* **1996**, *92*, 131–136.

70. Mai, H.; Zhang, D.; Shi, L.; Li, H. Highly active $Ce_{1-x}Cu_xO_2$ nanocomposite catalysts for the low temperature oxidation of CO. *Appl. Surf. Sci.* **2011**, *257*, 7551–7559.

71. Rao, K.N.; Venkataswamy, P.; Reddy, B.M. Structural Characterization and Catalytic Evaluation of Supported Copper–Ceria Catalysts for Soot Oxidation. *Ind. Eng. Chem. Res.* **2011**, *50*, 11960–11969.

72. Santos, V.P.; Carabineiro, S.A.C.; Bakker, J.J.W.; Soares, O.S.G.P.; Chen, X.; Pereira, M.F.R.; Órfão, J.J.M.; Figueiredo, J.L.; Gascon, J.; Kapteijn, F. Stabilized gold on cerium-modified cryptomelane: Highly Active in Low-Temperature CO Oxidation. *J. Catal.* **2014**, *309*, 58–65.

73. Varga, E.; Pusztai, P.; Óvári, L.; Oszkó, A.; Erdőhelyi, A.; Papp, C.; Steinrück, H.-P.; Kónya, Z.; Kiss, J. Probing the interaction of Rh, Co and bimetallic Rh–Co nanoparticles with the CeO₂ support: Catalytic Materials for Alternative Energy Generation. *Phys. Chem. Chem. Phys.* **2015**, *17*, 27154–27166.

74. Vári, G.; Óvári, L.; Papp, C.; Steinrück, H.-P.; Kiss, J.; Kónya, Z. The Interaction of Cobalt with CeO₂(111) Prepared on Cu(111). *J. Phys. Chem. C* **2015**, *119*, 9324–9333.

75. Ocampo, F.; Louis, B.; Roger, A.C. Methanation of carbon dioxide over nickel-based Ce₀.₇₂Zr₀.₂₈O₂ mixed oxide catalysts prepared by sol–gel method. *Appl. Catal. A* **2009**, *369*, 90–96.

76. Palma, V.; Castaldo, F.; Ciambelli, P.; Iaquaniello, G. CeO₂-supported Pt/Ni catalyst for the renewable and clean H₂ production via ethanol steam reforming. *Appl. Catal. B Environ.* **2013**, *145*, 73–84.

77. Carrero, A.; Calles, J.A.; Vizcaino, A.J. Hydrogen production by ethanol steam reforming over Cu-Ni/SBA-15 supported catalysts prepared by direct synthesis and impregnation. *Appl. Catal. A* **2007**, *327*, 82–94.

78. Pang, X.; Chen, Y.; Dai, R.; Cui, P. Co/CeO₂ Catalysts Prepared Using Citric Acid Complexing for Ethanol Steam Reforming. *Chin. J. Catal.* **2012**, *33*, 281–289.

79. Fatsikostas, A.N.; Verykios, X.E. Reaction network of steam reforming of ethanol over Ni-based catalysts. *J. Catal.* **2004**, *225*, 439–452.

80. Fierro, V.; Akdim, O.; Mirodatos, C. On-board hydrogen production in a hybrid electric vehicle by bio-ethanol oxidative steam reforming over Ni and noble metal based catalysts. *Green Chem.* **2003**, *5*, 20–24.

81. Mariño, F.; Boveri, M.; Baronetti, G.; Laborde, M. Hydrogen production from steam reforming of bioethanol using Cu/Ni/K/γ-Al₂O₃ catalysts. Effect of Ni. *Int. J. Hydrogen Energy* **2001**, *26*, 665–668.

82. Melnick, J.G.; Radosevich, A.T.; Villagrán, D.; Nocera, D.G. Decarbonylation of ethanol to methane, carbon monoxide and hydrogen by a [PNP]Ir complex. *Chem. Commun.* **2010**, *46*, 79–81.

83. Agus, H.; Sandun, F.; Naveen, M.; Sushil, A. Current Status of Hydrogen Production Techniques by Steam Reforming of Ethanol: A Review. *Energy Fuels* **2005**, *19*, 2098–2106.

84. Sadykov, V.; Mezentseva, N.; Alikina, G.; Bunina, R.; Pelipenko, V.; Lukashevich, A.; Vostrikov, Z.; Rogov, V.; Krieger, T.; Ishchenko, A.; *et al.* Nanocomposite Catalysts for Steam Reforming of Methane and Biofuels: Design and Performance. In *Advances in Nanocomposites—Synthesis, Characterization and Industrial Applications*; Reddy, B.S.R., Ed.; In Tech: Rijeka, Croatia, 2011; pp. 909–946.

85. De Lima, S.M.; da Silva, A.M.; da Costa, L.O.O.; Graham, U.M.; Jacobs, G.; Davis, B.H.; Mattos, L.V.; Noronha, F.B. Study of catalyst deactivation and reaction mechanism of steam reforming, partial oxidation, and oxidative steam reforming of ethanol over Co/CeO$_2$ catalyst. *J. Catal.* **2009**, *268*, 268–281.

86. Galetti, A.E.; Gomez, M.F.; Arrua, L.A.; Marchi, A.J.; Abello, M.C. Study of CuCoZnAl oxide as catalyst for the hydrogen production from ethanol reforming. *Catal. Commun.* **2008**, *9*, 1201–1208.

87. Wang, H.; Liu, Y.; Wang, L.; Qin, Y.N. Study on the carbon deposition in steam reforming of ethanol over Co/CeO$_2$ catalyst. *Chem. Eng. J.* **2008**, *145*, 25–31.

88. Yu, S.W.; Huang, H.H.; Tang, C.W.; Wang, C.B. The effect of accessible oxygen over Co$_3$O$_4$-CeO$_2$ catalysts on the steam reforming of ethanol. *Int. J. Hydrogen Energy* **2014**, *39*, 20700–20711.

89. Moon, D.J. Hydrogen production by catalytic reforming of liquid hydrocarbons. *Catal. Surv. Asia* **2011**, *15*, 25–36.

90. Finocchio, E.; Rossetti, I.; Ramis, G. Redox properties of Co- and Cu-based catalysts for the steam reforming of ethanol. *Int. J. Hydrogen Energy* **2013**, *38*, 3213–3225.

91. Da Silva, A.M.; de Souza, K.R.; Mattos, L.V.; Jacobs, G.; Davis, B.H.; Noronha, F.B. The effect of support reducibility on the stability of Co/CeO$_2$ for the oxidative steam reforming of ethanol. *Catal. Today* **2011**, *164*, 234–239.

92. Mosqueda, B.; Toyir, J.; Kaddouri, A.; Gelin, P. Steam reforming of methane under water deficient conditions over gadolinium doped ceria. *Appl. Catal. B* **2009**, *88*, 361–367.

93. Al-Musa, A.A.; Ioakeimidis, Z.S.; Al-Saleh, M.S.; Al-Zahrany, A.; Marnellos, G.E.; Konsolakis, M. Steam reforming of iso-octane toward hydrogen production over mono- and bi-metallic Cu-Co/CeO$_2$ catalysts: Structure-Activity Correlations. *Int. J. Hydrogen Energy* **2014**, *39*, 19541–19554.

94. Cai, W.; Wang, F.; Daniel, C.; van Veen, A.C.; Schuurman, Y.; Descorme, C.; Provendier, H.; Shen, W.; Mirodatos, C. Oxidative steam reforming of ethanol over Ir/CeO$_2$ catalysts: A Structure Sensitivity Analysis. *J. Catal.* **2012**, *286*, 137–152.

95. Singhto, W.; Laosiripojana, N.; Assabumrungrat, S.; Charojrochkul, S. Steam reforming of bio-ethanol over Ni on Ce-ZrO$_2$ support: Influence of Redox Properties on the Catalyst Reactivity. *J. Sci. Technol.* **2006**, *28*, 1251–1264.

96. Laosiripojana, N.; Kiatkittipong, W.; Charojrochkul, S.; Assabumrungrat, S. Effects of support and co-fed elements on steam reforming of palm fatty acid distillate (PFAD) over Rh-based catalysts. *Appl. Catal. A* **2010**, *383*, 50–57.

97. Salazar-Villalpando, M.D.; Berry, D.A.; Cugini, A. Role of lattice oxygen in the partial oxidation of methane over Rh/zirconia-doped ceria. Isotopic studies. *Int. J. Hydrogen Energy* **2010**, *35*, 1998–2003.

98. Huang, T.J.; Wang, C.H. Roles of Surface and Bulk Lattice Oxygen in Forming CO$_2$ and CO during Methane Reaction over Gadolinia-Doped Ceria. *Catal. Lett.* **2007**, *118*, 103–108.

Ni Catalysts Supported on Modified Alumina for Diesel Steam Reforming

Antonios Tribalis, George D. Panagiotou, Kyriakos Bourikas, Labrini Sygellou, Stella Kennou, Spyridon Ladas, Alexis Lycourghiotis and Christos Kordulis

Abstract: Nickel catalysts are the most popular for steam reforming, however, they have a number of drawbacks, such as high propensity toward coke formation and intolerance to sulfur. In an effort to improve their behavior, a series of Ni-catalysts supported on pure and La-, Ba-, (La+Ba)- and Ce-doped γ-alumina has been prepared. The doped supports and the catalysts have been extensively characterized. The catalysts performance was evaluated for steam reforming of *n*-hexadecane pure or doped with dibenzothiophene as surrogate for sulphur-free or commercial diesel, respectively. The undoped catalyst lost its activity after 1.5 h on stream. Doping of the support with La improved the initial catalyst activity. However, this catalyst was completely deactivated after 2 h on stream. Doping with Ba or La+Ba improved the stability of the catalysts. This improvement is attributed to the increase of the dispersion of the nickel phase, the decrease of the support acidity and the increase of Ni-phase reducibility. The best catalyst of the series doped with La+Ba proved to be sulphur tolerant and stable for more than 160 h on stream. Doping of the support with Ce also improved the catalytic performance of the corresponding catalyst, but more work is needed to explain this behavior.

Reprinted from *Catalysts*. Cite as: Tribalis, A.; Panagiotou, G.D.; Bourikas, K.; Sygellou, L.; Kennou, S.; Ladas, S.; Lycourghiotis, A.; Kordulis, C. Ni Catalysts Supported on Modified Alumina for Diesel Steam Reforming. *Catalysts* **2016**, *6*, 11.

1. Introduction

PEM (proton-exchange membrane) fuel cell technology promises to be an efficient and clean alternative with respect to fuel combustion for primary power generation in stationary and mobile source applications. All fuel cells currently being developed for near term use in electric vehicles require hydrogen as a fuel. Hydrogen can be stored directly or produced onboard the vehicle by reforming methanol, or hydrocarbon fuels derived from crude oil—e.g., gasoline, diesel, or middle distillates. The vehicle design is simpler with direct hydrogen storage, but requires the development of a more complex refuelling infrastructure [1]. Thus the H_2 production onboard by the catalytic reforming of liquid fuels is very attractive [2]. Hence, there is great interest in converting hydrocarbons and oxygenated hydrocarbons into hydrogen. The process of converting petroleum fuels to hydrogen-rich gas products that have been developed in the past generally fall into one of the following classes—steam reforming (SR), partial oxidation (POX), autothermal reforming (ATR), dry reforming (DR) or a combination of two or more of the above. Despite their advantages, each of these processes has drawbacks with respect to design, fuel, and operating temperature [3]. The choice of reforming processes for producing fuel-cell feeds of the necessary quality has been thermodynamically analyzed using different fuels [4]. Similarly, the challenges and opportunities of various fuel reforming technologies for application in low and high-temperature fuel cells have been thoroughly reviewed [5].

Fuels produced by pure steam-reforming contain ~70%–80% hydrogen [6] proving this process the most productive one. Recently, Remiro *et al.* [7] have reported on the possibility to obtain an almost pure H_2 stream (99% H_2) by in situ capturing of CO_2 produced in the steam reforming process. Fuels, such as natural gas, methanol, propane, gasoline, or aqueous bio-oil/bio-ethanol mixtures have been widely studied for their reforming characteristics [8–12]. Due to the complexity of diesel fuel, its conversion by steam reforming is a complicated problem [13].

The design of a fuel reforming catalyst is a difficult undertaking. Weight, size, activity, cost, transient operations, versatility to reform different fuels/compositions, catalyst durability, and fuel processor efficiency are critical considerations for both stationary and automotive fuel cell applications. A desirable catalyst is one which catalyzes the reaction at low temperatures and is resistant to coke formation and tolerant for various concentrations of poison (e.g. sulfur, halogens, heavy metals, *etc.*) for extended periods of time [2].

Generally, reforming catalysts consist of a base metal (Ni, Co, *etc.*) promoted by a noble metal (Pt, Pd, Rh, Ru, *etc.*) supported on stabilized supports (alumina, ceria, ceria promoted alumina, zeolite, *etc.*) [14]. Steam reforming Ni-based catalysts are the most popular. Although these catalysts are cost effective and commercially available, they have a number of drawbacks when considered for fuel cell applications. For instance, they have a high propensity toward coke formation, whereas in their active state they are pyrophoric and intolerant to sulfur. Diesel is a complex hydrocarbon mixture characterized by the presence of sulphur. Even though desulfurization can be used to remove most of the sulphur from a traditional fuel, the small quantities that remain (10 ppm in EU since 2010) can lead to deactivation after many hours on stream.

In the present work, we have tried to minimize the above mentioned drawbacks of the Ni-based steam reforming catalysts by preparing such catalysts supported on modified alumina. Alumina has been chosen, as it has several advantages in comparison with other ceramic materials: it is inexpensive, quite highly refractory, and supposedly relatively inert to water in the range of the SR process conditions [15]. BaO, La_2O_3, and CeO_2 have been used as modifiers, as these materials have already been tried in similar processes, namely catalytic

129

partial oxidation of hydrocarbons [16]. Although La and Ce-promoted Ni/γ-Al$_2$O$_3$ catalysts have been reported in the steam reforming of light hydrocarbons [17–19], there are few studies on the rare-earth promoted catalysts in the diesel steam reforming [20]. The prepared catalysts have been characterized by using a combination of N$_2$ adsorption-desorption, X-ray diffraction (XRD), UV-Vis diffuse reflectance spectroscopy (DRS), temperature programmed reduction (TPR), X-ray photoelectron spectroscopy (XPS), and microelectrophoresis techniques.

The catalysts' performance was evaluated in a fixed bed micro-reactor using *n*-hexadecane with or without dibenzothiophene doping as surrogate for desulphurized commercial (10 ppm S) or sulphur free diesel, respectively. Thus, we attempted to gain insight concerning the influence of support modification on carbon deposition and sulfur poisoning during steam reforming.

2. Results and Discussion

2.1. Physicochemical Characteristics of the Supports

Five supports, pure alumina as well as La-, Ba-, (La+Ba)-, Ce-doped alumina, were used for the preparation of nickel catalysts studied in the present work. The doped supports were prepared by wet impregnation of pure alumina with aqueous solutions of the nitrate salts of the above ions. The composition, specific surface areas, and isoelectric points of these supports are shown in Table 1.

Table 1. Dopant content, specific surface areas calculated per gram of support (SSA$_S$) and per gram of alumina (SSA$_{Al}$), and isoelectric points (IEP) of the supports used.

Supports	Dopant Content (% mole)			SSA$_S$ (m^2/g)	SSA$_{Al}$ (m^2/g) of Al$_2$O$_3$	IEP
	La	Ba	Ce			
Al	-	-	-	152	152	7.98
Al-La	3	-	-	145	159	8.91
Al-Ba	-	10	-	128	149	8.98
Al-(La+Ba)	3	10	-	117	150	9.70
Al-Ce	-	-	3	162	170	7.88

Inspection of this table shows that the doping of alumina by La, Ba or (La+Ba) results in doped supports with specific surface areas (S$_{BET}$) lower than that of the parent alumina. However, the calculation of the surface area per gram of alumina (S$_{BET/Al}$) results in values equal to that of undoped alumina. This indicates no pore blockage of the support and eventually quite good dispersion of the dopant on the support surface. In contrast, doping with Ce provoked a slight increase in the specific surface area due to the porous nature of the corresponding oxide itself.

Figure 1 presents the pore size distributions of the prepared supports. One can observe a very small narrowing of the distributions along with a small shift towards lower pore diameters. This picture corroborates that no pore blockage takes place upon doping.

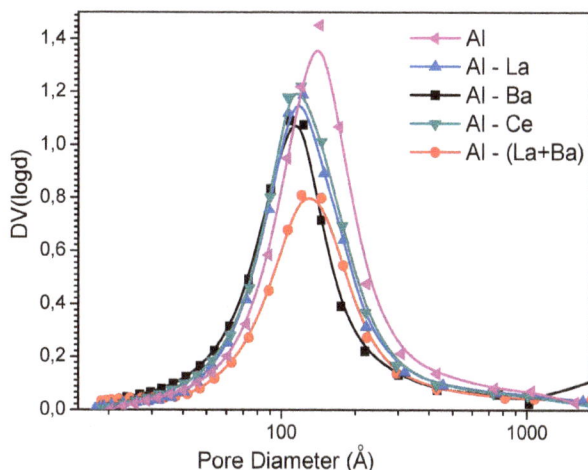

Figure 1. Pore size distributions of the supports used.

Figure 2 illustrates the XRD patterns of the supports used for the preparation of the corresponding Ni-catalysts. These patterns have been recorded after calcination at 850 °C. Diffraction peaks at $2\theta = 37.2$, 45.8, and 66.7° correspond to γ-Al_2O_3 (JCPDS 86-1410). The most intensive of these three peaks are visible in the patterns of all the supports studied.

Diffraction peaks assigned to crystalline species of either La_2O_3 (JCPDS 83-1355) or $LaAlO_3$ (JCPDS 85-1071) were not observed in the Al-La support. However, the corresponding pattern shows that doping with La decreases the alumina crystallinity broadening the corresponding peaks. Doping with Ba provoked a similar effect on the alumina crystallinity, while additional diffraction peaks, with the most intensive at $2\theta = 28.3°$, indicate the formation of $BaAl_2O_4$ phase (JCPDS 82–2001). The formation of the latter is much more pronounced in the support doped with La+Ba. Doping with Ce had a little influence on the alumina crystallinity, while a diffraction peak arising at $2\theta = 28.6°$ suggests the segregation of a CeO_2-phase (JCPDS 081-0792) over the γ-Al_2O_3 surface [21]. The formation of a porous CeO_2 phase explains the increase of the specific surface area of the Al-Ce support discussed previously (see Table 1).

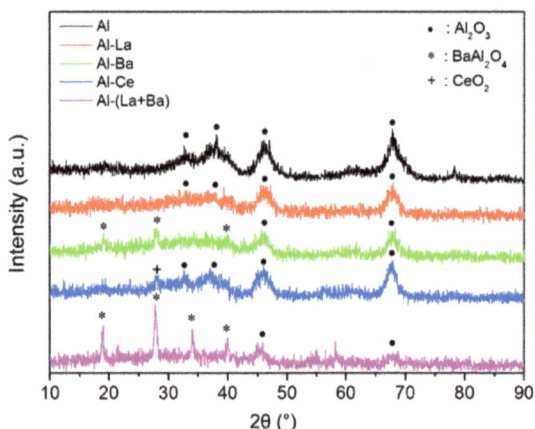

Figure 2. XRD patterns of pure (Al) and doped with La (Al-La), Ba (Al-Ba), Ce (Al-Ce), and La+Ba (Al-(La+Ba)) alumina after calcination at 850 °C.

The isoelectric point (IEP) values in Table 1 show that the doping with La and/or Ba increases the basicity of alumina, while the doping with Ce has a slightly opposite action. It is well known that the acid-base properties of the support influence the deposition mechanism of the active phase [22] and, thus, the active species formed on the support surface, as well as the resistance of the final catalyst to coke formation. It has been shown that the rate of carbon deposition on catalyst surfaces is reduced when the active metal is supported on a more basic carrier [23,24].

2.2. Physicochemical Characteristics of the Ni-Catalysts

The Ni-catalysts studied were prepared by wet impregnation of the support with a $Ni(NO_3)_2$ aqueous solution of appropriate concentration to obtain catalysts with Ni loading equal to 15wt. %. Comparing the specific surface areas of the final Ni-catalysts (SSA_C) presented in Table 2 with those of the corresponding supports (SSA_S) compiled in Table 1, we observe that the deposition of Ni caused a decrease of the specific surface area of ~20% in the case of the doped samples. This decrease was smaller in the undoped catalyst (~13%), which might indicate a higher dispersion of the Ni-phase in the undoped catalyst or a relatively high extent of $NiAl_2O_4$ formation by incorporation of Ni^{2+} ions into the alumina lattice.

Figure 3 shows the pore size distributions of the prepared catalysts. Similar distributions can be observed for all catalysts. However, in the Ni/Al sample a slightly higher contribution of small pores can be observed, in agreement with our previous comments for the Ni deposited on the surface of un-doped alumina.

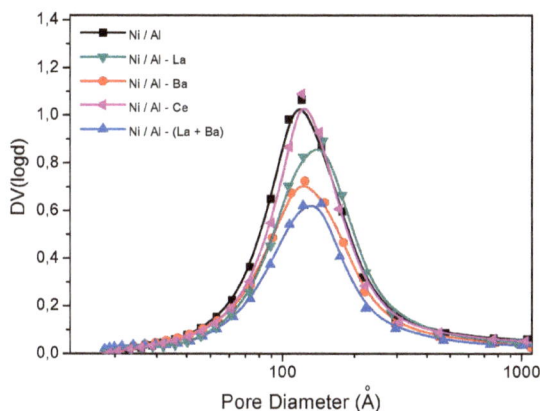

Figure 3. Pore size distributions of the catalysts studied.

Table 2. Specific surface areas of the catalysts (SSA_C), their crystal phases detected by XRD, their $NiAl_2O_4/NiO$ ratios determined by DRS (R_{DRS}) and TPR (R_{TPR}) and percentage of Ni reduced upon TPR measurements (% Ni^0).

Catalysts	SSA_C (m²/g)	Crystal Phases Detected by XRD	R_{DRS}	R_{TPR}	% Ni^0
Ni/Al	132	$NiAl_2O_4$, γ-Al_2O_3	4.86	3.75	79.9
Ni/Al-La	112	$NiAl_2O_4$, γ-Al_2O_3	4.67	1.94	87.4
Ni/Al-Ba	105	$BaAl_2O_4$, $NiAl_2O_4$, γ-Al_2O_3	3.11	1.67	86.1
Ni/Al-(La+Ba)	91	$BaAl_2O_4$, $NiAl_2O_4$, γ-Al_2O_3	2.63	1.50	98.1
Ni/Al-Ce	124	CeO_2, $NiAl_2O_4$, γ-Al_2O_3	4.82	3.80	77.3

Figure 4 shows the XRD patterns of the Ni-catalysts studied after calcination at 850 °C. The diffraction peaks observed in the XRD patterns of the corresponding supports (see Figure 2) can be also found in these patterns. A careful inspection of Figure 4 reveals that there is no evidence of a NiO phase (37.3, 43.3, and 62.9 JCPDS 78-0643). On the other hand, diffraction peaks attributed to $NiAl_2O_4$ crystallites can be found in the XRD patterns of all catalysts studied. Although, it is difficult to discriminate between $NiAl_2O_4$ and γ-Al_2O_3 from their XRD patterns, because almost all the diffraction peaks of these phases overlap, they can be distinguished from each other by the peak intensities of the main diffractions. Table 2 summarizes the crystal phases detected in the calcined Ni-catalysts.

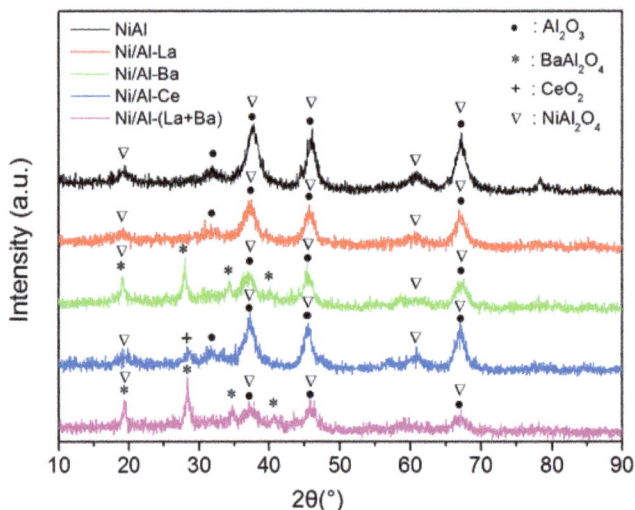

Figure 4. XRD patterns of Ni-catalysts supported on pure (Ni/Al) and doped with La (Ni/Al-La), Ba (Ni/Al-Ba), Ce (Ni/Al-Ce), and La+Ba (Ni/Al-(La+Ba)) alumina after calcination at 850 °C.

The formation of $NiAl_2O_4$ crystallites is attributed to the fact that Ni^{2+} ions preferentially incorporate into the tetrahedral vacancies of γ-Al_2O_3 at low nickel loading [25]. However, detailed analysis of the XRD patterns revealed that the $NiAl_2O_4$ content of the catalysts prepared depends on the alumina doping. More precisely, $NiAl_2O_4$ seems to be the predominant nickel phase in the Ni/Al sample. Its content decreases in the catalyst prepared using La doped alumina. La-phase has not been detected in the latter sample confirming its very good dispersion suggested by the SSA results discussed previously (Table 1).

Doping of alumina with Ba caused, in addition, the formation of $BaAl_2O_4$ as new peaks appeared in the corresponding XRD pattern at 2θ: 19.08°, 27.85°, 34.24°, and 40.09° [26]. The formation of this crystal phase seems to take place by insertion of Ba^{2+} ions into γ-Al_2O_3 lattice maintaining intact the corresponding SSA_{Al} (see Table 1). In such a case the Ba^{2+} insertion probably competes with the insertion of the Ni^{2+} ions for the same alumina lattice vacancies. This is corroborated by the observation that the intensities of the peaks assigned to $NiAl_2O_4$ are lower in the XRD pattern of Ni/Al-Ba sample than those recorded in the previous samples. Thus, the doping of alumina with Ba leads to a decrease of $NiAl_2O_4$ content in the corresponding catalyst. The XRD pattern of Ni/Al-(La+Ba) catalyst was similar to that of Ni/Al-Ba one. However, the $NiAl_2O_4$ content of the doubly doped catalyst seems to be the lowest detected among the catalysts studied in the present work.

The XRD pattern of Ni/Al-Ce catalyst showed the formation of CeO_2 crystallites in addition to those of γ-Al_2O_3 and $NiAl_2O_4$.

Figure 5 illustrates the diffuse reflectance (DR) spectra of the prepared catalysts recorded in the range 500–800 nm. An intense doublet with maxima at 595 and 635 nm appears in these spectra. This doublet is assigned to Ni^{2+} ions in tetrahedral symmetry and thus it is associated with the formation of the $NiAl_2O_4$ phase detected by XRD. Two shoulders appearing at 561 and 715 nm are assigned to Ni^{2+} ions in octahedral symmetry associated with the NiO phase [27,28].

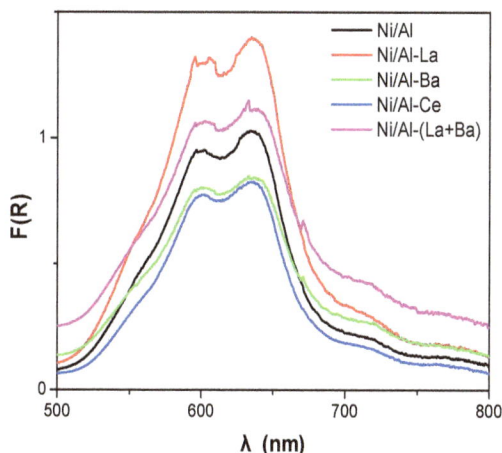

Figure 5. Diffuse reflectance spectra of the catalysts recorded in the range 500–800 nm.

To study the effect of doping on the catalysts, it is necessary to analyze the peaks of various Ni^{2+} species. However, the quantitative analysis of these species based on the DR spectra is relatively difficult because of the overlapping of the absorption domains of various Ni^{2+} ions. Thus, for obtaining a rough indication of the relative content of $NiAl_2O_4$ to NiO we used the ratio (R_{DRS}) of the height of the peak at 635 nm to the height of the peak at 715 nm [29].

The R_{DRS} values for the catalysts studied are included in Table 2. Taking into account these values one can conclude that doping of alumina with La and Ba decreases the $NiAl_2O_4$ content of the catalysts in accordance with XRD results mentioned before. More precisely, the $NiAl_2O_4$ content of the catalysts follows the order Ni/Al > Ni/Al-La > Ni/Al-Ba > Ni/Al-(La+Ba). On the other hand, doping of alumina with Ce does not influence the relative amount of $NiAl_2O_4$/NiO.

The reducibility of the Ni-species and their interaction with the support were studied via TPR experiments. It is evident from the TPR profiles (Figure 6) that the catalysts consume hydrogen in three temperature ranges. The first and minor peak

135

appears at 350 °C but only in the TPR profile of Ni/Al-Ce catalyst. It is assigned to the reduction of loosely attached NiO crystals. The small size of this peak shows that the relative amount of these crystals is very low. According to the literature [30,31], doping of Ni-catalysts with Ce promotes the NiO reduction due to action of the Ce^{3+}/Ce^{4+} redox pairs, which facilitate electron transfer.

The second and relatively more intense event characterized by a broad peak appearing in all TPR profiles, with maxima in the range 400–650 °C, arises from the reduction of NiO species exhibiting stronger interaction with the support. The fact that the peak is broad and asymmetric indicates the presence of a multitude of such NiO species with different strength of binding to the support [32,33]. The third and most important reduction event appears at 817 ± 10 °C. The corresponding well resolved peak should be assigned to the reduction of the $NiAl_2O_4$ phase [34] detected by XRD and DRS.

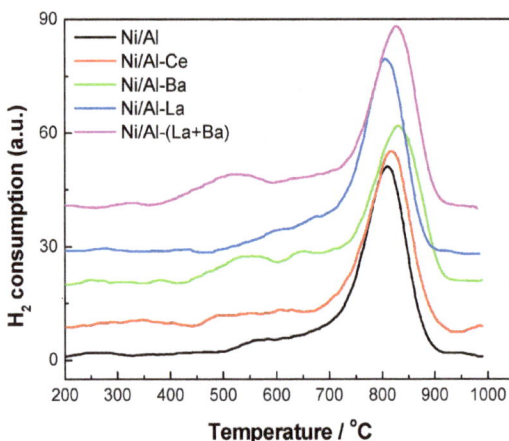

Figure 6. TPR profiles of the prepared catalysts recorded in the temperature range 150–1000 °C.

Deconvolution of the TPR curves has been performed in order to calculate the $NiAl_2O_4/NiO$ ratio (R_{TPR}). The corresponding values are given in Table 2. These values follow the same trend as those calculated from the DR spectra.

Taking into account the amount of hydrogen required for complete reduction of the supported Ni ($Ni^{2+} \rightarrow Ni^0$) and the total amount of hydrogen experimentally consumed, the percentage of Ni-phase reduced upon TPR measurements (experimental/theoretical amount of hydrogen consumed) is obtained. The percentage values of this parameter ($\%Ni^0$) are illustrated in the last column of Table 2. An inspection of these values reveals that doping with La and Ba and especially the combined doping with La and Ba increases the reducibility of the

Ni-phase. These results are in good agreement with the decrease of the $NiAl_2O_4$ content in the corresponding catalysts indicated by the XRD and DRS results. In contrast, doping with Ce seems to have a slight, if any, negative effect on the total reducibility of Ni-phase, although, as already mentioned, it facilitated the reduction of NiO loosely bound on the support surface. Taking into account the corresponding R value appearing in Table 2, which indicates relatively higher $NiAl_2O_4$ content in the Ni/Al-Ce catalyst, this behavior could be easily explained.

The surface analysis of the catalysts prepared was performed by X-ray photoelectron spectroscopy (XPS). XPS resulted in the expected detection of Ni, Al, and O on the surfaces of all catalysts studied, as well as of La in Ni/Al-La and Ni/Al-(La+Ba) catalysts, Ba in Ni/Al-Ba and Ni/Al-(La+Ba) catalyst and Ce in the Ni/Al-Ce one. Organic carbon from a surface contamination layer, which consists mainly of CH_x with the dominant C1s peak set at 284.8 eV as a binding energy (BE) reference and some organic oxygen of the CO_x type, was found in all specimens due to the atmospheric exposure. The quantitative surface analysis of the catalysts was based on the following peaks: $Ni2p_{3/2}$ (855.9 eV), O1s (531.0 eV), Al2s (118.9 eV), $Ba3d_{5/2}$ (780.2 eV), $La3d_{5/2}$ (835.2 eV), and $Ce3d_{5/2}$ (~885 eV).

Table 3 summarizes the results obtained from the quantitative analysis of the XPS data, based on measured peak areas, one for each detected element as mentioned above, and taking into account the respective relative sensitivity factors (RSF).

Table 3. Quantitative surface analysis of the prepared catalysts based on XPS results.

Catalysts	Average Surface Composition (atoms %)					
	Ni	Al	O	La	Ba	Ce
Ni/Al	2.5	32.5	65.0	-	-	-
Ni/Al-La	2.8	28.2	67.9	1.1	-	-
Ni/Al-Ba	5.1	28.6	64.7	-	1.6	-
Ni/Al-(La+Ba)	5.4	24.4	67.3	1.4	1.5	-
Ni/Al-Ce	2.6	32.3	64.9	-	-	0.2

Inspection of Table 3 reveals that doping of the γ-Al_2O_3 carrier with La leads to only a slight surface enrichment of the corresponding catalyst with Ni-species. This effect becomes more intense by doping with Ba and even more intense when combined doping with La and Ba takes place. This surface enrichment with Ni-species is in good accordance with the increase of reducibility of these catalysts and the decrease in the R value discussed previously.

Although we do not have direct experimental measurement for Ni particle size (e.g., TEM), the XPS data (Table 3) offer strong evidence of enhanced Ni dispersion over the Ba-doped support. The measured Ni/Al atomic ratio on the Ba-doped alumina is more than two times larger than in the case of undoped alumina, whereas Ba has a tendency to be incorporated in alumina and not to cover it. Indeed the surface atomic concentration of Ba in the Ni/Al-Ba and Ni/Al-(La+Ba) catalysts was found to be relatively low compared to the nominal Ba concentration. This suggests the easy insertion of Ba ions into the alumina lattice and is in very good agreement with our XRD results, which indicated the formation of $BaAl_2O_4$ crystals (see Table 2). Doping with Ce seems to have a slight, if any, positive influence on the surface enrichment of Ni-species, while the surface concentration of Ce atoms was found to be extremely low. The latter effect can be explained by the formation of relatively large CeO_2 crystallites, as suggested also by XRD.

2.3. Catalyst Evaluation

The catalytic performance of the catalysts studied was evaluated in a fixed bed micro-reactor for the steam reforming of hexadecane (surrogate sulphur-free diesel) and hexadecane doped with dibenzothiophene corresponding to surrogate commercial diesel with 10 ppm sulphur. The only products detected in the gas stream at the reactor outlet were H_2, CO, CO_2 and CH_4. Figure 7 shows the production rate of H_2 measured over the Ni-catalysts supported on pure, La-, Ba-, and (La+Ba)-doped γ-Al_2O_3. Complete conversion of n-hexadecane was observed at 795 °C over all of the catalysts studied in accordance with the results described recently by Kaynar et al. [35].

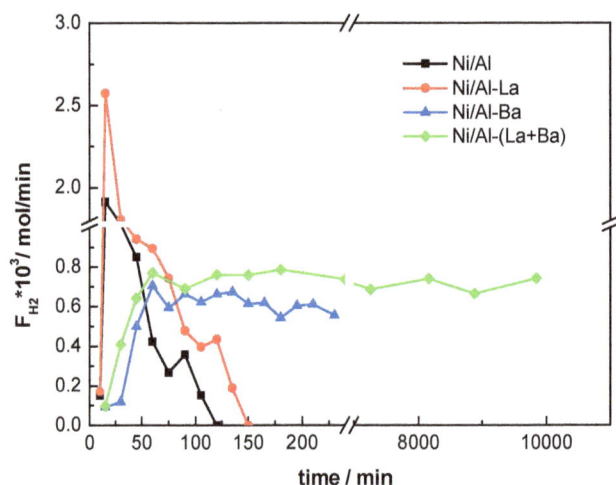

Figure 7. Hydrogen production rates measured over the Ni-catalysts supported on pure (■), La- (•), Ba- (▲), and (La+Ba)- (♦) modified γ-Al_2O_3 with the time on stream. (Reaction conditions: Steam to Carbon ratio = 2.67, Reaction temperature: 795 °C, catalyst mass = 0.05 g, GHSV= 20000 h^{-1})

In all catalytic runs the H_2 mass flow rates (F) measured at the reactor outlet increased initially and reached their maximum value after 0.5–1 h on stream. This is in accordance with the fact that metallic nickel is the active phase of the catalysts studied, while the reactor was loaded with their oxide form. Thus, the initial increase of H_2 production rate should be attributed to the in situ activation (reduction) of the catalyst by the H_2 produced (see Experimental section). The Ni/Al and Ni/Al-La catalysts exhibited important initial activities (after ~0.5 h on stream) but they were fully deactivated after 2 and 2.5 h on stream, respectively. However, the beneficial action of doping with La was clear resulting in higher initial activity and longer deactivation time compared to that of un-doped catalyst. The enhanced initial H_2 production observed over the Ni/Al and Ni/Al-La samples can be attributed to decomposition of the hydrocarbon, which is expected to increase the coke formation over the catalysts, thus explaining the simultaneous rapid deactivation of these samples [35]. Coke formation in the form of both amorphous and graphitic filamentous carbon was identified as the main reason for deactivation of rich in $NiAl_2O_4$ catalysts studied by Jiménez-González et al. [11] for steam reforming of isooctane.

Doping of the catalyst support with Ba or (Ba+La) further improved the catalytic performance of the corresponding samples (Figure 7). This improvement could be attributed to the following effects caused by doping with Ba or (Ba+La): surface enrichment of the nickel phase as observed by the XPS analysis, decrease of the support acidity measured by microelectrophoresis and increase of the Ni-phase reducibility determined by TPR. Taking into account that all the catalysts studied in the present work had the same Ni loading, their surface enrichment with nickel phase caused by doping of the support with Ba or Ba+La indicates an increase of the dispersion of the active phase. Christensen et al. [36] reported that lower coking rates are possible for Ni-catalysts with higher metallic dispersion, and Quitete et al. [37] found that coke formation is favored by increasing nickel particle size. On the other hand, it is well known that low tendency to coke formation is observed when the Ni active phase is dispersed on less acidic supports [38]. As to the influence of Ni-phase reducibility on catalytic performance it is well known that metallic Ni atoms are active sites for the steam reforming of hydrocarbons [39].

Figure 8 shows the production rates of H_2, CO, CO_2, and CH_4 measured at the outlet of the reactor on a dry basis after 4 h on stream over all catalysts under sulfur free conditions and over the Ni/Al-(La+Ba) catalyst, which is the most promising one, in the presence of sulphur in the feed. Comparing the Ni/Al-Ba and Ni/Al-(La+Ba) catalysts one can observe that the enhanced H_2 production rate observed in the latter is accompanied by lower CH_4 and higher CO and CO_2 production rates. This probably shows that doping hinders the methanation reactions.

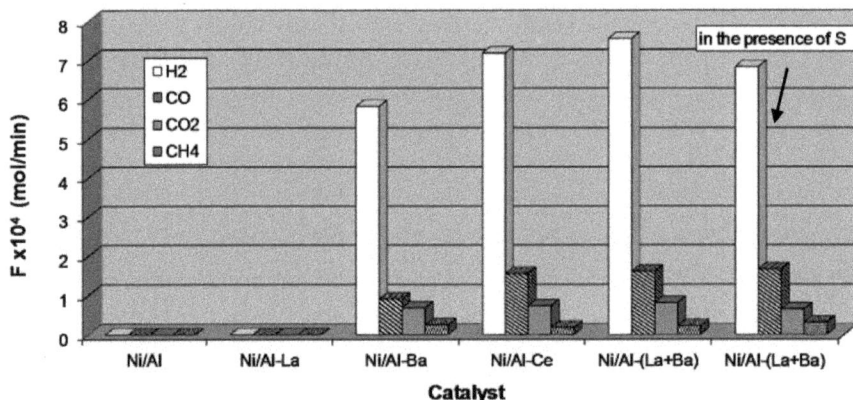

Figure 8. Production rates of H_2, CO, CO_2, and CH_4 after 4 h on stream over the catalysts under sulphur free conditions and over the Ni/Al-(La+Ba) catalyst in the presence of 10 ppm sulphur in the feed (Steam/Carbon ratio: 2.67, Reaction temperature: 795 °C, catalyst mass: 0.05 g, GHSV= 20000 h^{-1}).

The Ni/Al-(La+Ba) catalyst proved to be quite active even in the presence of sulphur in the stream because only a small decrease (<10%) in its activity has been observed when the reactor was fed with surrogate diesel containing 10 ppm sulphur. The activity of this most promising Ni/Al-(La+Ba) catalyst was tested for long times under sulphur free conditions and proved to be quite stable for more than 168 h on stream (see Figure 7).

Doping of the support with Ce caused also significant increase in the activity of the corresponding catalyst, which could not be explained taking into account the physicochemical characteristics of the Ni-phase examined in the present work. According to the literature, the beneficial action of Ce could be attributed to the scavenging of the coke from the surface of the Ni phase via oxygen provided by the CeO_x phase [40]. Indeed, this phase has been detected in our Ni/Al-Ce sample by XRD. However, more work is necessary in order to explain the important catalytic behavior of the Ni/Al-Ce sample.

3. Experimental Section

Alumina extrudates (AKZO, HDS-000-1,5mmE, SSA = 264 m^2/g, pore volume: 0.65 cm^3/g) were crushed and sieved to obtain a support powder with particle size 90–105 μm. It was wet impregnated with aqueous nitrate solutions of $La(NO_3)_3 \cdot 6H_2O$ (Sigma-Aldrich, Taufkirchen, Germany, 99.999%), $Ce(NO_3)_3 \cdot 6H_2O$ (Alfa Aesar, Lancashire, United Kingdom, 99.5%), and $Ba(NO_3)_2$ (Sigma-Aldrich, 99.999%) of volume equal to 20 times the total pore volume of the support used. The solution contained the appropriate amount of salt to obtain the desired concentration

140

of dopant in the doped support and the final catalyst. After the evaporation of water in a rotary evaporator (BUCHI, Flawil, Switzerland) impregnates were dried overnight at 120 °C in air, then they were calcined by increasing the temperature initially up to 400 °C (3 °C/min) and then up to 850 °C (10 °C/min), where they remained for 3 h. A series of Ni catalysts supported on pure and modified alumina containing 15% w/w Ni, was prepared following the above described procedure for the deposition of the nickel phase using aqueous solution of $Ni(NO_3)_2$ $6H_2O$ (Sigma-Aldrich, Taufkirchen, Germany 98%). The final impregnates were submitted to the same thermal treatment (drying and calcination) with that described above for the preparation of the supports.

The determination of the specific surface area (S_{BET}) of the samples was based on the nitrogen adsorption–desorption isotherms recorded using a Micromeritics apparatus (Tristar 3000 porosimeter, Micromeritics, Aachen, Germany) and the corresponding software. Pore size distributions have been determined using the BJH method and the N_2 desorption curve.

XRD measurements were carried out with a D8 Advance X-Ray diffractometer, Bruker, Leiderdorp, The Netherlands, equipped with nickel-filtered Cu Ka (0.15418 nm) radiation source. The step size and the time per step were respectively fixed at 0.02° and 0.5 s in the range of $20° \leqslant 2\theta \leqslant 80°$.

The diffuse reflectance spectra of the calcined samples were recorded in the range 200–800 nm at room temperature. A UV-Vis spectrophotometer (Varian Cary 3, Agilent Technologies, Santa Clara, CA, USA) equipped with an integration sphere was used. The corresponding support was used as reference. The samples were mounted in a quartz cell. This provided a sample thickness greater than 3 mm to guarantee the "infinite" sample thickness.

The isoelectric points of the pure and modified alumina supports were determined using a Zetasizer Nano-ZS apparatus (Malvern, United Kingdom) equipped with a DTS 1060C-Clear disposable zeta cell (Malvern, United Kingdom).

XPS measurements were carried out on catalyst powders prepared as described above and pressed in a hollow aluminum metal receptacle. The experiments took place in a MAX200 (LEYBOLD/SPECS, Berlin, Germany) Electron Spectrometer using non-monochromatic MgK_α radiation (1253.6 eV) and an EA200-MCD analyzer (SPECS, Berlin, Germany) operated at a constant pass energy (PE) of 140 eV in order to maximize signal intensity while maintaining a resolution of the order of 1.1 eV. The analysis was performed along the surface normal and the analyzed specimen area was a $\sim 4 \times 7$ mm^2 rectangle near the specimen center with a depth of analysis of the order of 10nm. The binding energy (BE) scale was corrected for electrostatic charging via the main C1s peak from surface contamination set at 284.8 eV BE, the uncertainty in the corrected values being of the order of ± 0.15 eV. The relative sensitivity factor (RSF) database used for quantitative analysis has been adapted to the MAX200 spectrometer (SPECS, Berlin, Germany) operating conditions from the empirical collection of Wagner *et al.* [41], whereas the assignment of the BE for the various chemical states of the elements was made using an appropriate database [42].

The TPR experiments were performed in a laboratory-constructed equipment described elsewhere [43]. An amount of sample, 0.1 g, was placed in a quartz reactor and the reducing gas mixture (H_2/Ar: 5/95 v/v) was passed through it for 2 h with a flow rate of 40 mL min^{-1} at room temperature. Then the temperature was increased to 1000 °C with a constant rate of 10 °C min^{-1}. Reduction leads to a decrease of the hydrogen concentration of the gas mixture, which was detected by a thermal conductivity detector (TCD) (Shimadzu, Duisburg, Germany). The reducing gas mixture was dried in a cold trap (-95 °C) before reaching the TCD.

Catalytic performance of the prepared samples was evaluated using a stainless steel fixed-bed reactor (diameter: 1/4 inch) loaded with 0.05 g of catalyst in its oxide form diluted with 0.05 g of quartz sand. Helium from a cylinder was fed through mass flow-meter (Brooks) into a saturator containing hexadecane or hexadecane/dibenzothiophen to simulate sulfur-free diesel and a diesel containing 10 ppm of sulfur, respectively. The temperature of the saturator was set at 153 °C for obtaining constant partial pressure of surrogate diesel in the He stream. Triply distilled water was fed to the latter stream with a syringe pump (Braintree Scientific Inc., Braintree, MA, USA). A steam/carbon ratio of 2.67 was ensured. The line from the saturator to the reactor entrance was maintained at 350 °C in order to avoid any condensation. Thus, a GHSV= 20000 h^{-1} was obtained. The reactor temperature was set at 795 $\pm$ 1 °C. The outlet stream of the reactor was analyzed after cooling and gas-liquid separation (at 0–5 °C). Two gas chromatographs (GC-8A and GC-14B, Shimadzu, Duisburg, Germany) equipped with thermal conductivity detectors and SUPELCO columns (100/120 Carbosieve SII 10 ft $\times$ 1/8IN SS and 60/80 Carboxen 1000 15 ft $\times$ 1/8IN SS) were used for the dry-base analysis of the gas products. Sampling was performed using six port valves every 15 minutes for the first four hours of each catalytic test and then every four hours for the long term test performed over the Ni/Al-(La+Ba) catalyst. A delay was observed in the evolution of H_2 in the outlet of the reactor. Thus the maximum H_2 production was observed after 0.5 h on stream over the Ni/Al and Ni/Al-La samples and after 1 h over the rest of the catalysts. The aforementioned delay of H_2 evolution is attributed to its consumption for the in situ activation (reduction) of the catalysts by the H_2 initially produced.

4. Conclusions

The main conclusions drawn from the present work can be summarized as follows:

Doping of the γ-alumina support with Ba or (La+Ba) resulted in Ni catalysts with improved activity and stability for surrogate diesel steam reforming. This improvement can be attributed to the increase of the dispersion of the nickel phase, the decrease of the support acidity and the increase of Ni-phase reducibility.

Doping of the support with Ce caused also significant increase in the activity of the corresponding catalyst.

The doubly doped (La+Ba) catalyst proved also to be sulphur tolerant and stable for more than 160 h on stream.

Author Contributions: A.T. performed the preparation, evaluation, N_2-physisorption, and XRD study of the catalysts. G.D.P. performed TPR experiments and data analysis. K.B. performed DRS and IEP measurements and data analysis. L.S., S.K., and S.L. performed XPS measurements and data analysis. A.L. and C.K. supervised the work and wrote the manuscript. All authors contributed to the revision of the manuscript.

Conflicts of Interest: The authors declare no conflict of interest.

References

1. Ogden, J.M.; Steinbugler, M.M.; Kreutz, T.G. A comparison of hydrogen, methanal and gasoline as fuels for fuel cell vehicles: implications for vehicle design and infrastructure development. *J. Power Sources* **1999**, *79*, 143–168.
2. Farrauto, R.; Hwang, S.; Shore, L.; Ruettinger, W.; Lampert, J.; Giroux, T.; Liu, Y.; Ilinich, O. New material needs for hydrocarbon fuel processing: Generating hydrogen for the PEM fuel cell. *Annu. Rev. Mater. Res.* **2003**, *33*, 1–27.
3. Cheekatamarla, P.K.; Finnerty, C.M. Reforming catalysts for hydrogen generation in fuel cell applications. *J. Power Sources* **2006**, *160*, 490–499.
4. Semelsberger, T.A.; Brown, L.F.; Borup, R.L.; Inbody, M.A. Equilibrium products from autothermal processes for generating hydrogen-rich fuel-cell feeds. *Int. J. Hydrogen Energy* **2004**, *29*, 1047–1064.
5. Song, C.S. Fuel processing for low-temperature and high-temperature fuel cells—Challenges, and opportunities for sustainable development in the 21st century. *Catal. Today* **2002**, *77*, 17–49.
6. Brown, L.F. A comparative study of fuels for on-board hydrogen production for fuel-cell-powered automobiles. *Int. J. Hydrogen Energy* **2001**, *26*, 381–397.
7. Remiro, A.; Valle, B.; Aramburu, B.; Aguayo, A.T.; Bilbao, J.; Gayubo, A.G. Steam reforming of the bio-oil aqueous fraction in a fluidized bed reactor with in situ CO_2 capture. *Ind. Eng. Chem. Res.* **2013**, *52*, 17087–17098.
8. Zhang, Y.; Wang, W.; Wang, Z.Y.; Zhou, X.T.; Wang, Z.; Liu, C.-J. Steam reforming of methane over Ni/SiO_2 catalyst with enhanced coke resistance at low steam to methane ratio. *Catal. Today* **2015**, *256*, 130–136.
9. Barrios, C.E.; Bosco, M.V.; Baltanas, M.A.; Bonivardi, A.L. Hydrogen production by methanol steam reforming: Catalytic performance of supported-Pd on zinc-cerium oxides' nanocomposites. *Appl. Catal. B* **2015**, *179*, 262–275.

10. Hou, T.F.; Yu, B.; Zhang, S.Y.; Zhang, J.H.; Wang, D.Z.; Xu, T.K.; Cui, L.; Cai, W.J. Hydrogen production from propane steam reforming over Ir/Ce$_{0.75}$Zr$_{0.25}$O$_2$ catalyst. *Appl. Catal. B* **2015**, *168*, 524–530.

11. Jimenez-Gonzalez, C.; Boukha, Z.; de Rivas, B.; Ramon Gonzalez-Velasco, J.; Ignacio Gutierrez-Ortiz, J.; Lopez-Fonseca, R. Behaviour of nickel-alumina spinel (NiAl$_2$O$_4$) catalysts for isooctane steam reforming. *Int. J. Hydrogen Energy* **2015**, *40*, 5281–5288.

12. Remiro, A.; Valle, B.; Oar-Arteta, L.; Aguayo, A.T.; Bilbao, J.; Gayubo, A.G. Hydrogen production by steam reforming of bio-oil/bio-ethanol mixtures in a continuous thermal-catalytic process. *Int. J. Hydrogen Energy* **2014**, *39*, 6889–6898.

13. Thormann, J.; Pfeifer, P.; Schubert, K.; Kunz, U. Reforming of diesel fuel in a micro reactor for APU systems. *Chem. Eng. J.* **2008**, *135*, S74–S81.

14. Lakhapatri, S.L.; Abraham, M.A. Deactivation due to sulfur poisoning and carbon deposition on Rh-Ni/Al$_2$O$_3$ catalyst during steam reforming of sulfur-doped n-hexadecane. *Appl. Catal. A* **2009**, *364*, 113–121.

15. Faure, R.; Rossignol, F.; Chartier, T.; Bonhomme, C.; Maître, A.; Etchegoyen, G.; del Gallo, P.; Gary, D. Alumina foam catalyst supports for industrial steam reforming processes. *J. Eur. Ceram. Soc.* **2011**, *31*, 303–312.

16. Haynes, D.J.; Campos, A.; Smith, M.W.; Berry, D.A.; Shekhawat, D.; Spive, J.J. Reducing the deactivation of Ni-metal during the catalytic partial oxidation of a surrogate diesel fuel mixture. *Catal. Today* **2010**, *154*, 210–216.

17. Sanchez-Sanchez, M.C.; Navarro, R.M.; Fierro, J.L.G. Ethanol steam reforming over Ni/La-Al$_2$O$_3$ catalysts: Influence of lanthanum loading. *Catal. Today* **2007**, *129*, 336–345.

18. Sanchez-Sanchez, M.C.; Navarro, R.M.; Fierro, J.L.G. Ethanol steam reforming over Ni/M$_x$O$_y$-Al$_2$O$_3$ (M = Ce, La, Zr and Mg) catalysts: Influence of support on the hydrogen production. *Int. J. Hydrogen Energy* **2007**, *32*, 1462–1471.

19. Lucredio, A.F.; Filho, G.T.; Assaf, E.M. Co/Mg/Al hydrotalcite-type precursor, promoted with La and Ce, studied by XPS and applied to methane steam reforming reactions. *Appl. Surf. Sci.* **2009**, *255*, 5851–5856.

20. Xu, L.; Mi, W.; Su, Q. Hydrogen production through diesel steam reforming over rare-erth promoted Ni/γ-Al$_2$O$_3$ catalysts. *J. Nat. Gas Chem.* **2011**, *20*, 287–293.

21. Campos, C.H.; Osorio-Vargas, P.; Flores-Gonzalez, N.; Fierro, J.L.G.; Reyes, P. Effect of Ni Loading on Lanthanide (La and Ce) Promoted γ-Al$_2$O$_3$ Catalysts Applied to Ethanol Steam Reforming. *Catal. Lett.* **2015**.

22. Bourikas, K.; Kordulis, C.; Lycourghiotis, A. The Role of the Liquid-Solid Interface in the Preparation of Supported Catalysts. *Catal. Rev.-Sci. Eng.* **2006**, *48*, 363–444.

23. Cavallaro, S.; Mondello, N.; Freni, S. Hydrogen produced from ethanol for internal reforming molten carbonate fuel cell. *J. Power Sources* **2001**, *102*, 198–204.

24. Basagiannis, A.C.; Verykios, X.E. Catalytic steam reforming of acetic acid for hydrogen production. *Int. J. Hydrogen Energy* **2007**, *32*, 3343–3355.

25. Wang, R.; Li, Y.; Shi, R.; Yang, M. Effect of metal-support interaction on the catalytic performance of Ni/Al$_2$O$_3$ for selective hydrogenation of isoprene. *J. Mol. Catal. A* **2011**, *344*, 122–127.

26. Machida, M.; Eguchi, K.; Arai, H. Effect of Additives on the Surface Area of Oxide Supports for Catalytic Combustion. *J. Catal.* **1987**, *103*, 385–393.

27. Wang, Y.; Wang, L.; Gan, N.; Lim, Z.Y.; Wu, C.Z.; Peng, J.; Wang, W.G. Evaluation of $Ni/Y_2O_3/Al_2O_3$ catalysts for hydrogen production by autothermal reforming of methane. *Int. J. Hydrogen Energy* **2014**, *39*, 10971–10979.

28. Boukha, Z.; Jiménez González, C.; de Rivas, B.; GonzálezVelasco, J.R.; Gutiérrez Ortiz, J.I.; López Fonseca, R. Synthesis, characterisation and performance evaluation of spinel-derived Ni/Al_2O_3 catalysts for various methane reforming reactions. *Appl. Catal. B* **2014**, *158–159*, 190–201.

29. Yang, R.; Zhang, Z.; Wu, J.; Li, X.; Wang, L. Hydrotreating Performance of La-Modified Ni/Al_2O_3 Catalysts Prepared by Hydrothermal Impregnation Method. *Kinet. Catal.* **2015**, *56*, 222–225.

30. Daza, C.E.; Gallego, J.; Mondragon, F.; Moreno, S.; Molina, R. High stability of Ce-promoted Ni/Mg-Al catalysts derived from hydrotalcites in dry reforming of methane. *Fuel* **2010**, *89*, 592–603.

31. Debek, R.; Radlik, M.; Motak, M.; Galvez, M.E.; Turek, W.; da Costa, P.; Grzybek, T. Ni-containing Ce-promoted hydrotalcite derived materials as catalysts for methane reforming with carbon dioxide at low temperature—On the effect of basicity. *Catal. Today* **2015**, *257*, 59–65.

32. Damyanova, S.; Pawelec, B.; Arishtirova, K.; Fierro, J.L.G. Ni-based catalysts for reforming of methane with CO_2. *Int. J. Hydrogen Energy* **2012**, *37*, 15966–15975.

33. Kraleva, E.; Pohl, M.-M.; Jurgensen, A.; Ehrich, H. Hydrogen production by bioethanol partial oxidation over Ni based catalysts. *Appl. Catal. B* **2015**, *179*, 509–520.

34. Zaungouei, M.; Moghaddam, A.Z.; Arasteh, M. The Influence of Nickel Loading on Reducibility of NiO/Al_2O_3 Catalysts Synthesized by Sol-Gel Method. *Chem. Eng. Res. Bull.* **2010**, *14*, 97–102.

35. Kaynar, A.D.D.; Dogu, D.; Nail, Y. Hydrogen production and coke minimization through reforming of kerosene over bi-metallic ceria-alumina supported Ru-Ni catalysts. *Fuel Process. Technol.* **2015**, *140*, 96–103.

36. Christensen, K.O.; Chen, D.; Lodeng, R.; Holmen, A. Effect of supports and Ni crystal size on carbon formation and sintering during steam methane reforming. *Appl. Catal. A* **2006**, *314*, 9–22.

37. Quitete, C.P.B.; Bittencourt, R.C.P.; Souza, M.M.V.M. Coking resistance evaluation of tar removal catalysts. *Catal. Commun.* **2015**, *71*, 79–83.

38. Fajardo, H.V.; Longo, E.; Mezalira, D.Z.; Nuernberg, G.B.; Almerindo, G.I.; Collasiol, A.; Probst, L.F.D.; Garcia, I.T.S.; Carreño, N.L.V. Influence of support on catalytic behavior of nickel catalysts in the steam reforming of ethanol for hydrogen production. *Environ. Chem. Lett.* **2010**, *8*, 79–85.

39. Zou, X.; Wang, X.; Li, L.; Shen, K.; Lu, X.; Ding, W. Development of highly effective supported nickel catalysts for pre-reforming of liquefied petroleum gas under low steam to carbon molar ratios. *Int. J. Hydrogen Energy* **2010**, *35*, 12191–12200.

145

40. Shekhawat, D.; Gardner, T.H.; Berry, D.A.; Salazar, M.; Haynes, D.J.; Spivey, J.J. Catalytic partial oxidation of *n*-tetradecane in the presence of sulfur or polynuclear aromatics: Effects of support and metal. *Appl.Catal. A* **2006**, *311*, 8–16.
41. Wagner, C.D.; Davis, L.E.; Zeller, M.V.; Taylor, J.A.; Raymond, R.H.; Gale, L.H. Empirical atomic sensitivity factors for quantitative analysis by electron spectroscopy for chemical analysis. *Surf. Interface Anal.* **1981**, *3*, 211–225.
42. Wagner, C.D.; Riggs, W.M.; Davis, L.E.; Moulder, J.F.; Muilenberg, G.E. *Handbook of X-ray Photoelectron Spectroscopy*; Perkin-Elmer Corp: Eden Prairie, MN, USA, 1979.
43. Spanos, N.; Matralis, H.K.; Kordulis, C.; Lycourghiotis, A. Molybdenum-oxo Species Deposited on Titania by Adsorption and Characterization of the Calcined Samples. *J. Catal.* **1992**, *136*, 432–445.

Synthesis of Ethanol from Syngas over Rh/MCM-41 Catalyst: Effect of Water on Product Selectivity

Luis Lopez, Jorge Velasco, Vicente Montes, Alberto Marinas, Saul Cabrera, Magali Boutonnet and Sven Järås

Abstract: The thermochemical processing of biomass is an alternative route for the manufacture of fuel-grade ethanol, in which the catalytic conversion of syngas to ethanol is a key step. The search for novel catalyst formulations, active sites and types of support is of current interest. In this work, the catalytic performance of an Rh/MCM-41 catalyst has been evaluated and compared with a typical Rh/SiO$_2$ catalyst. They have been compared at identical reaction conditions (280 °C and 20 bar), at low syngas conversion (2.8%) and at same metal dispersion (H/Rh = 22%). Under these conditions, the catalysts showed different product selectivities. The differences have been attributed to the concentration of water vapor in the pores of Rh/MCM-41. The concentration of water vapor could promote the water-gas-shift-reaction generating some extra carbon dioxide and hydrogen, which in turn can induce side reactions and change the product selectivity. The extra hydrogen generated could facilitate the hydrogenation of a C$_2$-oxygenated intermediate to ethanol, thus resulting in a higher ethanol selectivity over the Rh/MCM-41 catalyst as compared to the typical Rh/SiO$_2$ catalyst; 24% and 8%, respectively. The catalysts have been characterized, before and after reaction, by N$_2$-physisorption, X-ray photoelectron spectroscopy, X-ray diffraction, H$_2$-chemisorption, transmission electron microscopy and temperature programmed reduction.

Reprinted from *Catalysts*. Cite as: Lopez, L.; Velasco, J.; Montes, V.; Marinas, A.; Cabrera, S.; Boutonnet, M.; Järås, S. Synthesis of Ethanol from Syngas over Rh/MCM-41 Catalyst: Effect of Water on Product Selectivity. *Catalysts* **2015**, 5, 1737–1755.

1. Introduction

At present, fuel-grade ethanol is utilized as a renewable component in gasoline or as a pure fuel in flex-fuel vehicles [1,2]. In 2013, about 70 million tons of fuel-grade ethanol were produced worldwide [3]. Most of the production technologies use food-related raw materials, such as corn in the USA or sugar cane in Brazil. Non-food related resources such as forest and agricultural biomass (known as cellulosic biomass) are being considered as alternative raw materials. Indeed, several public and private institutions have started R&D programs in order to produce fuel-grade ethanol from cellulosic biomass, at competitive cost [4].

In principle, any kind of biomass can be converted into fuels and chemicals thermochemically [5,6]. The thermochemical process is divided in two stages; in the first, biomass is converted to an intermediate mixture of gases known as "synthesis gas" or "syngas" typically via gasification, in the second, syngas is catalytically converted to the final product. In the latter stage, the performance of the selected catalyst is of key importance for the overall process. Much effort has been put into designing a selective catalyst for the synthesis of ethanol from syngas. The rhodium-based catalysts are among the most selective catalysts reported in the literature [7,8]. However, few reports are found using mesoporous silica as catalyst support [9–11], although various mesoporous materials have been applied in other catalytic reactions showing interesting results [12,13]. MCM-41 is a mesoporous silica that has 1D-hexagonal porous arrangement with a pore diameter between 1.6–10 nm and a wall thickness of around 0.8 nm [14,15]. The large surface area of MCM-41, usually 1000 m^2/g or more, can be of great utility for dispersing the active sites and hence boost the catalyst activity per unit of mass. To the best of our knowledge, there is no comparison between a mesoporous Rh/MCM-41 catalyst and a typical Rh/SiO$_2$ catalyst. Ma *et al.* [9] have compared Rh-Mn/MCM-41 with Rh-Mn/SiO$_2$, however, the presence of manganese may considerably affect the reactivity of the catalysts and no direct information about the effect of MCM-41 could be inferred. Chen *et al.* [10,11] have studied the effect of metal promoters (Mn and Fe) using another type of mesoporous silica (SBA-15).

In order to evaluate the catalytic performance of Rh/MCM-41 and Rh/SiO$_2$, some considerations regarding the metal loading, degree of metal dispersion (H/Rh) and syngas conversion level must be taken. Arakawa *et al.* [16] and Underwood and Bell [17] have studied the effect of metal dispersion, from 10% to 82%, which was obtained by increasing the metal loading from 0.1% to 30% Rh. Both studies showed a large effect of the metal dispersion on the product selectivity. However, the experiments carried out by Arakawa *et al.* [16] might have been affected by secondary reactions since the syngas conversion level was not the same in all the experiments (it varied from 0.4% to 27.9% depending on the metal loading). Underwood and Bell [17] kept the syngas conversion level below 0.1%, thus reducing the risk of secondary reactions, however, the various extents of metal loading could have affected their results. Tago *et al.* [18] showed that the extent of metal loading affects the product selectivity, according to their experiments carried out at constant H/Rh = 25% with the metal loading varied from 0.6% to 3.5% Rh. In a more recent work, Zhou *et al.* [19] used Rh/SiO$_2$ catalysts with different metal dispersions while keeping the same metal loading (3% Rh) testing them at low syngas conversion level (0.7%–1.5%). They observed that the degree of metal dispersion directly affects the product selectivity. Therefore, in order to avoid the effects of (i) secondary reactions; (ii) degree of metal dispersion and (iii) extent of metal loading over the product selectivity, the catalytic testing of Rh/MCM-41 and Rh/SiO$_2$ should be carried out at (1) low syngas conversion level; (2) equal degree of metal dispersion and (3) equal metal loading.

In the present work, we have evaluated the catalytic performance of a mesoporous Rh/MCM-41 catalyst and compared it with a typical Rh/SiO$_2$ catalyst. Both catalysts were tested for the synthesis of ethanol from syngas at low syngas conversion level (2.8%), same metal dispersion (H/Rh = 22%) and equal Rh loading (3 wt. %). Different product selectivities were found over Rh/MCM-41 and Rh/SiO$_2$. Additional experiments have been made in order to clarify the obtained results: (a) addition of water to the syngas feed-stream and (b) lowering of the syngas ratio (H$_2$/CO). The results from these experiments together with the catalyst characterization (BET, XPS, XRD, TEM, TPR), before and after reaction, indicate that the differences in the product selectivities can be attributed to the concentration of water vapor in the pores of Rh/MCM-41, which promote the water-gas-shift-reaction (WGSR) and produce extra CO$_2$ and H$_2$. These results confirm a previous study where high selectivity to CO$_2$ was observed over Rh/MCM-41 at various levels of syngas conversion as well as at different catalyst reduction temperatures [20].

2. Results and Discussion

The results obtained in the present study are divided into two sections. The first, Section 2.1, describes the catalytic performances of Rh/SiO$_2$ and Rh/MCM-41, which includes the effect of water addition and different syngas ratios (H$_2$/CO) (Section 2.1.1.). The second, Section 2.2, describes the catalysts characterization, before and after reaction, through the following techniques: N$_2$-physisorption (Section 2.2.1.), X-ray photoelectron spectroscopy (Section 2.2.2.), powder X-ray diffraction (Section 2.2.3.), transmission electron microscopy (Section 2.2.4.) and temperature programmed reduction (Section 2.2.5.). Finally, an interpretation of the obtained results and a discussion are then presented (Section 2.3.).

2.1. Catalytic Performances of Rh/SiO$_2$ and Rh/MCM-41

In order to examine the activity of the Rh/SiO$_2$ and Rh/MCM-41 catalysts, similar reaction conditions were used, that is reaction temperature and pressure of 280 °C and 20 bar, respectively. The gas hourly space velocity (GHSV) was varied in order to obtain a syngas conversion equal to 2.8% and the syngas ratio was equal to H$_2$/CO = 2/1. Similar conversion levels were applied for studies regarding the metal dispersion on rhodium-based catalysts [17–19]. At low syngas conversion, the occurrence of secondary reactions can be diminished and, at the same time, a uniform catalyst bed temperature can be reached. In all the experiments, the axial temperature gradient through the catalyst bed (measured by a mobile thermocouple introduced in a thermowell inside the catalyst bed) was always less than 1 °C.

Different product selectivities are obtained over Rh/SiO$_2$ and Rh/MCM-41 catalysts, as shown in Figure 1. If the selectivity toward all hydrocarbon compounds (methane and higher hydrocarbons) and the selectivity toward all oxygenated compounds (methanol, carbon dioxide, ethanol, acetaldehyde, acetic acid, methyl acetate and ethyl acetate) are considered in Figure 1, the following results are obtained: 58.2% hydrocarbon compounds and 41.8% oxygenated compounds for Rh/SiO$_2$, while for Rh/MCM-41 results 43.4% hydrocarbon compounds and 56.6% oxygenated compounds. This indicates that more hydrocarbon compounds are formed in Rh/SiO$_2$ than in Rh/MCM-41. Underwood and Bell have used a Rh/SiO$_2$ catalyst with a similar metal dispersion to the one used in our study (H/Rh = 10%) and also found a higher selectivity to hydrocarbons than to oxygenates [17].

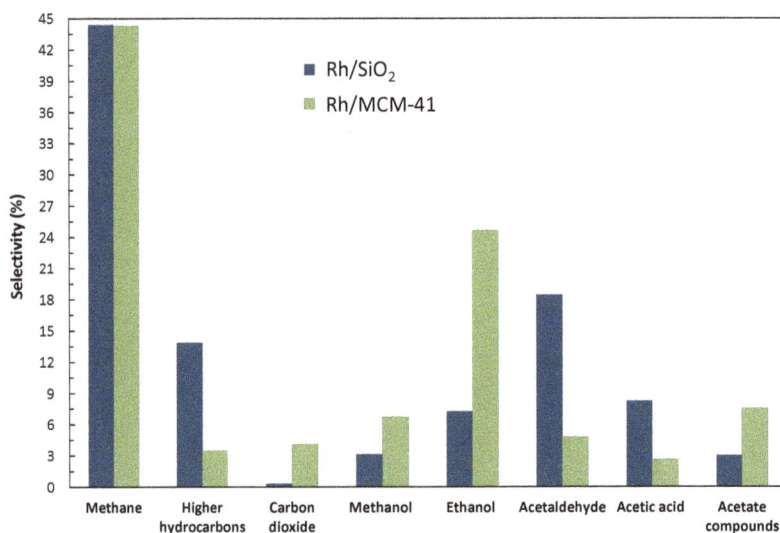

Figure 1. Product selectivities obtained from the conversion of syngas over Rh/SiO$_2$ and Rh/MCM-41. Higher hydrocarbons: ethane, propane and butane. Acetate compounds: methyl acetate and ethyl acetate. Reaction conditions: 280 °C, 20 bar, GHSV = 12,000 mL$_{syngas}$/g$_{cat}$·h for Rh/SiO$_2$ and GHSV = 3000 mL$_{syngas}$/g$_{cat}$·h for Rh/MCM-41.

Regarding the formation of the C$_2$-oxygenated compounds (ethanol, acetaldehyde and acetic acid), it can be seen from Table 1 (syngas ratio H$_2$/CO = 2/1) that the selectivity trend for Rh/SiO$_2$ decreases in the following order: acetaldehyde > acetic acid > ethanol. While for Rh/MCM-41 it decreases as: ethanol > acetaldehyde >

acetic acid. Interestingly, the total selectivity to C_2-oxygenated compounds is similar in both catalysts; 34% for Rh/SiO$_2$ and 32% for Rh/MCM-41.

Table 1. Selectivity to C_2-oxygenated compounds (ethanol, acetaldehyde and acetic acid) at different syngas ratios (H_2/CO) and with/without addition of water. Temperature of 280 °C and pressure of 20 bar.

Catalyst	Syngas Ratio H$_2$/CO	GHSV (mL/g·h)	Conversion (%)	TOF (s^{-1})	Total Selectivity to C$_2$-oxygenated (%)	Selectivity between C$_2$-oxygenated. (%)		
						Ethanol (%)	Acetaldehyde (%)	Acetic Acid (%)
	2/1	12000	2.8	0.039	34	21	54	24
Rh/SiO$_2$	2/1*	6000	1.6	0.011	18	8	83	9
	1/1	6000	2.1	0.022	32	16	51	34
	2/1	3000	2.8	0.010	32	77	15	8
Rh/MCM-41	2/1*	3000	1.0	0.005	1	65	35	0
	1/1	3000	0.4	0.002	35	69	10	21

* Water added.

2.1.1. Addition of Water Vapor and Lower Syngas Ratio (H_2/CO = 1/1)

Figure 2 shows the effect of the addition of water to the syngas feed-stream for Rh/SiO$_2$ and Rh/MCM-41 catalysts. The addition of water vapor decreases the syngas conversion from 2.8% to 1.0% for Rh/MCM-41 at GHSV = 3000 mL$_{syngas}$/g$_{cat}$·h. For Rh/SiO$_2$, since a very low syngas conversion was found at GHSV = 12,000 mL$_{syngas}$/g$_{cat}$·h, it was necessary to decrease the space velocity to GHSV = 3000 mL$_{syngas}$/g$_{cat}$·h to achieve a syngas conversion of 1.6%, which indicates a considerable decrease of the activity for Rh/SiO$_2$ in the presence of water vapor. It can also be observed that the addition of water notably increases the selectivity to CO$_2$ in both catalysts; from 0.3% to 18.5% for Rh/SiO$_2$ and from 4.1% to 90.4% for Rh/MCM-41. Likewise, while the selectivity to the rest of the products is reduced in both catalysts, the selectivity to methanol is considerably increased over Rh/SiO$_2$, but it is somewhat reduced over Rh/MCM-41. Finally, the total selectivity to C_2-oxygenated compounds is notably reduced in both catalysts, as can be seen in Table 1 (syngas ratio H_2/CO = 2/1 *), being more marked in Rh/MCM-41. Moreover, the selectivity trends are kept in the same order for Rh/SiO$_2$ (acetaldehyde > acetic acid > ethanol) and for Rh/MCM-41 (ethanol > acetaldehyde > acetic acid) as in the experiments without water addition (Table 1).

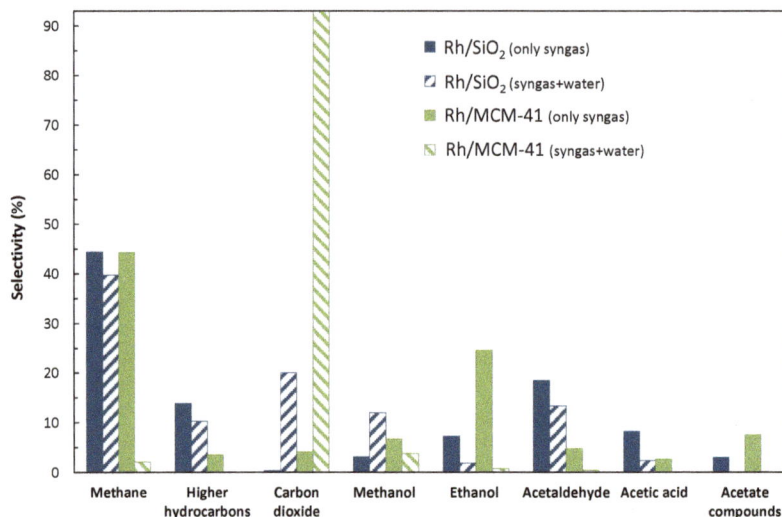

Figure 2. Comparison of the product selectivities obtained from the conversion of syngas with and without addition of water over Rh/SiO_2 and Rh/MCM-41. Reaction conditions (with addition of water): 280 °C, 20 bar, GHSV = 6000 mL_{syngas}/g_{cat}·h (H_2:CO:H_2O = 2:1:2.7) for Rh/SiO_2 and GHSV = 3000 mL_{syngas}/g_{cat}·h (H_2:CO:H_2O = 2:1:1.5) for Rh/MCM-41. Reaction conditions for the experiments without addition of water as indicated in Figure 1.

Figure 3 shows the effect of changing the syngas ratio from H_2/CO = 2/1 to H_2/CO = 1/1 for Rh/SiO_2 and Rh/MCM-41 catalysts. The lowering of syngas ratio from H_2/CO = 2 to H_2/CO = 1 decreases the syngas conversion in both catalysts; from 2.8% to 0.4% for Rh/MCM-41 at GHSV = 3000 mL_{syngas}/g_{cat}·h and from 2.8% (GHSV = 12,000 mL_{syngas}/g_{cat}·h) to 2.1% (GHSV = 6000 mL_{syngas}/g_{cat}·h) for Rh/SiO_2. The reduction of the syngas conversion by the lowering the syngas ratio has also reported in the literature, which is attributable to a lesser extent of the hydrogenation reactions (hydrocarbon formation) [7,8].

In both catalysts, the lowering of the syngas ratio results in less production of methane and methanol while, in contrast, more higher hydrocarbons, carbon dioxide, and acetate compounds are obtained (Figure 3). When the syngas H_2/CO ratio is lowered, the partial pressure of hydrogen is expected to be decreased. This suggests that the products with high H/C ratios, such as CH_4 and CH_3OH, could be favored at high partial pressure of hydrogen (high syngas H_2/CO ratio). Likewise, the products with lower H/C ratios, such as the higher hydrocarbons and oxygenated compounds could be favored at low partial pressure of hydrogen (low syngas H_2/CO ratio). These results are in agreement with the literature regarding the effect of the syngas ratio for rhodium-based catalysts [7,8]. In addition, among the C_2-oxygenated compounds it can be observed that the selectivity to acetic acid is favored at low syngas H_2/CO ratio, while the selectivities to the more hydrogenated compounds (ethanol and acetaldehyde) are favored at higher syngas H_2/CO ratio, as is observed in Table 1 (see syngas ratio $H_2/CO = 2/1$ and $H_2/CO = 1/1$).

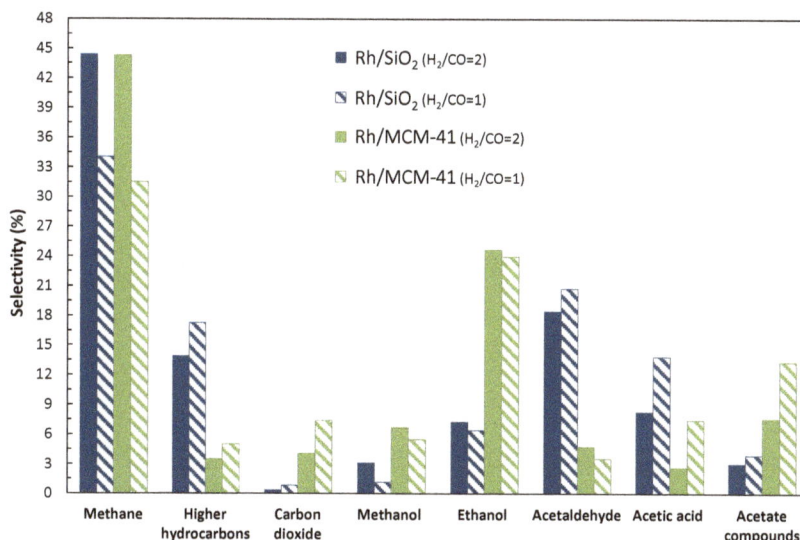

Figure 3. Comparison of the product selectivities obtained from the conversion of syngas with $H_2/CO = 2$ and $H_2/CO = 1$ over Rh/SiO$_2$ and Rh/MCM-41. Reaction conditions (syngas $H_2/CO = 1$): 280 °C, 20 bar, GHSV = 6000 mL$_{syngas}$/g$_{cat}$·h for Rh/SiO$_2$ and GHSV = 3000 mL$_{syngas}$/g$_{cat}$·h for Rh/MCM-41. Reaction conditions for the experiments with syngas $H_2/CO = 2$ as indicated in Figure 1.

In a previous study a high selectivity to CO_2 in the Rh/MCM-41 catalyst was also observed, it was found to be independent of the catalyst reduction temperature and the syngas conversion level [20]. Figures 4 and 5 show the product selectivities and the syngas conversion with time on stream at different reaction conditions for the Rh/SiO$_2$ and MCM-41 catalysts, respectively. It can be noted that almost no CO_2 is formed over the Rh/SiO$_2$ catalyst (Figure 4, Period A–F) whatever the reaction condition (T, P, GHSV) or catalyst reduction temperature (200, 370, 500 °C and non-reduced) applied. On the other hand, a high selectivity to CO_2 is observed over the MCM-41 catalyst at all reaction conditions studied and at different catalyst reduction temperatures (Figure 5, Period A′–G′). Thus, the product selectivity is notably affected by the MCM-41 support in a large range of reaction conditions and at different catalyst reduction temperatures. These results together with the catalysts characterization (Section 2.2) are discussed later on in Section 2.3.

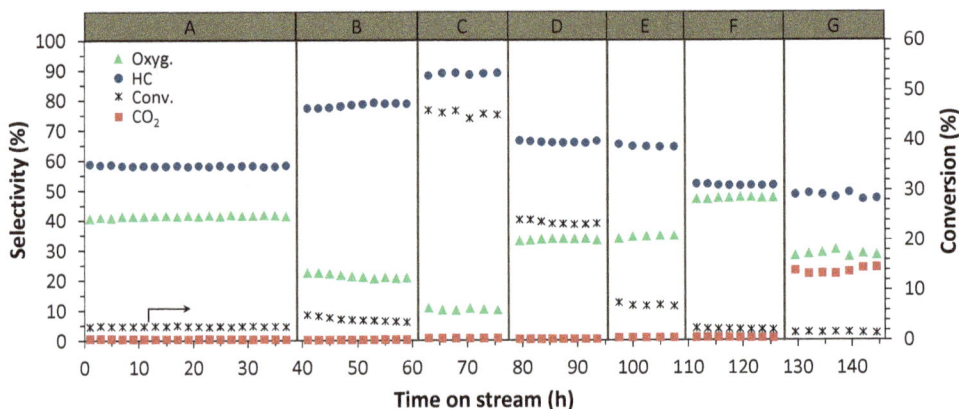

Figure 4. Selectivities and conversion with time on stream for the Rh/SiO$_2$ catalyst. Oxyg.: alcohols, acetaldehyde, acetic acid and esters. HC: methane and higher hydrocarbons. See Table S1 in supporting information for details about the reaction conditions in periods A–G.

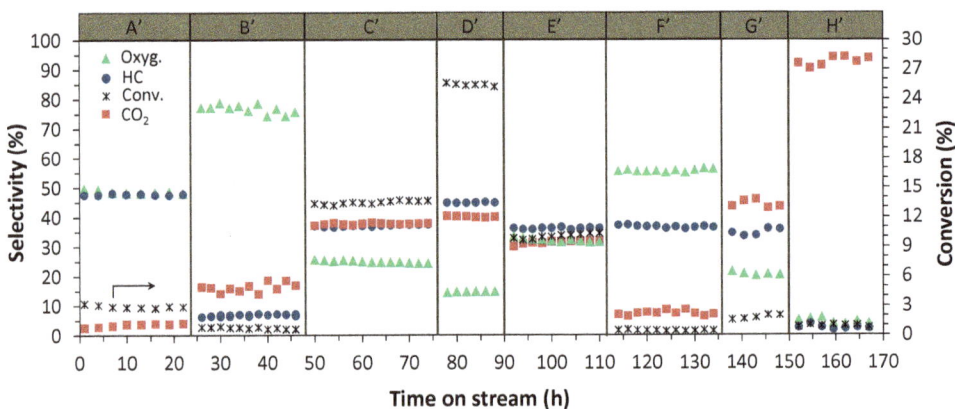

Figure 5. Selectivities and conversion with time on stream for the Rh/MCM-41 catalyst. Oxyg.: alcohols, acetaldehyde, acetic acid and esters. HC: methane and higher hydrocarbons. See Table S1 in supporting information for details about the reaction conditions in periods A′–H′.

2.2. Catalyst Characterization

2.2.1. N$_2$-Physisorption

In Table 2 the results obtained from N$_2$-physisorption are reported. When comparing the pure supports, the surface area of MCM-41 is about four times larger than the surface area of SiO$_2$. A similar relation is observed when 3 wt. %

Rh is impregnated in both supports. This indicates that the process of rhodium incorporation (aqueous impregnation, drying and calcination) affects to a similar extent the surface area of MCM-41 and the surface area of SiO_2. In some reported studies, the surface area of MCM-41 is drastically affected by metal incorporation: for example, the impregnation with 5wt. % Co reduced the original MCM-41 surface area by nearly 50% [21]. However, the incorporation of Rh into MCM-41 only reduces the original MCM-41 surface area by 0.9%, as shown in Table 2. After catalytic testing, the surface area was reduced by 3.7% for Rh/SiO_2 and 9.0% for Rh/MCM-41. More notable changes are observed when water is added to the syngas feed-stream, then the surface areas of Rh/SiO_2 and Rh/MCM-41 are reduced by 10.3% and 19.3%, respectively (Table 2). The more significant effect of water on the surface area of Rh/MCM-41 is consistent with the structure degradation of mesoporous materials proposed by Landau *et al.* [22]. Therefore, the hydration of the siloxane structure is followed by siloxane hydrolysis-hydroxylation and their rearrangement-redehydroxylation during calcination. This results in a few 1D-channels collapsing into a single one [22]. As a consequence, there is a loss of Bragg intensity (as will be evidenced by XRD analysis, Figure 6), decreasing the pore volume (as observed in Table 2), while retaining uniform pore size (as described below and further shown by TEM analysis, Figure 7).

Regarding the pore size distribution of Rh/SiO_2 and Rh/MCM-41 (see Figure S1 in supporting information) confirmed, a wide pore size distribution is observed in the pure SiO_2, which is narrowed after Rh impregnation. Pure MCM-41 presents a narrow pore size distribution and smaller pore sizes as compared to SiO_2; after Rh impregnation no significant changes are observed. From the adsorption isotherms, the average pore size of Rh/SiO_2 is much larger than in the case of Rh/MCM-41; 19 nm and 2.5 nm, respectively. Finally, after catalytic testing (with and without water) there is no significant change in the pore size distribution for any of the catalysts.

2.2.2. X-ray Photoelectron Spectroscopy (XPS)

In Table 2, the results from XPS analysis are summarized. Before the catalytic testing, we can observe that the oxidation state of Rh is similar in both Rh/SiO_2 and Rh/MCM-41 catalysts. After the catalytic testing (without addition of water), the fraction of Rh^{3+} is increased in Rh/SiO_2 while it is decreased in Rh/MCM-41. After the catalytic testing with addition of water, all Rh species are in the form of Rh^0 in both catalysts. Finally, the Rh/Si ratio indicates that Rh species migrate to the surface of the catalysts as a consequence of the catalytic testing. The latter phenomenon is more pronounced when water is added to the catalytic system.

Table 2. N_2-physisorption and XPS analyses of the Rh/SiO_2 and Rh/MCM-41 catalysts, before and after the catalytic testing.

Catalyst	Condition	Surface Area (m^2/g)	Change in Surface Area	Pore Volume (cm^3/g)	Rh^0 %	Rh^0 Binding energy (eV)	Rh^{3+} %	Rh^{3+} Binding energy (eV)	Rh/Si
	Pure support	238	-	0.87	-	-	-	-	-
	Before catalytic testing	236	0.5% *	0.90	86.5	307.3	13.4	309.7	0.0057
Rh/SiO_2	After catalytic testing (280 °C, 20 bar, 12,000 mL/g·h): *only syngas*	228	3.7% **	0.90	82.4	307.3	17.6	309.7	0.0064
	After catalytic testing (280 °C, 20 bar, 6000 mL/g h): *syngas and water*	212	10.3% **	0.88	100	307.2	-	-	0.0079
	Pure support	970	-	1.08	-	-	-	-	-
	Before catalytic testing	961	0.9% *	1.15	80.6	307.1	19.4	308.6	0.0079
Rh/ MCM-41	After catalytic testing (280 °C, 20 bar, 3000 mL/g·h): *only syngas*	875	9.0% **	1.04	90.8	307.4	9.2	309.3	0.0087
	After catalytic testing (280 °C, 20 bar, 3000 mL/g·h): *syngas and water*	776	19.3% **	0.94	100	307.1	-	-	0.0100

* Compared to the pure support. ** Compared to the catalyst before catalytic testing.

2.2.3. Powder X-ray Diffraction (XRD)

Mesoporous MCM-41 has characteristic signals at small-angle XRD, which are indicated in Figure 6 as (100), (110), and (200) reflections. These signals correspond to a hexagonal structure with unit cell parameter $a = 2d_{100}/\sqrt{3}$ [14]. An additional (210) reflection is reported in the literature, which gives a low intensity signal at around $2\theta = 5.9$ [14]. In qualitative terms, the intensity of the XRD signals can be attributed to the long-range periodic structure of MCM-41 [21]. Although the long-range ordering in Rh/MCM-41 decreases after impregnation with Rh (Figure 6A) and after catalytic testing (Figure 6B,C), it does not completely disappear. In addition, it can be observed that all signals appear at the same 2θ, *i.e.*, same d-spacing, which suggests that there is no significant lattice contraction either after Rh impregnation or after catalytic testing. Typical SiO_2 support does not present a pore ordering as mesoporous MCM-41, thus no signals are found at small-angle XRD (Figure 6).

The wide-angle XRD patterns in Figure 6 indicate that both catalysts Rh/SiO_2 and Rh/MCM-41 are amorphous materials. It also indicates that there is no segregation of rhodium oxides (or at least they are not large enough to be detected by the XRD technique) after impregnation with Rh (Figure 6A) and after catalytic testing without addition of water (Figure 6B). After catalytic testing with addition of water (Figure 6C) a low-intensity and broad signal appears at around $2\theta = 35\text{--}40$. This range of 2θ corresponds to the characteristic signals of Rh_2O_3 and Rh^0. According to the XPS analysis, all Rh species are converted to Rh^0 when water is added to the catalytic system. Therefore, we can conclude that the presence of water in the syngas feed-stream slightly favors the growth of Rh^0 clusters.

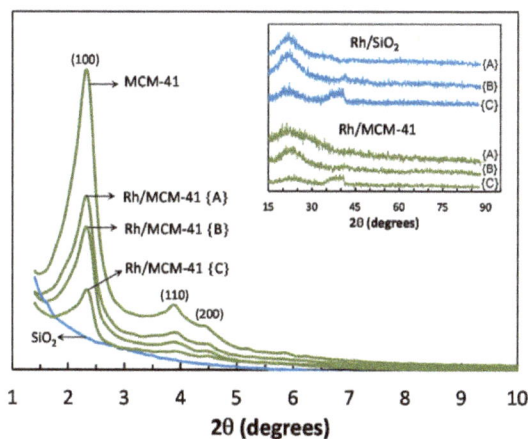

Figure 6. Small and wide angle XRD patterns of pure supports SiO_2 and MCM-41, impregnated with Rh (**A**); after catalytic testing (**B**); and after catalytic testing with addition of water (**C**). Catalytic testing at same conditions as indicated in Table 2.

2.2.4. Transmission Electron Microscopy (TEM)

In Figure 7, TEM images for Rh/SiO_2 and Rh/MCM-41 catalysts are shown. Before reaction, the average Rh particle size is similar in both catalysts; 4 nm for Rh/SiO_2 and 3 nm for Rh/MCM-41. Which agrees with the average metal particle size equal to 4 nm derived from the metal dispersion (H/Rh = 22%), assuming an icosahedral particle shape [23]. For the Rh/SiO_2 catalyst, the average Rh particle size has grown after the catalytic testing (with and without water). Less effect is observed for the Rh/MCM-41 catalyst, where the average particle size is kept to 3 nm after the catalytic testing, even with addition of water. It can be noted that the pore diameter of MCM-41 is in the same range as the Rh particle size in the Rh/MCM-41 catalyst (Figure 7A-Rh/MCM-41), it may suggest that the pore diameter of MCM-41 limits the growth of Rh particles. This is supported by Zhou et al. [19] who observed a relation between the Rh particle size and the pore diameter in various silica supports. If it is so, it could explain the narrow Rh particle size distribution in the mesoporous Rh/MCM-41 catalyst, even after the catalytic testing (Figure 7B,C-Rh/MCM-41), as compared to the particle size distribution in the Rh/SiO$_2$ catalyst (Figure 7A–C-Rh/SiO$_2$).

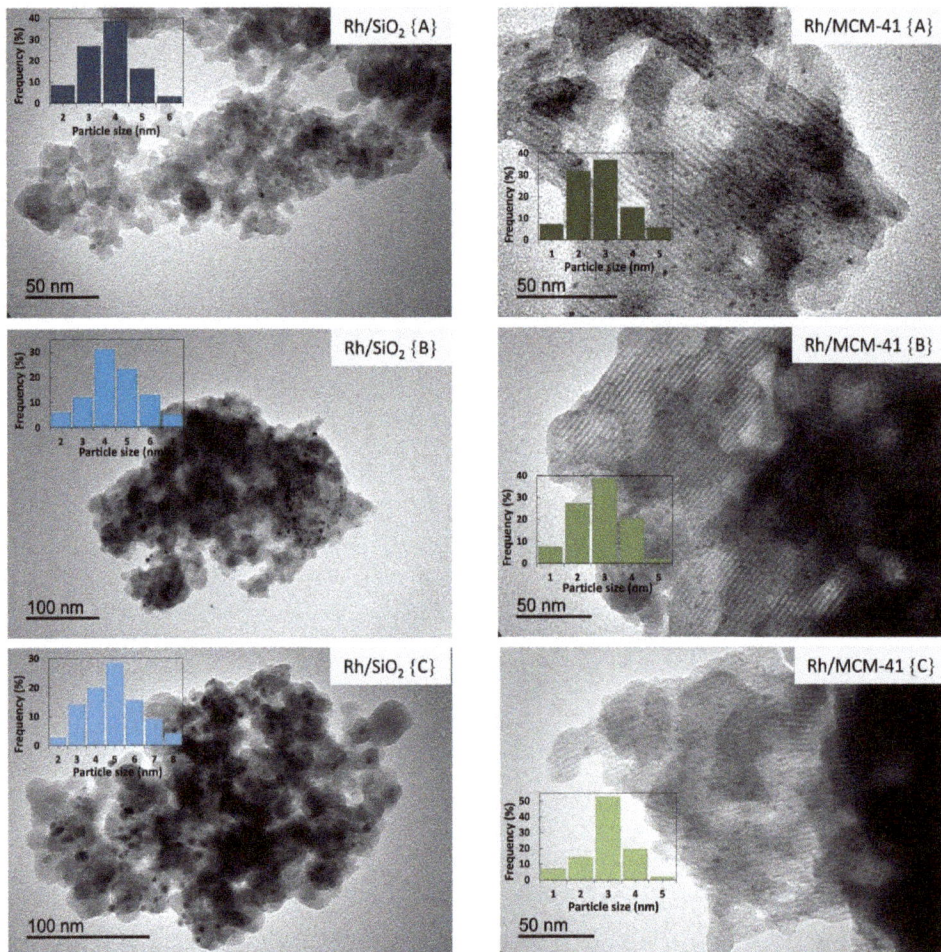

Figure 7. TEM images of Rh/SiO$_2$ and Rh/MCM-41 catalysts; reduced (**A**); after catalytic testing (**B**); and after catalytic testing with addition of water (**C**). Catalytic testing at same conditions as indicated in Table 2.

From XRD analysis (Figure 6), after the catalytic testing (with addition of water) a new and broad signal is observed in both catalysts. This signal corresponds to Rh^0 according to XPS (Table 2). For the Rh/SiO$_2$ catalyst, these results agree with the growth of the particle size observed in TEM images (Figure 7). For the Rh/MCM-41 catalyst, it seems that the pores of MCM-41 support limit the growth of Rh and thus the particle size distribution is little affected. Possibly, the signal detected in XRD for Rh/MCM-41 may be due mostly to an increased number of Rh^0 dispersed along the pores.

2.2.5. Temperature Programmed Reduction (TPR)

Figure 8 shows the TPR profiles for the Rh/SiO_2 and Rh/MCM-41 catalysts. It can be seen that the reduction of rhodium oxide species begins at low temperatures, in agreement with the literature [24,25]. All Rh species seem to be reduced at temperatures below 200 °C. The TPR profile of the Rh/MCM-41 catalyst suggests some metal-support interactions, which widen the reduction peak at higher temperatures. Metal-support interactions are observed in other mesoporous catalysts as well as in traditional catalysts [13,26]. Moreover, after the reduction treatment at 370 °C the surface of the Rh/SiO_2 and Rh/MCM-41 catalysts are composed of a similar proportion of Rh^0 and Rh^{3+}, as is indicated by XPS analysis (Section 2.2.2.).

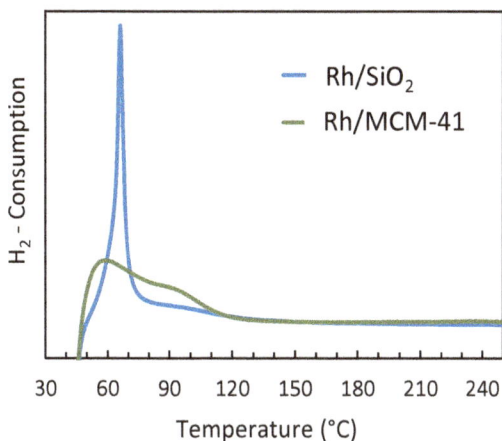

Figure 8. Temperature programmed reduction profiles of the Rh/SiO_2 and Rh/MCM-41 catalysts.

2.3. Interpretation of the Catalytic Performance of Rh/MCM-41 Compared to Rh/SiO₂

As observed in Figure 1, the product selectivity obtained for the conversion of syngas over Rh/MCM-41 is notably different compared to the product selectivity found over Rh/SiO_2. These differences can be due to the different characteristics of the supports: MCM-41 and SiO_2. It may be that the 1D-channels with small pore diameter of MCM-41, which seem to be retained in Rh/MCM-41 (as suggested from N_2-physisorption, XRD and TEM analyses), could affect the catalytic performance when comparing with a typical SiO_2 support which has large and non-ordered pore diameters. Another characteristic is the metal-support interactions in Rh/MCM-41,

159

which seems to be absent in Rh/SiO$_2$ (as suggested from TPR analysis). However, the later characteristic may not be important during the catalytic testing since the final surface composition is similar in both catalysts (as indicated from XPS analysis). The product selectivity could neither have been affected by the Rh particle size because, on the one hand, both catalysts have been pre-reduced at conditions to obtain similar metal dispersion (*i.e.*, similar metal particle size, which is confirmed by TEM images). On the other hand, we cannot relate the slight growth of the Rh particle size in the Rh/SiO$_2$ catalyst with the increased selectivity to CO$_2$ (approx. 20%), because, if this would be the case, the selectivity to CO$_2$ in the Rh/MCM-41 catalyst should be less than 20% since the growth of the Rh particle size is even lower for this catalyst (as shown the TEM images), but its selectivity to CO$_2$ is above 90%.

It has been suggested that a high concentration of water vapor can be formed in the pores of MCM-41 during catalyst reduction, which could promote metal-support interactions [13,27]. If so, water vapor can also be concentrated in the catalyst pores during catalytic testing, since most of the syngas reactions produce water as a main by-product (reactions 1–4) [7,8]. As a consequence, an environment rich in water vapor can be formed which may induce some reactions, such as the water-gas-shift-reaction (WGSR). WGSR generates extra carbon dioxide and hydrogen (reaction 5). In concordance, more carbon dioxide is produced by Rh/MCM-41 than by Rh/SiO$_2$ (see Figure 1). Hydrogen was unfortunately not measured due to analytical limitations since hydrogen has similar thermal conductivity to the carrier helium in the GC analysis. However, if we consider the selectivity towards the C$_2$-oxygenated compounds (Table 1), the selectivity to the more hydrogenated compounds, *i.e.*, ethanol and acetaldehyde, accounts for 92% over Rh/MCM-41, while it is only 75% over Rh/SiO$_2$. This may suggest that the extra hydrogen generated from the WGSR over Rh/MCM-41 would facilitate the hydrogenation of acetic acid (or a C$_2$-oxygenated intermediary, perhaps an acetyl intermediate [28]) to acetaldehyde and ethanol.

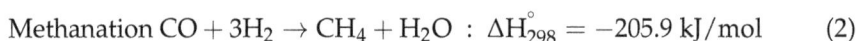

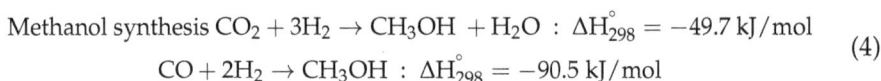

$$\text{Ethanol generation } 2CO + 4H_2 \rightarrow C_2H_5OH + H_2O \;:\; \Delta H^{\circ}_{298} = -253.6 \text{ kJ/mol} \qquad (1)$$

$$\text{Methanation } CO + 3H_2 \rightarrow CH_4 + H_2O \;:\; \Delta H^{\circ}_{298} = -205.9 \text{ kJ/mol} \qquad (2)$$

$$\text{Hydrocarbons formation } CO + 2H_2 \rightarrow -CH_2- + H_2O \;:\; \Delta H^{\circ}_{298} = -165 \text{ kJ/mol} \quad (3)$$

$$\text{Methanol synthesis } CO_2 + 3H_2 \rightarrow CH_3OH + H_2O \;:\; \Delta H^{\circ}_{298} = -49.7 \text{ kJ/mol}$$
$$CO + 2H_2 \rightarrow CH_3OH \;:\; \Delta H^{\circ}_{298} = -90.5 \text{ kJ/mol} \qquad (4)$$

$$\text{Water gas shift reaction } CO + H_2O \rightarrow CO_2 + H_2 \;:\; \Delta H^{\circ}_{298} = -41.1 \text{ kJ/mol} \quad (5)$$

In accordance with this view, the WGSR should also be promoted over the Rh/SiO$_2$ catalyst if water were to be concentrated in the catalyst particle. In effect, when water is added to the syngas feed-stream (Figures 2 and 4) the selectivity to CO$_2$ is enhanced from almost zero to approximately 20%. In the case of the Rh/MCM-41 catalyst, an even higher concentration of water is obtained in the 1-D channel pores, boosting the selectivity to CO$_2$ (>90%). This means that the WGSR is highly favored in the presence of water, whatever kind of silica support is used, SiO$_2$ or MCM-41. In the presence of water vapor, all Rh species are in the form of Rh^0 (as suggested by XPS analysis). Still there is no agreement in the literature on which Rh species (Rh^0, Rh^{n+}) is more active for the oxygenate formation. Chuang *et al.* [29] suggest that Rh^0 is less active than the oxidized form. Our results might support this idea since the lowest activities found in both catalysts occurred when both catalysts had 100% Rh^0. At the same time, as CO$_2$ is produced by the addition of water in both catalysts, the selectivity to methanol is increased in Rh/SiO$_2$ (Figure 2). For Rh/MCM-41, although the selectivity to methanol is somewhat reduced by the addition of water, it is not as drastically reduced as the selectivity to the rest of products (Figure 2). Therefore, it seems that the formation of CO$_2$ and methanol is favored in the presence of water, where the WGSR would play an important role. Indeed, it is generally accepted that the synthesis of methanol from syngas occurs primary via the hydrogenation of CO$_2$ [30]. This means that the occurrence of the WGSR, as a consequence of the concentration of water vapor in Rh/MCM-41, may contribute to the generation of methanol from CO$_2$, which may explain the higher selectivity to methanol observed in Rh/MCM-41 compared to Rh/SiO$_2$ (Figure 1).

Regarding the effect of the hydrogen partial pressure on the selectivity to the C$_2$-oxygenated compounds, when it decreases, the selectivity towards acetic acid should increase in both catalysts. It can be observed that when changing the syngas ratio from H$_2$/CO = 2/1 to H$_2$/CO = 1/1 (Table 1), more acetic acid is evidently formed in both catalysts. This may also support the idea that WGSR occurs in Rh/MCM-41 producing extra H$_2$, which may facilitate the hydrogenation of acetic acid (or a C$_2$-oxygenated intermediary) yielding more acetaldehyde and ethanol than in the typical Rh/SiO$_2$ catalyst (Table 1). Furthermore, the product distribution between the C$_2$-oxygenates in Rh/MCM-41 always keeps the relation according to the extent of oxidation of each compound: ethanol > acetaldehyde > acetic acid (Figures 1–3).

The reaction mechanism for the conversion of syngas to oxygenated compounds over rhodium-based catalyst, is not fully understood [7,8,28]. For example, it is suggested that acetate compounds can be formed by the reaction of CO$_2$ with surface intermediates such as CH$_x$* species [31]. If this reaction occurs over Rh/SiO$_2$ and Rh/MCM-41, the additional formation of CO$_2$ via WGSR in Rh/MCM-41 would consume much more CH$_x$* species than in Rh/SiO$_2$. This means that a lesser number

161

of $CH_x{}^*$ species would be available for conversion into hydrocarbon compounds and more acetate compounds should be expected in the Rh/MCM-41 catalyst. This is in accordance with our results, since lower selectivity to hydrocarbon compounds and higher selectivity to acetate compounds are found over Rh/MCM-41 as compared to Rh/SiO_2 (Figure 1). However, we must indicate that the discussion of the elementary steps or reaction mechanism for ethanol formation is not the objective of this work, we rather study the performance of a Rh/MCM-41 catalyst. Studies such as micro-kinetics together with theoretical calculations can be helpful for a fundamental understanding of the reaction mechanism. Very interesting reports on this topic have been published in recent years [32,33]. The inclusion of the WGSR and its effect on the product distribution can be interesting to study and compare with the results obtained in the present work.

From the above discussion, we believe that the differences in activity and selectivity observed for Rh/SiO_2 and Rh/MCM-41 are highly related to the concentration of water vapor in Rh/MCM-41. This is also supported by a previous investigation where high selectivity to CO_2 was found over the mesoporous Rh/MCM-41 catalyst at various levels of syngas conversion as well as at different catalyst reduction temperatures [20] (Figures 4 and 5). The metal-support interactions in Rh/MCM-41 might contribute to the observed differences, but probably to a much lower extent than the effect of water vapor. In summary, the high concentration of water vapor in the pores of Rh/MCM-41 may promote the occurrence of the WGSR, generating extra CO_2 and H_2, which in turn facilitate side reactions changing the product selectivity. Finally, it seems that, as a consequence of the extra hydrogen generated from the WGSR in Rh/MCM-41, more acetaldehyde and/or acetic acid (or a C_2-oxygenated intermediary) is hydrogenated yielding more ethanol than in the typical Rh/SiO_2 catalyst.

3. Experimental Section

3.1. Catalyst Preparation and Characterization Techniques

Hexagonal mesoporous silica (MCM-41) is usually obtained by the "atrane route" [34]. In this method, cetyltrimethylammonium bromide is used as the structural directing agent, and triethanolamine (TEA) is used as a hydrolysis retarding agent. Silatrane complexes are formed between tetraethyl orthosilicate and TEA, as metal precursors of Si. Further preparation procedure has been described elsewhere [34,35]. A commercial MCM-41 from Sigma Aldrich (Sigma Aldrich, St. Louis, MO, USA) was indistinctly used to the MCM-41 obtained from the "atrane route". Silica (SiO_2) catalyst support was purchased from Alfa Aesar (Alfa Aesar, Karlsruhe, Germany). The Rh/MCM-41 and Rh/SiO_2 catalysts were prepared by successive incipient wetness impregnation, using an aqueous solution of RhCl$_3 \cdot n$H$_2$O. After impregnation, the catalysts were dried and then calcined at 500 °C/5 h. The total metal loading was 3 wt. % Rh for both Rh/MCM-41 and Rh/SiO$_2$ catalysts.

Powder X-ray diffraction (XRD) was performed using a Siemens D5000 instrument (Siemens, Karlsruhe, Germany) with Cu K-α radiation (2θ = 10°–90°, step size = 0.04°) equipped with a Ni filter and operated at 40 kV and 30 mA. N$_2$-physisorption was carried out using a Micromeritics ASAP 2000 instrument (Micromeritics, Norcross, GA, USA). The Brunauer-Emmett-Teller (BET) method was used to calculate the surface area and the Barrett-Joyner-Halenda (BJH) method was used to calculate the pore size and pore volume from the desorption isotherm. Temperature programmed reduction (TPR) was carried out in a Micromeritics Autochem 2910 instrument (Micromeritics, Norcross, GA, USA), a reducing gas mixture (5% H$_2$ in Ar) at a flow of 50 mL/min passed through the catalyst sample while the temperature was increased by 5 °C/min up to 900 °C. Transmission electron microscopy (TEM) images were obtained using a JEOL JEM 1400 microscope (JEOL, Tokio, Japan). Samples were mounted on 3 mm holey carbon copper grids. Particle size and distribution were estimated after examination of more than 100 metal particles.

The metal dispersion was measured by H$_2$-chemisorption using a Micromeritics ASAP 2020 instrument (Micromeritics, Norcross, GA, USA). Prior to the analysis, the catalyst sample was reduced with hydrogen. After vacuum evacuation, a dynamic mode of hydrogen injection was performed at 40 °C [36]. Repeated analyses were made in order to discriminate between the amount of hydrogen adsorbed via physisorption or chemisorption [37]. The stoichiometry of hydrogen chemisorbed on metallic rhodium was considered to be 1:1. Different degrees of metal dispersion were obtained by changing the reduction temperature and the reduction time. In order to obtain the same metal dispersion, the catalysts were reduced at 370 °C during 1h for Rh/MCM-41 and 6 h for Rh/SiO$_2$, resulting in H/Rh = 22% in both catalysts.

X-ray photoelectron spectroscopy (SPECS, Berlin, Germany) data were recorded on 4 mm × 4 mm pellets of 0.5 mm thickness that were obtained by gently pressing the powdered materials, following outgassing to a pressure below 2 × 10^{-8} Torr at 150 °C in the instrument pre-chamber to remove chemisorbed volatile species. The main chamber of a Leybold-Heraeus LHS10 spectrometer was used, capable of operating down of 2 × 10^{-9} Torr, which was equipped with an EA-200MCD hemispherical electron analyzer with a dual X-ray source using AlKα (hv = 1486.6 eV) at 120 W, 30 mA, with C(1s) as energy reference (284.6 eV). The catalyst samples were analyzed as; (i) reduced at the same conditions as in the catalytic testing; (ii) after catalytic testing and; (iii) after catalytic testing with addition of water. The reaction conditions are indicated in Table 2.

3.2. Catalytic Testing

The experiments were carried out in a stainless-steel, down-flow fixed bed reactor. The system components and the on-line gas chromatograph (GC) analysis are similar to those described by Andersson *et al.* [38]. The internal diameter of the reactor was 8.3 mm, in which about 300 mg of catalyst was charged with a particle size between 160 and 250 μm. Before reaction, the catalysts were reduced following the same procedure as the H_2-chemisorption described above, in order to obtain a metal dispersion of H/Rh = 22%. After reduction, the reactor was cooled to 280 °C and pressurized with syngas up to 20 bar. The gas hourly space velocity (GHSV) was varied between 3000–19,000 mL_{syngas}/g_{cat}. Premixed syngas bottles (AGA Linde) with a H_2/CO ratio of 2:1 and 1:1 were used. The amount of water added to the system was regulated by means of a Gilson 307 pump, then the dosed water was evaporated by an external tape heater at 225 °C and then mixed with the syngas feed-stream.

An on-line GC Agilent 7890A (Agilent, Santa Clara, CA, USA) was used to quantify the reaction products. A detailed description of the GC configuration and the analytical procedure have been published previously by Andersson *et al.* [38]. N_2 added to the syngas mixture was used as internal standard to quantify CH_4, CO and CO_2 in a thermal conductivity detector [39]. The internal normalization of corrected peak areas was used to quantify the hydrocarbon and oxygenated compounds in a flame ionization detector [39]. The expressions used to calculate the syngas conversion, carbon mole selectivity and carbon balance are reported in [38]. In all the experiments the carbon balance was in the range of 99.1%–100%.

Preliminary estimations were carried out to ascertain transport effects on the Rh/MCM-41 catalyst. At the experimental conditions used in this study, intraparticle diffusion limitation might slightly initiate at high syngas conversion (>70% in term of CO conversion), according to Weisz-Prater's criterion [40]. However, a proper estimation of the diffusion limitation would require a careful determination of the morphology of the Rh/MCM-41 catalyst, which is outside the scope of the present study. The Koros-Nowak criterion indicates that the reaction rate, in the kinetic regime, is directly proportional to the concentration of the active material [40]. This means that the turnover frequency (TOF, s^{-1}) must be invariant as the concentration of the active material is changed. An experimental test was carried out at 230 °C, 20 bar and 40,000 mL_{syngas}/g_{cat} h (H_2/CO = 2/1) using a catalyst sample with a fixed number of active sites (3 wt. % Rh and metal dispersion of H/Rh = 22%). A fraction of this sample was diluted with pure MCM-41 support at a ratio of 1:3, resulting in a catalyst with nominal composition 0.75 wt.% Rh and a metal dispersion of H/Rh = 22%. A minor variation (<10%) was observed between the TOFs of the non-diluted catalyst and the diluted catalyst, suggesting that the reaction operates in the kinetic regime.

4. Conclusions

The catalytic performance of Rh/MCM-41 catalyst has been evaluated for the synthesis of ethanol from syngas and compared with a typical Rh/SiO$_2$ catalyst. Equal reaction conditions have been applied (280 °C and 20 bar), low syngas conversion (2.8%), the same metal dispersion (H/Rh = 22%) and the same Rh loading (3 wt. %). Under these conditions, different product selectivities have been obtained for Rh/SiO$_2$ and Rh/MCM-41 catalysts. In order to clarify the obtained results, additional experiments were conducted: addition of water to the syngas feed-stream and lowering of the syngas ratio (H$_2$/CO). The results from these experiments together with the catalyst characterization, before and after reaction, indicate that the differences in the catalytic performances of Rh/SiO$_2$ and Rh/MCM-41 can be attributed to a high concentration of water vapor in the pores of Rh/MCM-41, which seem not to occur in Rh/SiO$_2$. The concentration of water vapor in Rh/MCM-41 could essentially promote the occurrence of the water-gas-shift-reaction (WGSR) which generates some extra carbon dioxide and hydrogen. The extra carbon dioxide and hydrogen could induce side reactions and thus change the product selectivity, as compared with Rh/SiO$_2$. The extra hydrogen generated from the WGSR in Rh/MCM-41 could possibly facilitate the hydrogenation of acetic acid and/or acetaldehyde (or a C$_2$-oxygenated intermediary) to ethanol, which may be related to the higher selectivity to ethanol observed in Rh/MCM-41 (24%) compared to Rh/SiO$_2$ (8%).

Acknowledgments: The Swedish International Development Cooperation Agency (SIDA) and the Junta de Andalucia (P09-FQM-4781 project) are gratefully acknowledged for financial support. SCAI at the University of Córdoba is also acknowledged for the use of TEM.

Author Contributions: The experimental work and drafting of the manuscript was done by L.L., assisted by J.V., who carried out some catalyst characterization at KTH, and V.M., who performed the TEM measurements. M.B. and S.J. have supervised the experimental work and participated in the analysis and interpretation of the results. A.M., S.C., M.B. and S.J. supported the work and cooperation between KTH, UMSA and the University of Córdoba. The manuscript was written through the comments and contributions of all authors. All authors have approved for the final version of the manuscript.

Conflicts of Interest: The authors declare no conflicts of interest.

References

1. International Energy Agency. *Renewable Energy: Medium-Term Market (Market Trends and Projections to 2018)*; International Energy Agency—IEA: Paris, France, 2013.
2. Du, X.; Carriquiry, M.A. Flex-fuel vehicle adoption and dynamics of ethanol prices: Lessons from Brazil. *Energy Policy* **2013**, *59*, 507–512.
3. Licht, F.O. Ethanol Industry Outlook 2008–2013 Reports. Available online: http://www. afdc.energy.gov/data/ (accessed on 9 September 2014).

4. Chabrelie, M.F.; Gruson, J.F.; Sagnes, C. *Overview of Second-Generation Biofuel Projects*; IFP Energies Nouvelles: Paris, Fance, 2014.

5. Brown, R.C. *Thermochemical Processing of Biomass Conversion into Fuels, Chemicals and Power*; Wiley: Chichester, West Sussex, UK, 2011.

6. Suarez Paris, R.; Lopez, L.; Barrientos, J.; Pardo, F.; Boutonnet, M.; Jaras, S. Chapter 3 catalytic conversion of biomass-derived synthesis gas to fuels. In *Catalysis: Volume 27*; The Royal Society of Chemistry: London, UK, 2015; Volume 27, pp. 62–143.

7. Spivey, J.J.; Egbebi, A. Heterogeneous catalytic synthesis of ethanol from biomass-derived syngas. *Chem. Soc. Rev.* **2007**, *36*, 1514–1528.

8. Subramani, V.; Gangwal, S.K. A review of recent literature to search for an efficient catalytic process for the conversion of syngas to ethanol. *Energy Fuels* **2008**, *22*, 814–839.

9. Ma, H.T.; Yuan, Z.Y.; Wang, Y.; Bao, X.H. Temperature-programmed surface reaction study on C_2-oxygenate synthesis over sio_2 and nanoporous zeolitic material supported Rh-Mn catalysts. *Surf. Interface Anal.* **2001**, *32*, 224–227.

10. Chen, G.; Guo, C.Y.; Zhang, X.; Huang, Z.; Yuan, G. Direct conversion of syngas to ethanol over Rh/Mn-supported on modified SBA-15 molecular sieves: Effect of supports. *Fuel Process. Technol.* **2011**, *92*, 456–461.

11. Chen, G.; Guo, C.-Y.; Huang, Z.; Yuan, G. Synthesis of ethanol from syngas over iron-promoted Rh immobilized on modified SBA-15 molecular sieve: Effect of iron loading. *Chem. Eng. Res. Des.* **2011**, *89*, 249–253.

12. Fechete, I.; Wang, Y.; Védrine, J.C. The past, present and future of heterogeneous catalysis. *Catal. Today* **2012**, *189*, 2–27.

13. Martínez, A.; Prieto, G. The application of zeolites and periodic mesoporous silicas in the catalytic conversion of synthesis gas. *Top. Catal.* **2009**, *52*, 75–90.

14. Kresge, C.T.; Leonowicz, M.E.; Roth, W.J.; Vartuli, J.C.; Beck, J.S. Ordered mesoporous molecular sieves synthesized by a liquid-crystal template mechanism. *Nature* **1992**, *359*, 710–712.

15. Biz, S.; Occelli, M.L. Synthesis and characterization of mesostructured materials. *Catal. Rev.* **1998**, *40*, 329–407.

16. Arakawa, H.; Takeuchi, K.; Matsuzaki, T.; Sugi, Y. Effect of metal dispersion on the activity and selectivity of Rh/SiO_2 catalyst for high pressure co hydrogenation. *Chem. Lett.* **1984**, *13*, 1607–1610.

17. Underwood, R.P.; Bell, A.T. Influence of particle size on carbon monoxide hydrogenation over silica- and lanthana-supported rhodium. *Appl. Catal.* **1987**, *34*, 289–310.

18. Tago, T.; Hanaoka, T.; Dhupatemiya, P.; Hayashi, H.; Kishida, M.; Wakabayashi, K. Effects of Rh content on catalytic behavior in CO hydrogenation with Rh-silica catalysts prepared using microemulsion. *Catal. Lett.* **2000**, *64*, 27–31.

19. Zhou, S.T.; Zhao, H.; Ma, D.; Miao, S.J.; Cheng, M.J.; Bao, X.H. The effect of Rh particle size on the catalytic performance of porous silica supported rhodium catalysts for co hydrogenation. *Z. Phys. Chem.* **2005**, *219*, 949–961.

20. Lopez, L.; Velasco, J.; Cabrera, S.; Boutonnet, M.; Järås, S. Effect of syngas conversion and catalyst reduction temperature in the synthesis of ethanol: Concentration of water vapor in mesoporous Rh/MCM-41 catalyst. *Catal. Commun.* **2015**, *69*, 183–187.

21. Khodakov, A.Y.; Zholobenko, V.L.; Bechara, R.; Durand, D. Impact of aqueous impregnation on the long-range ordering and mesoporous structure of cobalt containing MCM-41 and SBA-15 materials. *Microporous Mesoporous Mater.* **2005**, *79*, 29–39.

22. Landau, M.V.; Varkey, S.P.; Herskowitz, M.; Regev, O.; Pevzner, S.; Sen, T.; Luz, Z. Wetting stability of Si-MCM-41 mesoporous material in neutral, acidic and basic aqueous solutions. *Microporous Mesoporous Mater.* **1999**, *33*, 149–163.

23. Borodziński, A.; Bonarowska, M. Relation between crystallite size and dispersion on supported metal catalysts. *Langmuir* **1997**, *13*, 5613–5620.

24. Ehwald, H.; Ewald, H.; Gutschick, D.; Hermann, M.; Miessner, H.; Ohlmann, G.; Schierhorn, E. A bicomponent catalyst for the selective formation of ethanol from synthesis gas. *Appl. Catal.* **1991**, *76*, 153–169.

25. Wong, C.; McCabe, R.W. Effects of oxidation/reduction treatments on the morphology of silica-supported rhodium catalysts. *J. Catal.* **1987**, *107*, 535–547.

26. Robertson, S.D.; McNicol, B.D.; de Baas, J.H.; Kloet, S.C.; Jenkins, J.W. Determination of reducibility and identification of alloying in copper-nickel-on-silica catalysts by temperature-programmed reduction. *J. Catal.* **1975**, *37*, 424–431.

27. Panpranot, J.; Goodwin, J.G., Jr.; Sayari, A. Synthesis and characteristics of MCM-41 supported coru catalysts. *Catal. Today* **2002**, *77*, 269–284.

28. Mei, D.; Rousseau, R.; Kathmann, S.M.; Glezakou, V.-A.; Engelhard, M.H.; Jiang, W.; Wang, C.; Gerber, M.A.; White, J.F.; Stevens, D.J. Ethanol synthesis from syngas over Rh-based/SiO$_2$ catalysts: A combined experimental and theoretical modeling study. *J. Catal.* **2010**, *271*, 325–342.

29. Chuang, S.C.; Stevens, R., Jr.; Khatri, R. Mechanism of C$_{2+}$ oxygenate synthesis on Rh catalysts. *Top. Catal.* **2005**, *32*, 225–232.

30. Hansen, J.B.; Højlund Nielsen, P.E. Methanol synthesis. In *Handbook of Heterogeneous Catalysis*; Ertl, G., Knözinger, H., Weitkamp, J., Eds.; VCH: Weinheim, Germany, 2008.

31. Bowker, M. On the mechanism of ethanol synthesis on rhodium. *Catal. Today* **1992**, *15*, 77–100.

32. Medford, A.J.; Lausche, A.C.; Abild-Pedersen, F.; Temel, B.; Schjodt, N.C.; Norskov, J.K.; Studt, F. Activity and selectivity trends in synthesis gas conversion to higher alcohols. *Top. Catal.* **2014**, *57*, 135–142.

33. Choi, Y.M.; Liu, P. Mechanism of ethanol synthesis from syngas on Rh(111). *J. Am. Chem. Soc.* **2009**, *131*, 13054–13061.

34. Cabrera, S.; El Haskouri, J.; Guillem, C.; Latorre, J.; Beltrán-Porter, A.; Beltrán-Porter, D.; Marcos, M.D.; Amorós, P. Generalised syntheses of ordered mesoporous oxides: The atrane route. *Solid State Sci.* **2000**, *2*, 405–420.

35. El Haskouri, J.; Cabrera, S.; Guillem, C.; Latorre, J.; Beltrán, A.; Beltrán, D.; Marcos, M.D.; Amorós, P. Atrane precursors in the one-pot surfactant-assisted synthesis of high zirconium content porous silicas. *Chem. Mater.* **2002**, *14*, 5015–5022.

36. Bartholomew, C.H.; Farrauto, R.J. *Fundamentals of Industrial Catalytic Processes*, 2nd ed.; Wiley-Interscience: Hoboken, NJ, USA, 2006.

37. Webb, P. *Introduction to Chemical Adsorption Analytical Techniques and Their Applications to Catalysis*; Micromeritics Instrument Corp.: Norcross, GA, USA, 2003.

38. Andersson, R.; Boutonnet, M.; Järås, S. On-line gas chromatographic analysis of higher alcohol synthesis products from syngas. *J. Chromatogr. A* **2012**, *1247*, 134–145.

39. Elsevier. Chapter 15 quantitative analysis by gas chromatography measurement of peak area and derivation of sample composition. In *Journal of Chromatography Library*; Georges, G., Claude, L.G., Eds.; Elsevier: Amsterdam, The Netherlands; Oxford, UK; New York, NY, USA; Tokyo, Japan, 1988; Volume 42, pp. 629–659.

40. Madon, R.J.; Boudart, M. Experimental criterion for the absence of artifacts in the measurement of rates of heterogeneous catalytic reactions. *Ind. Eng. Chem. Fund.* **1982**, *21*, 438–447.

Influence of Cobalt Precursor on Efficient Production of Commercial Fuels over FTS Co/SiC Catalyst

Ana Raquel de la Osa, Amaya Romero, Fernando Dorado, José Luis Valverde and Paula Sánchez

Abstract: β-SiC-supported cobalt catalysts have been prepared from nitrate, acetate, chloride and citrate salts to study the dependence of Fischer–Tropsch synthesis (FTS) on the type of precursor. Co*m*/SiC catalysts were synthetized by vacuum-assisted impregnation while N_2 adsorption/desorption, XRD, TEM, TPR, O_2 pulses and acid/base titrations were used as characterization techniques. FTS catalytic performance was carried out at 220 °C and 250 °C while keeping constant the pressure (20 bar), space velocity (6000 Ncm^3/g·h) and syngas composition (H_2/CO:2). The nature of cobalt precursor was found to influence basic behavior, extent of reduction and metallic particle size. For β-SiC-supported catalysts, the use of cobalt nitrate resulted in big Co crystallites, an enhanced degree of reduction and higher basicity compared to acetate, chloride and citrate-based catalysts. Consequently, cobalt nitrate provided a better activity and selectivity to C_5^+ (less than 10% methane was formed), which was centered in kerosene-diesel fraction ($\alpha = 0.90$). On the contrary, catalyst from cobalt citrate, characterized by the highest viscosity and acidity values, presented a highly dispersed distribution of Co nanoparticles leading to a lower reducibility. Therefore, a lower FTS activity was obtained and chain growth probability was shortened as observed from methane and gasoline-kerosene ($\alpha = 0.76$) production when using cobalt citrate.

Reprinted from *Catalysts*. Cite as: de la Osa, A.R.; Romero, A.; Dorado, F.; Valverde, J.L.; Sánchez, P. Influence of Cobalt Precursor on Efficient Production of Commercial Fuels over FTS Co/SiC Catalyst. *Catalysts* **2016**, *6*, 98.

1. Introduction

In general, there are several factors that affect the performance of Fischer–Tropsch synthesis (FTS) catalysts: media type, nature of the precursor, nature and quantity of promoters as well as dispersion of the active phase [1–3]. Regarding active metal, the production of commercial fuels from syngas is mostly based on supported Co catalysts. Iglesia et al. established that neither the nature of the support nor the dispersion of Co particles influenced turnover frequency (TOF) of these catalysts, but it was proportional to the concentration of active sites on the surface [4–6]. Johnson et al. [7] agree with the fact that TOF was not affected by dispersion, although CO conversion seemed to be strongly dependent on Co reducibility. However, Reuel and Bartholomew [8] or Bessell [9] exposed that both low metal–support interactions and high dispersion are required parameters to obtain a really active cobalt based catalyst. Furthermore, catalysts with high surface density favor the formation of n-paraffins of high molecular weight [4], which is desirable to maximize the production of hydrocarbons in the range of diesel fraction. Hence, at a certain Co loading and established catalyst support, the number of active centers would mainly depend on two parameters, dispersion and degree of reduction of cobalt oxide supported particles [5,7]. Key in this regard is the strength of the metal–support interaction, which depends largely on the preparation method [10–12] and nature of metal precursor [12–15].

The methods most commonly employed in the preparation of cobalt nitrate based catalysts are vacuum assisted and incipient wetness impregnation (IWI) [5,9]. However, the use of nitrate often results in a catalyst characterized by large Co particles and therefore, a lesser active surface area [16]. On the other hand, a large particle size would favor chain growth probability and low metal–support interactions and therefore, reducibility would be enhanced. For that reason, several authors [13,14,16,17] have studied the influence of the nature of precursor on particle size showing that the use of different salts other than nitrate (acetate, chloride, carbonyl, citrate, EDTA, etc.) results in particles of different size by changing Co^0 dispersion and hence, metal–support interaction. It has been further shown that the final dispersion of metallic cobalt usually depends on the dispersion of Co_3O_4 in the oxidized precursor [18,19].

β-SiC is a ceramic material that despite of a low-to-medium surface area, presents potential properties for its use as support in heterogeneous catalytic processes. These include chemical inertness, excellent mechanical strength and thermal conductivity. In this sense, β-SiC has been tested in several highly endo-/exothermic reactions, including Fischer–Tropsch synthesis [20–26]. Its superior thermal conductivity (one hundred times higher than that of SiO_2 [27]) is of practical interest for FTS since allows hindering hot spots inside the reactor, reducing thermal gradients and consequently, improving plant security and selectivity to heavy hydrocarbons. Nature of precursor of nickel or iron [28] supported over SiC has been investigated showing important differences in catalytic activity. Nonetheless, literature related to the nature of Co/SiC is still scarce and more research is required to improve Co/SiC composition for a highly selective FTS catalyst. Then, the present work was conducted to study the dependence of FTS catalytic performance on the nature of Co metal precursors (nitrate, acetate, chloride and citrate) supported on porous β-SiC.

2. Results and Discussion

Results related to the phase and surface composition, extent of reduction and dispersion of cobalt oxide precursor along with the catalytic activity of Com/β-SiC catalysts are presented and discussed. "m" denotes cobalt precursor as N (nitrate), A (acetate), Cl (chloride) and Cit (citrate).

2.1. Thermal Analysis

The decomposition of cobalt precursor is a fundamental stage in the preparation of FTS catalyst since the heat released during this step may affect the structure of cobalt species in the final catalyst and consequently, change reducibility and dispersion [29]. Therefore, in order to determine a proper calcination temperature, thermogravimetric analysis (TG/DTG) of the corresponding parent salts were performed (Figure 1).

Nitrate and cobalt acetate are the most common precursors in the preparation of FTS catalysts. From Figure 1a, it can be observed that in case of the nitrate precursor, there is a progressive weight loss between 25 °C and 210 °C attributable to residual water desorption, while the oxidation of the nitrate ligands occurred at 210–260 °C. The decomposition of cobalt acetate proceeded at slightly higher temperatures, being the main weight loss at 285 °C (Figure 1b). However, from Figure 1c, it can be noted that the decomposition of cobalt chloride took place in several stages, losing water from room temperature to 200 °C and producing dechlorination over 500 °C [30]. Figure 1d shows that cobalt citrate decomposed between 212 °C and 307 °C. It is important to mention that these characterization experiments were run until 600 °C since β-SiC support was reported to be stable under oxidant atmosphere up to this temperature [22,31,32].

Considering the above, and taking into account that supported cobalt catalysts are known to need higher reduction temperatures due to the precursor–support interaction (as demonstrated later by TPR analysis, Section 2.4), higher calcination temperatures are required. In this sense, 550 °C has been recently reported as the optimum calcination temperature for FTS on Co/β-SiC since prevents deactivation by formation Co_2C species. In addition, the resulting metal–support interaction was found to stabilize cobalt particles against sintering [33]. Therefore, 550 °C was selected as a suitable calcination temperature to completely decompose the precursor into Co_3O_4 as well as to ensure the stability of cobalt species supported on porous β-SiC. In agreement, after calcination step, water appeared to be removed and all ligands were decomposed to the corresponding cobalt oxide phase, as confirmed by XRD measurements (Section 2.3).

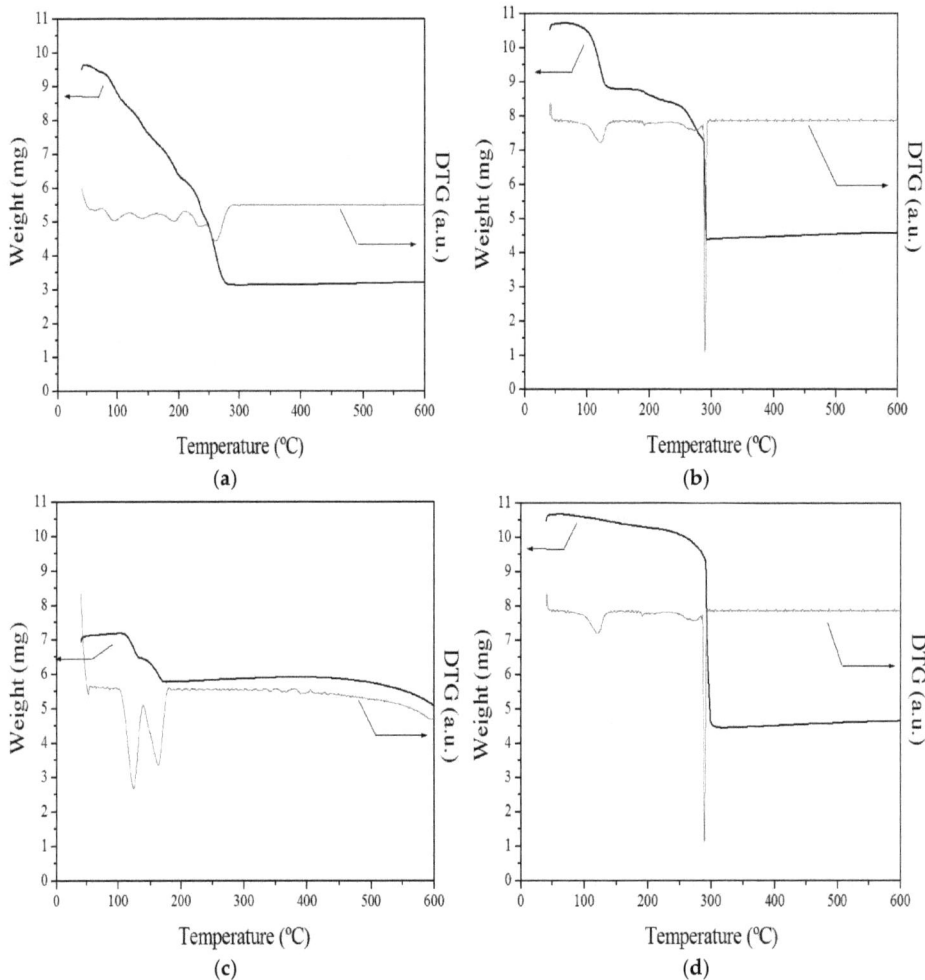

Figure 1. TG/DTG Co precursors: (**a**) CoN; (**b**) CoA; (**c**) CoCl; and (**d**) CoCit.

2.2. Nitrogen Adsorption/Desorption Measurements

Table 1 summarizes textural properties of each oxidized catalyst, estimated from N_2 adsorption–desorption isotherms and BJH method, as described in Section 3.2.3.

Table 1. Physicochemical properties of FTS *Com*/SiC catalysts.

Catalyst	Wt % Co ± 0.1	dCo$_3$O$_4$ (nm) [1]	dCo^0 (nm) [11]	D (%) [12]	D (%) [22]	dCo^0 (nm) [2]	Degree of Reduction (%) [3] ± 2	BET Area (m^2/g) [4] ± 1.1	Pore Diameter (nm) [5] ± 0.3	Total Pore Volume (cm^3/g) [6]	Basicity (cm^3/g) [7] ± 0.03
CoN/SiC	9.9	45.0 ± 5.8	33.8	2.8	1.7	56.3 ± 10.1	74.1	21.5	7.2	0.039	0.26
CoCl/SiC	7.2	75.9 ± 8.7	56.9	1.7	2.6	37.1 ± 5.2	41.5	22.1	7.3	0.055	−0.53
CoA/SiC	10.9	34.4 ± 5.2	25.8	3.7	3.6	27.0 ± 5.5	58.5	24.8	8.6	0.106	−1.21
CoCit/SiC	8.9	21.4 ± 4.6	16.1	5.9	5.7	17.0 ± 2.1	41.8	34.9	8.6	0.151	−4.19

[1] Co$_3$O$_4$ crystallite size calculated from XRD at 2θ: 36.9°; [11] Co^0 crystallize size from dCo$_3$O$_4$ (XRD), Equation (2); [12] Dispersion calculated from XRD Equation (1); [2] Co^0 crystallite size from TEM imaging, Equation (3); [22] Dispersion from TEM imaging Equation (1); [3] Obtained from O$_2$ Pulses at 400 °C; [4] Calculated by multipoint BET method; [5] Estimated by BJH desorption branch; [6] Determined from a single point of adsorption at P/P$_0$ = 0.998; [7] HCl added (pKi ≤ 7).

N$_2$ adsorption/desorption isotherms associated to all of these samples (Figure 2a) show as a combination of isotherms II and IV, as referenced by IUPAC designation. Type II corresponded to the beginning of the isotherm and was caused by adsorption of monolayer/multilayer [34]. Type H3 hysteresis loop observed at higher partial pressures fitted type IV isotherm, characteristic of capillary condensation in presence of mesoporous [35], as denoted by pore distribution (Figure 2b).

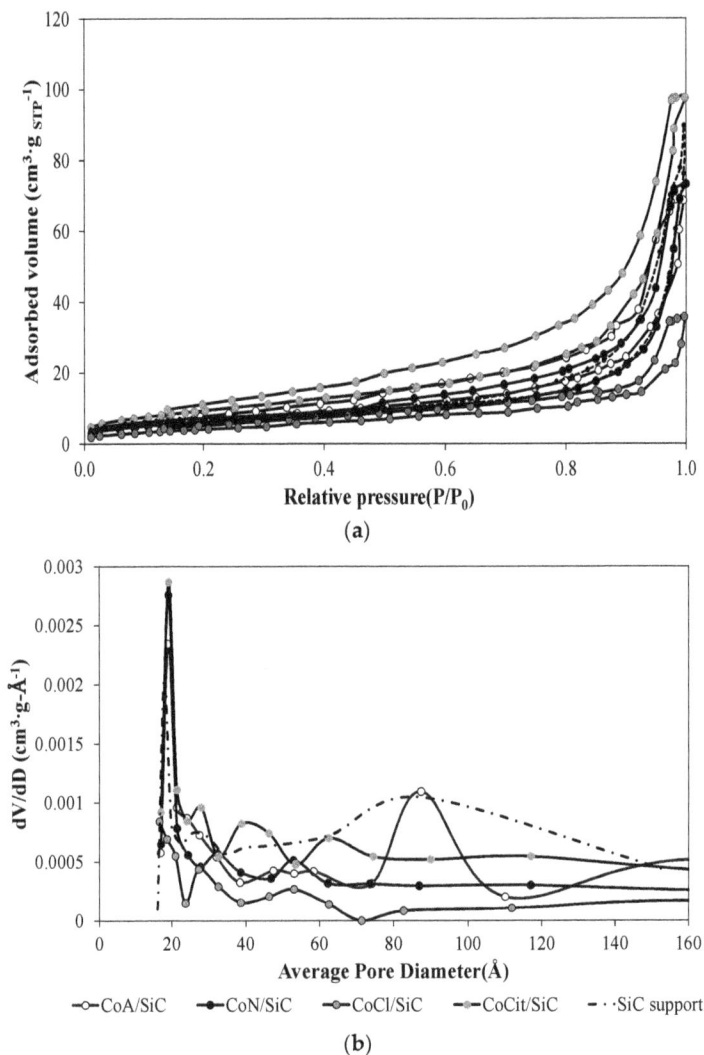

Figure 2. (a) N$_2$ adsorption/desorption isotherms; and (b) average pore size distribution.

Moreover, in agreement with isotherms type, pore sizes values shown in Table 1 are within the range of mesopores (2–50 nm) [35]. Particularly, a very similar pore size distribution was found for the four samples, as depicted in Figure 2b, exhibiting a maximum peak ca. 1.9 nm and a second one ca. 8.7 nm, typical of β-SiC support. It should be noted that all samples presented an average Co_3O_4 crystallite size larger than support pore size, as detailed later by means of XRD and TEM imaging, demonstrating that the location of cobalt crystallites on the external surface of β-SiC and not inside its pores may be considered.

From Table 1, it can be also observed that specific surface areas (SA) were similar, 21.5–34.9 m^2/g, within the measurement error range. Upon impregnation of cobalt solutions over β-SiC, a decrease of specific surface area occurred. This effect was significant on CoCl- and CoN-based catalysts, probably resulting from a partial pore overlay by aggregate particles. However, BET surface area diminished in a lesser extent when cobalt acetate was used as precursor. In addition, pore volume was found to increase probably due to a smaller crystallite size and the slight acid nature of this precursor. It seemed that, for a similar cobalt loading, the more acidic the precursor is, the more amount of metal impurities blocking SiC pores would be eliminated while a lower particle size would be deposited on the support surface, increasing SA and porosity. Accordingly, CoCit/SiC presented the highest total pore volume and SA as detected after acid treatment of silicon carbide catalyst surface [22].

2.3. XRD and TEM

X-ray diffractograms related to β-SiC support, bulk Co_3O_4 and calcined Co*m*/SiC catalysts under study are presented in Figure 3.

β-SiC support presented characteristic spectra of the two main SiC polytypes, namely α and β. Reflections at 2θ: 35.5° [111] and 33.7° [0001] are ascribed to hexagonal α-SiC, being the first one (C-SiC) caused by heaping defects generated during the formation of the material [36]. However, those at 2θ: 41.4° [002], 59.9° [202] and 71.7° [113] are assigned to face-centered cubic, β-SiC [37,38]. As depicted in Figure 3, catalyst preparation procedures were confirmed to result in the formation of a pure cobalt oxide spinel phase, Co_3O_4, regardless of the type of precursor. Different to that reported by Cook et al. [39], no evidence of CoO, whose crystal phase arises at 36.5°, 42.4° and 61.5° (JCPDS 75-0393), was found in our samples. Thus, only crystalline Co_3O_4 and β-SiC were detected after calcination.

According to XRD, average Co_3O_4 size related to each synthetized catalyst, deduced from the Scherrer equation at 2θ:36.9°, is listed in Table 1. It was found to increase in the following order: CoCit/SiC < CoA/SiC < CoN/SiC < CoCl/SiC. According to Panpranot et al. [40], cobalt chloride provided the largest particle size and, therefore, the lowest dispersion. It was also noted that, for all samples, particle size was bigger than support pore diameter, which predicts that crystallites

175

of Co oxide were deposited on the outer surface, rather than in the pores of the β-SiC support.

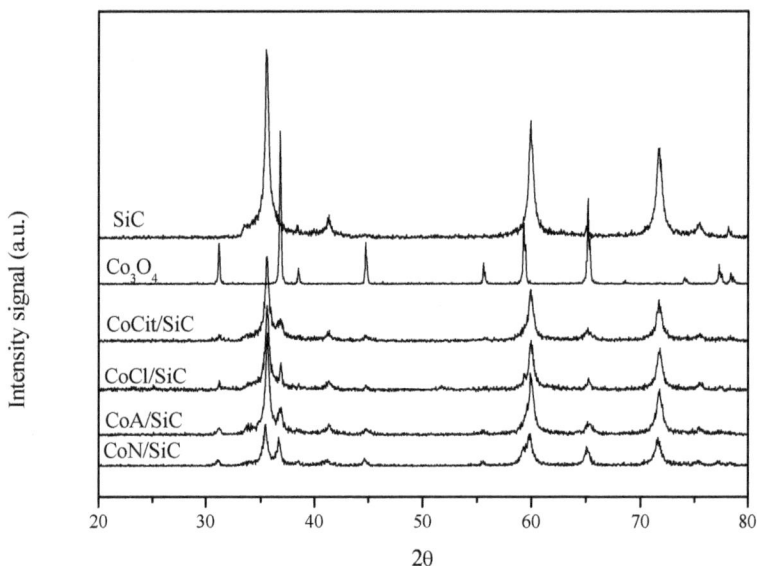

Figure 3. XRD patterns of bulk Co_3O_4, β-SiC support and catalysts Co*m*/SiC.

In order to verify the accuracy of XRD characterization results, several TEM micrographs of all the catalysts were collected. Figure 4 shows some examples of TEM images related to Co*m*/SiC catalysts revealing that Co particles disposition on β-SiC support was well determined by the nature of precursor.

TEM image of Figure 4a (CoN/SiC) shows several agglomerates (nm) that turned out to form large Co nanoparticles, even bigger than those expected from XRD. Figure 4b presents a better dispersion of cobalt for CoA/SiC with an average Co^0 particle size of 27 nm, in accordance with XRD data. CoCl/SiC (Figure 5c) displays a broaden size distribution since 15–35 nm crystallites were found to merge into coarse particles. In this case, particle size was proved to be lesser than that of CoN/SiC. In contrast, Figure 4d exhibits a high dispersion of quite fine grains that clumped together in 15–25 nm agglomerates. This also confirms XRD estimation of CoCit/SiC (16–17 nm) crystallite diameter.

Dispersion percentage, D (%), was calculated by assuming spherical cobalt crystallites of uniform diameter (dCo^0) with a site density of 14.6 at/nm^2, using the formula reported by Jones and Bartholomew [41], i.e.,

$$[dp = 96/D] \tag{1}$$

where dp refers to metallic cobalt particle diameter in nm (i.e., dCo^0). Consistent with TEM micrographs in which small crystallites are observed, the oxide produced from decomposition of Co citrate is shown by XRD to be more highly dispersed than the oxides produced from other Co precursors. Considering that both nature of support and average Co^0 size are known to strongly influence cobalt–support interactions, large crystallites of cobalt-nitrate based catalyst were supposed to result in a weaker interaction Co-β-SiC. This fact would lead to a high degree of reduction of cobalt species. It is important to remark that cobalt crystallites on β-SiC support presented globular form, as observed from TEM images, corroborating the proposed hypotheses on spherical particles.

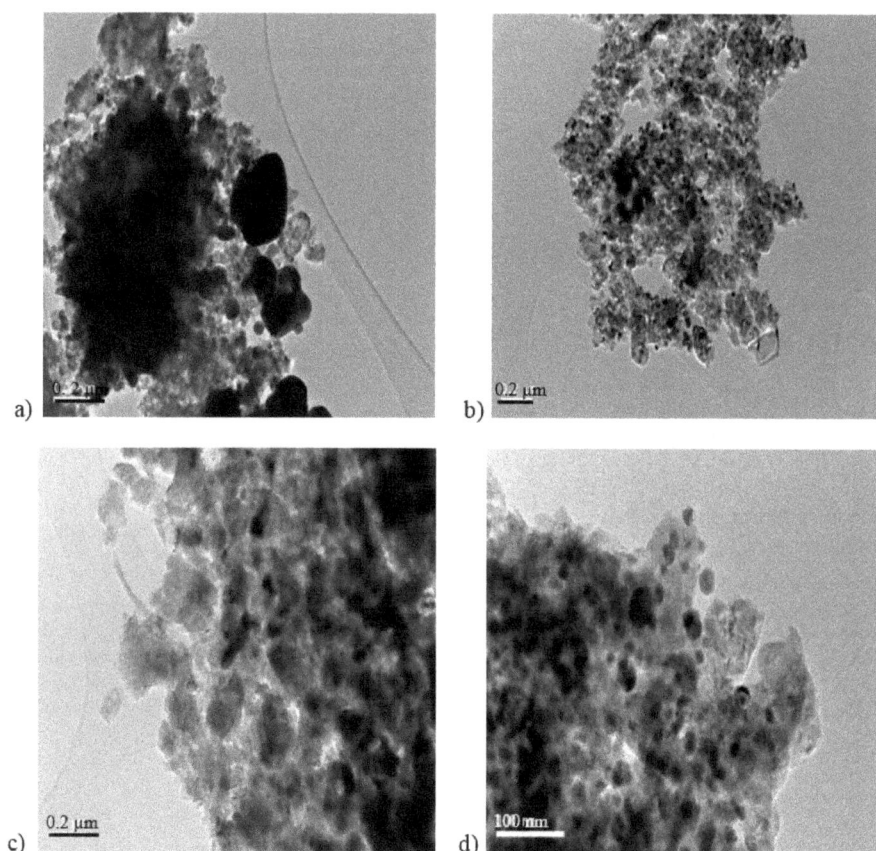

Figure 4. TEM Images: **(a)** CoN/SiC; **(b)** CoA/SiC; **(c)** CoCl/SiC; and **(d)** CoCit/SiC.

As displayed in Figure 5 average particle size distribution obeyed the series: CoCit/SiC < CoA/SiC < CoCl/SiC < CoN/SiC. The use of citrate as cobalt precursor was confirmed to provide the smallest particle size and a most uniform and narrow disposition (15–20 nm). In case of cobalt acetate and cobalt chloride precursors, particle size distribution was wider, specifically centered on 20 to 30 nm and 30 to 50 nm, verifying estimation from XRD. However, size classification provided by cobalt nitrate was much more heterogeneous, exhibiting a distribution mainly based on medium (35–60 nm) crystallites although the formation of large particles (even higher than 130 nm) from crystalline aggregates was also observed. These data are consistent with XRD diffraction peaks broadening and intensity, indicating heterogeneously distributed large cobalt oxide crystallites, and results reported by Martinez et al. [3] on Co/SBA-15, wherein particle size of cobalt nitrate was bigger than those coming from acetate or acetylacetonate that lead to a higher dispersion and a stronger metal–support interaction. Accordingly, the reducibility of CoN/SiC catalyst should be favored and therefore, its catalytic performance [1,5,42].

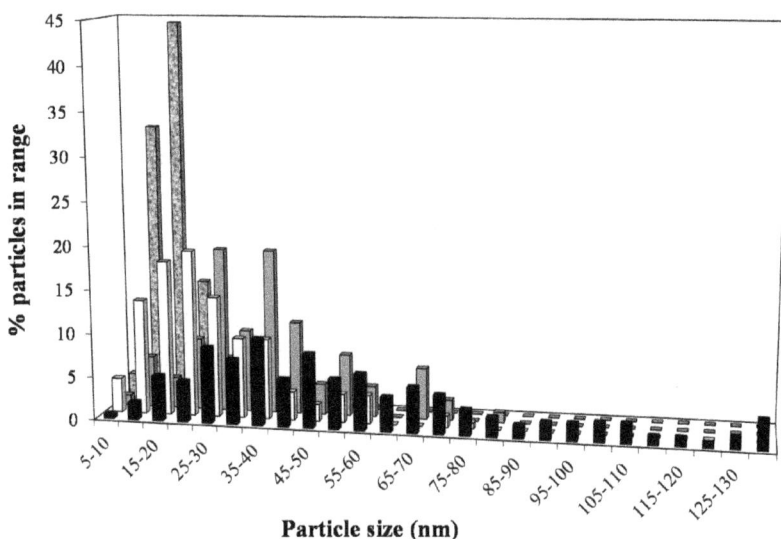

Figure 5. TEM particle size distribution: ■ CoN/SiC; □ CoA/SiC; ■ CoCl/SiC; and ◨ CoCit/SiC.

As evidenced by XRD and TEM, Co particle size distribution on β-SiC support is attributed to the nature of precursor salts. Cobalt citrate, chloride and acetate proved to be much smaller than that of Co-nitrate based catalyst. Terörde et al. [43] reported that both the viscosity of dissolved salts and their interactions with the support surface are critical parameters in the preparation of iron catalysts by impregnation technique. Thus, citrate precursor was shown to result in a smaller particle size of iron oxide than that of the other precursors used in their study, mainly due to its higher viscosity. Considering this argument, as a result of the drying step, cobalt nitrate crystallizes on β-SiC outer surface, whereas cobalt citrate, which presented a high viscosity, produces a gel that provides well dispersed and scarcely-crystallized species interacting with the support [28].

2.4. Cobalt Oxide Reducibility

Fischer–Tropsch synthesis exclusively proceeds on Co^0 sites. As shown from XRD patterns, Co_3O_4 is the main phase in the prepared catalysts, thus requiring a pretreatment. Despite most of the cobalt is commonly referred to exist as Co^0 phase under reaction conditions, some parameters may hinder reduction step. Co particles would be agglomerated under a high calcination temperature, consequently diminishing the active surface and catalytic performance despite providing a good degree of reduction. On the other hand, small particles frequently lead to a partial degree of reduction [44]. Co particle size and dispersion were found to depend on parent compounds, modifying the corresponding interaction of active sites with the support and therefore, reducibility. Since the degree of reduction of cobalt species seems to play a key role on FTS catalyst activity, the corresponding extent was studied by pulses re-oxidation and TPR techniques.

Figure 6a–e outlines TPR profiles of bulk Co_3O_4 and Com/SiC catalysts, respectively. It is important to mention that a blank test was carried out on fresh β-SiC with no evidence of hydrogen consumption. Hence, TPR peaks can only be attributed to cobalt oxide species reduction.

It can be seen in Figure 6a that bulk Co_3O_4 was completely reduced to Co^0 at temperatures lower than 350 °C. However, supported cobalt catalysts were expected to show a displacement to higher reduction temperatures. For conventional Co/SiO_2, which is the most similar catalyst to establish a comparison, four TPR reduction peaks are usually reported and identified as: (i) peak α, assigned to Co_3O_4 to CoO reduction; (ii) peak β, related to successive reduction of CoO to Co, at lower temperatures than 400 °C [19]; (iii) peak γ, between 400 °C and 600 °C, referred to mild interaction Co-SiO_2 to Co°; and (iv) peak δ, at temperatures higher than 600 °C, corresponding to Co-silicate/hydrosilicate species to Co^0 [45]. It should be noted that subsequent impregnations could also result in a significant increase in the intensity of high temperature maxima of cobalt silicate structures [46].

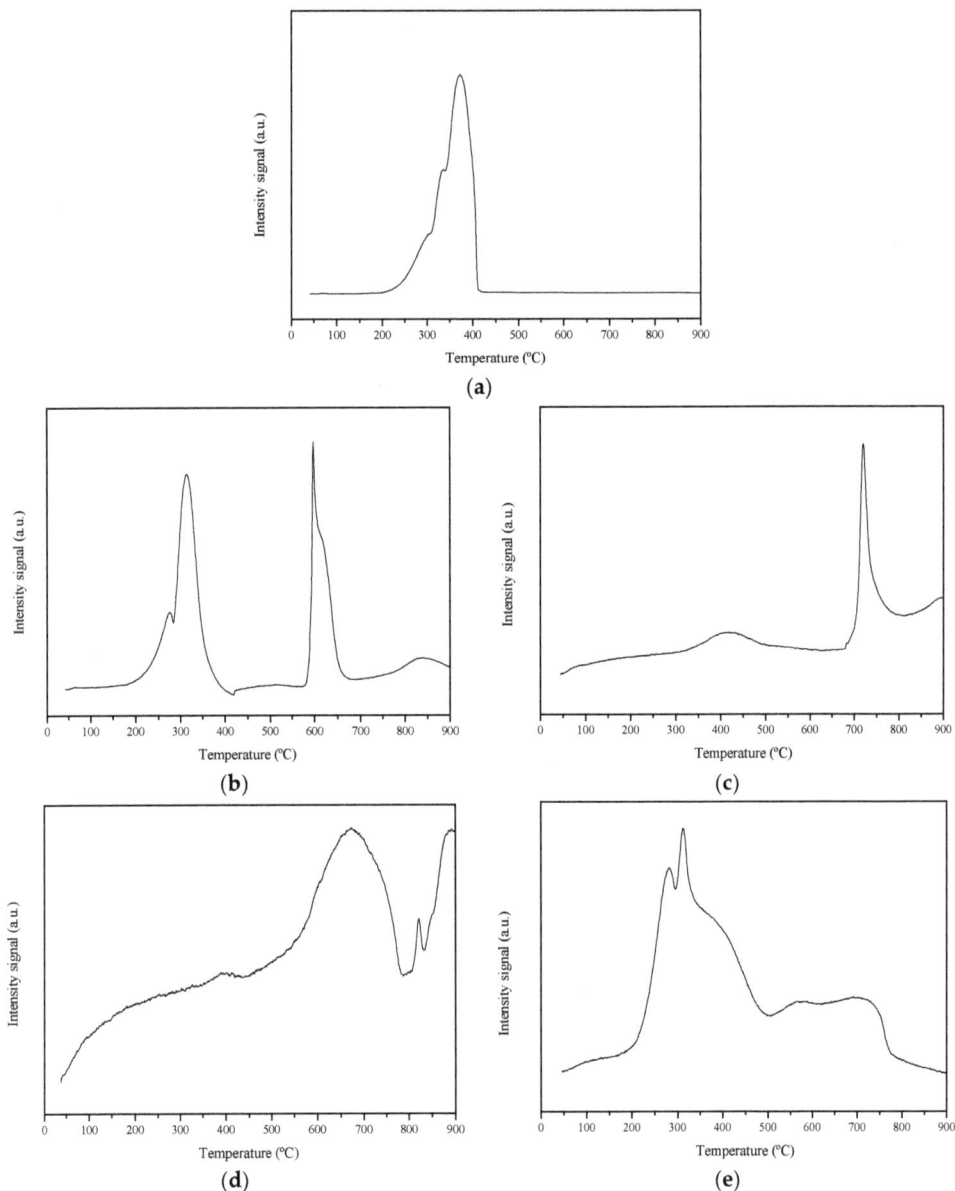

Figure 6. TPR profiles: (**a**) Bulk Co$_3$O$_4$; (**b**) CoN/SiC; (**c**) CoA/SiC; (**d**) CoCl/SiC; and (**e**) CoCit/SiC.

In Figure 6b,c, it can be observed that TPR-spectra of catalysts CoN/SiC and CoA/SiC consisted of two main reduction peaks. In agreement with Schanke et al. [19] and Rodrigues et al. [47], and as described above, the first maximum of CoN/SiC at 350 °C could be attributed to the reduction of Co$_3$O$_4$

180

agglomerates to Co^0 (α and β peaks). The second peak (550–650 °C) was indicative of Co^0 formation from mild-interaction of surface multilayer Co species (CoO_x) [46]. Nevertheless, the presence of immobilized cobalt ions (silicate and hydrosilicates Co-SiO_x species) may not be excluded as revealed by a third small, diffuse maximum between 800 and 850 °C. In the case of catalyst CoA/SiC (Figure 6c), the first maximum of H_2 consumption was observed to be quite lower compared to that of CoN/SiC. Moreover, TPR profile of CoA/SiC was shifted to higher reduction temperatures (420 °C and 720 °C), confirming a lesser degree of reduction due to a stronger interaction between Co species and β-SiC support. This fact would corroborate that catalysts with larger particles are more easily reduced than those with smaller ones, which lead to a higher concentration of hardly reducible cobalt species (silicate-type) [48–50]. TPR profiles associated to CoCl/SiC (Figure 6d) or CoCit/SiC (Figure 6e) resulted to be more complex, showing different Co-species formation upon preparation. Broad maxima indicated the presence of several species reducing at approximately the same temperature, whereas the sharp peaks indicated the existence of a single species. Despite its particle size, a poorer degree of reduction was supposed for catalyst CoCl/SiC compared to CoN/SiC or even CoA/SiC, since its broader profile was found to be widely shifted to higher temperatures. Catalyst CoCit/SiC, presenting the smallest particle size and higher dispersion, also showed a set of overlapped peaks indicating, apart from cobalt oxide reduction (200–450 °C), superposition of mild to strongly immobilized Co species with degrees of different order at temperatures between 500 and 800 °C. Therefore, citrate precursor would provide a diminished degree of reduction to the active phase than that of CoN/SiC.

In order to quantify the reducibility of cobalt oxide precursors, a series of pulse oxidation experiments of reduced samples was developed. The corresponding results, sorted in Table 1, determined that the extent of reduction increased accordingly to: CoCl/SiC~CoCit/SiC < CoA/SiC < CoN/SiC. This trend confirmed that concluded from TPR reduction profiles. It should be mentioned that some authors [51] have found that the degree of reduction may be underestimated for large Co crystallites (10–12 nm) under present pulse oxidation conditions. Titration is then proposed to be conducted at higher O_2 partial pressure (1–2 atm), temperature (450 °C instead of 400 °C), and exposure times (longer than 1 h) to completely oxidize Co^0 to Co_3O_4. In this sense, it should be specified that despite authors have not found evidence of that commented above, large crystallites in this work may subject to the same problem. Therefore, working under suggested conditions may improve the extent of reduction results. In any case, catalyst CoN/SiC would be expected to provide the best FTS catalytic behavior, since it showed a higher extent of reduction. A greater selectivity towards C_5^+ hydrocarbons would be also favored due to its large particle size that would promote chain growth probability [52]. Conversely, theoretically, acid-based (acetate, citrate and especially chloride) samples showed a lower re-oxidation performance, which indicated that a higher amount of Co species were anchored to the support, remarkably decreasing their reducibility.

2.5. Acid-Base Titrations

Finally, since basicity was established to modify C_5^+ hydrocarbon product distribution [52,53], Figure 7 presents the acid-base titration curves of the different catalysts under study.

Titration curves evolved towards higher HCl consumption values as follows: CoCit/SiC < CoA/SiC < CoCl/SiC < CoN/SiC. Table 1 reports surface basicity data, expressed as spent HCl (cm^3) per gram of catalyst sample to reach neutral pH level, confirming differences depending on cobalt precursor. As expected, CoN/SiC was the only catalyst that showed certain basicity with respect to acetate, chloride and citrate, which are postulated as acid ligands. Consequently, in addition to a higher extent of reduction and particle size, as basicity was reported to favor chain growth probability [52,54], this catalyst should result in a shift of C_5^+ hydrocarbon product distribution from gasoline towards higher molecular weights valuable products.

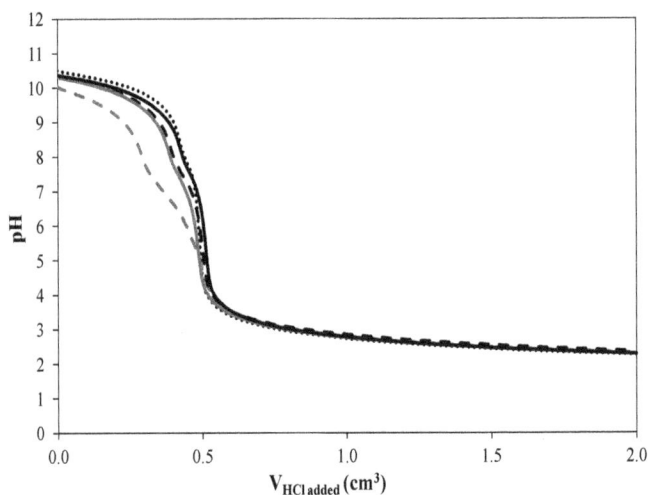

Figure 7. Acid/base titration measurements: ····· Reference; ▬▬ CoN/SiC; ▬▬ CoA/SiC; ▬ ▪ CoCl/SiC; and ▬ ▪ CoCit/SiC.

2.6. Fischer–Tropsch Synthesis

In order to study the differences on FTS catalytic performance attributed to the nature of cobalt precursor, a series of catalytic tests were carried out over those synthetized catalysts whose physicochemical characteristics were found to be more diverse, i.e., CoN/SiC and CoCit/SiC. Table 2 collects steady state performance parameters.

Nature of precursor was observed to significantly modify CO conversion and selectivity to C_5^+ hydrocarbons, improving both of them when cobalt nitrate-based catalyst was used. Despite providing a higher CO conversion in the selected range of temperature, CoCit/SiC exhibited a lower catalytic activity when referred to FTS rate. Moreover, this catalyst favored the production of light hydrocarbons (mainly CH_4) and CO_2, compared to CoN/SiC. According to several authors [3,8] and in agreement with characterization results, the increased selectivity to methane can be due to the formation of well dispersed and hardly reducible compounds, which could also catalyze Water Gas Shift reaction, as observed from the increase of CO_2 formation. CoCit/SiC, with a significant acidity, registered a strong Co-β-SiC interaction and deducible low reducibility, then providing low FTS activity. Zhang et al. [55] found that the presence of a relevant amount of acid sites enhanced cobalt–support interactions leading to a lower degree of reduction and a poorer catalytic activity. The chelating ligand facilitated the formation of coordinate species, where Co^{2+} is fixed on the support surface and cobalt-silicate compounds hindered FTS activity. Compared with Co-citrate, CoN/SiC presented Co nanocrystallites whose size, higher than those from the citrate precursor, provided a compromise between an enhanced extent of reduction and a suitable density of active sites, thus improving the performance for FT synthesis. A lower CO conversion was observed at 220 °C for this catalyst, which may occurred due to support pore filling with HC products. It can be seen that, at this temperature, a high selectivity toward C_5^+ hydrocarbons is pointed out and hence, further CO hydrogenation could be prevented by blockage of the active cobalt sites. It is worth it to mention that, under the fixed operating condition, CoN/SiC did not result in CO_2 formation and produced less than 10% of methane, the main FTS side reaction. Moreover, an increase in the reaction temperature was found to only enhance catalytic activity, whereas selectivity to C_5^+ was barely modified when using this catalyst under the fixed range. It should be mentioned that no evidence of catalytic performance deactivation or carbon deposition was found for at least 24 h of reaction, indicating that metal sintering or Boudouard reaction were prevented, probably due to the relatively moderate interaction of Co nanoparticles with β-SiC support, as recently reported by Koo et al. [56].

Regarding FTS selectivity to C_5^+, Table 3 and Figure 8 detailed liquid hydrocarbon product distribution obtained for each catalyst at 250 °C.

Table 2. Influence of cobalt precursor on FTS: 20 bar; 6000 Ncm³/g·h; H₂/CO: 1.8.

Catalyst	T (°C)	FTS Rate (mol/mol$_{Co}$·h) ± 0.4	WGS Rate (mol/mol$_{Co}$·h) ± 0.03	Conversion (%) ± 0.5		CO_2 ± 0.03	Selectivity (%)			
				CO	H₂		C_1–C_4 ± 0.4	C_2 OR[1]	C_3 OR[1]	C_5^+ ± 0.4
CoN/SiC	220.00	3.47	0.02	7.40	24.10	0.50	6.20	0.36	1.00	93.30
	235.00	31.50	0.21	67.20	80.20	0.70	5.30	0.09	0.73	94.10
	250.00	34.81	0.41	74.90	85.70	0.70	9.30	0.10	0.72	90.00
CoCit/SiC	220.00	5.69	0.04	24.50	29.80	0.70	9.60	0.09	0.74	89.80
	235.00	17.22	0.74	76.50	89.20	4.10	25.00	0.01	0.19	70.90
	250.00	19.85	1.69	92.30	97.50	7.80	26.50	0.00	0.11	65.70

[1] OR: Olefins ratio calculated as O/O+P where O: Olefins and P: Paraffins; Confidence level 95% (α: 0.05).

Table 3. C_5^+ hydrocarbon distribution: 20 bar; 250 °C; 6000 Ncm³/g·h; H₂/CO: 1.8.

Catalyst	C_5^+ Hydrocarbon Distribution (wt %)					Diesel (vol %)	Diesel Yield (%)	α
	Gasoline (C_7–C_{10})	Kerosene (C_{11}–C_{14})	Diesel (C_{15}–C_{18})	Lubricants (C_{19}–C_{20})	Waxes (C_{20+})			
CoN/SiC	13.3	56.8	23.5	3.2	3.2	17.2	12.9	0.90
CoCit/SiC	56.3	21.1	14.1	4.1	4.8	5.9	5.4	0.76

Confidence level 95% (α: 0.05).

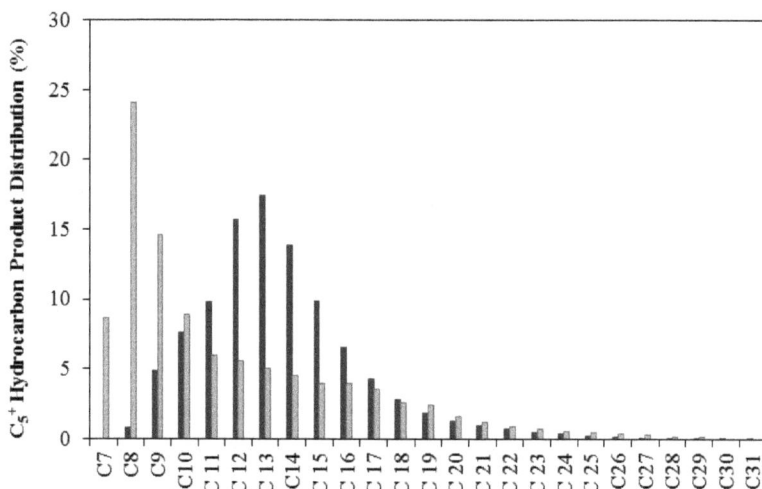

Figure 8. C_5^+ hydrocarbon product distribution: ☐ CoCit/SiC; and ■ CoN/SiC. Reaction conditions: 20 bar; 250 °C; 6000 Ncm^3/g·h; H_2/CO: 2.

FTS selectivity is known to depend on cobalt particle size [1] and basicity [53]. Accordingly, the distribution of C_5^+ hydrocarbons was also modified by the nature of the precursor. Larger cobalt particles are more selective for higher molecular weight hydrocarbons in FTS reaction because of dissociative adsorption of CO, which leads to the formation of -CH_2- required for chain growth. Basicity was also reported to promote chain growth [52,53,57]. It can be observed that catalyst CoCit/SiC, with the smallest particle size and the highest acidity, enhanced the production of gasoline fraction while catalyst CoN/SiC mainly shifted C_5^+ distribution towards the production of kerosene and diesel. It is important to note that although in a lesser extent, the latter reached a diesel yield twice (17%) than that of CoCit/SiC (6%). In agreement, alpha value (α) was set to 0.90 instead of 0.76. Therefore, cobalt nitrate was demonstrated to be the most suitable precursor for the production of synthetic diesel via Fischer–Tropsch synthesis over β-SiC-based catalyst.

3. Experimental Section

3.1. Synthesis of Com/SiC Catalyst

Extrudates of porous β-SiC (1 mm diameter and length, SICAT Catalyst) were used as catalytic support in this study. Medium specific surface area (35 m^2/g) β-SiC was successively vacuum-assisted impregnated with aqueous solutions of: (a) cobalt nitrate (CoN) [Co(NO$_3$)$_2$·6H$_2$O] (Merck KGaA, Darmstadt, Germany); (b) cobalt chloride (CoCl) [CoCl$_2$.6H$_2$O] (Panreac Química S.A.U., Barcelona, Spain) or (c) cobalt acetate (CoA) [C$_4$H$_6$CoO$_4$.4H$_2$O] (Panreac Química S.A.U., Barcelona,

Spain), in order to prepare a final catalyst with nominal 10 wt % Co loading. β-SiC was precalcined in a SELECTA muffle furnace (224129 model), under static air atmosphere at 550 °C (at a rate of 5 °C/min from ambient temperature) for 6 h before impregnation. Since pore volume of β-silicon carbide support is quite low, synthesis procedure was not based on incipient wetness impregnation technique. In order to obtain the specified Co content, usually, 15 g of β-SiC support were repeatedly impregnated (with intermediate drying) with aqueous solutions comprising the minimum amount of deionized water to dissolve 6.7–8.2 g of cobalt chloride, cobalt acetate or cobalt nitrate. Cobalt concentrations were determined from precursor weight change. After impregnation, the as-prepared mixture was dried in a SELECTA oven (240099 model) under static air atmosphere at 120 °C for 12 h. In addition, another 10 wt % cobalt-based catalyst was prepared by the citrate method as described elsewhere [58]. Cobalt nitrate was dissolved in an appropriate volume of deionized water to give a 0.1 M solution and then, citric acid [$C_6H_8O_7$] (Panreac Química S.A.U., Barcelona, Spain) was added in 5 wt % excess to guarantee entire complexation. Water was removed at 40 °C, under vacuum, to form a gel that was subsequently dried overnight at 70 °C. The resulting precursor presented a highly hygroscopic nature.

After impregnation and drying steps, all samples were calcined, in the same muffle furnace described above, at 550 °C for 6 h (5 °C/min heating rate) under static air atmosphere, to transform chloride, nitrate, acetate and citrate-ligands into the oxidized phase and then reduced to the Co^0 (Sections 2.4 and 3.3). Each catalyst was denoted as Com/SiC where m refers to the cobalt precursor. Physicochemical properties of synthetized Com/SiC are shown in Table 1.

3.2. Catalyst Characterization

3.2.1. Thermal Analysis (TG/DTG)

Thermal analysis over bulk salt precursors (7–12 mg) was conducted in a METTLER TOLEDO model TGA-DSC 1 STAR$_e$ SYSTEM (Mettler-Toledo AG, Schwerzenbach, Switzerland), under oxidant atmosphere. TG/DTG curves were registered from room temperature to 600 °C (5 °C/min heating rate), with an experimental error of ±0.5% in the weight loss measurement and ±2 °C in the temperature measurement.

3.2.2. Atomic Absorption (AA)

The final composition, referred to as wt % Co of calcined catalysts, was determined in a SPECTRAA 220FS (Varian Australia Pty Ltd., Mulgrave, Victoria, Australia), comprising a simple beam and background correction (±1% error). The samples were dissolved in HF before analysis.

3.2.3. Textural Characteristics

A Micromeritics ASAP 2010 sorptometer (Micromeritics, Norcross, GA, USA) was employed to measure porosity and surface area, with an error of $\pm 3\%$. Catalysts were outgassed under vacuum at 433 K for 16 h before measurement. Total specific surface areas were estimated by the multi-point Brunauer-Emmett-Teller (BET) [59] method using liquid N_2 (77 K) as sorbate, while pore size distributions were determined using Barrett–Joyner–Halenda (BJH) method [60].

3.2.4. X-ray Powder Diffraction

XRD spectra were collected using a Philips model X'Pert MPD (Philips, Eindhoven, the Netherlands) with Co-filtered Cu Kα radiation ($\lambda = 1.54056$ Å). Patterns were recorded from $2\theta = 3°$ to $90°$ with $0.04°$ step using a 0.4 s acquisition time per step. Crystalline phase's identification after drying and calcination steps was made by comparison to Joint Committee on Powder Diffraction Standards. Co_3O_4 average particle size (dCo_3O_4, nm) was estimated by Scherrer's equation at $2\theta{:}36.9°$ [61,62]. Then, metallic Co crystallite size (dCo^0, nm) was estimated by Equation (2) [63]:

$$[dCo^0 = 0.75 \cdot dCo_3O_4] \tag{2}$$

3.2.5. Transmission Electron Microscopy

Morphology and structure of the catalysts were characterized by a JEOL 2100 Transmission electron microscope (JEOL USA, Inc.), operated at 200 kV with a resolution of 0.194 nm. The catalysts were reduced ex situ under the same conditions as for FTS testing and then cooled to room temperature and passivated with low oxygen containing mixture (1 vol %) before performing the TEM analysis. Both Co dispersion and diameter distribution were calculated by counting a minimum of 400 particles. Average surface Co particle diameter was evaluated according to Equation (3) [64,65]:

$$[dCo^0 = \sum_i n_i \cdot d_i^3 / \sum_i n_i \cdot d_i^2] \tag{3}$$

where n_i refers to the number of particles with diameter d_i. The standard deviation (σ) of the obtained particle diameter was calculated with the following formula (Equation (4)):

$$[\sigma = \sqrt{\Sigma_i (d_i - \overline{d})^2 / n}] \tag{4}$$

where $\overline{d}$ is the mean Co particle diameter and n the total number of particles.

3.2.6. Temperature Programmed Reduction

Reducibility of Co_3O_4 particles was examined by TPR experiments performed in an Autochem HP 2950 analyzer (Micromeritics, Norcross, GA, USA). Calcined samples were firstly outgassed under flowing argon at 250 °C (10 °C/min heating rate) for 30 min. TPR was then performed using Ar forming gas with 17 v/v % H_2 as reductant (50 mL/min). Data of temperature and detector signal was monitored from room temperature to 900 °C (5 °C/min heating rate). TPR profiles were consistent, standard deviations for the temperature of the peak maxima being ±2%.

3.2.7. O_2 Pulse

Degree of reduction was calculated in an Autochem HP 2950 (Micromeritics, Norcross, GA, USA) by oxygen titration of reduced samples (±2% average error). After reduction at the same conditions fixed for FTS testing [550 °C (5 °C/min heating rate) for 2 h under pure H_2 (⩾99.9990%) flow], the flow was switched to helium for 1 h to desorb any chemisorbed H_2. Then, O_2 was injected at 400 °C, by means of calibrated pulses in helium carrier, until complete saturation of the catalyst surface (no further O_2 is detected to react with the sample by the TCD) [66]. Calculation of dispersion was based assuming that the H/Co stoichiometric ratio was 1/1 [19], while for the degree of reduction, stoichiometric re-oxidation of Co^0 to Co_3O_4 [67,68] was supposed.

3.2.8. Titrations

A Metrohm 686 apparatus (Metrohm Ltd., Herisau, Switzerland) was used to conduct acid/base titrations. 25 mg of catalyst were dissolved in 50 cm^3 of a NaCl solution (0.1 M, also used as a reference) and basified to pH = 10.3 (0.1 M NaOH). Then, titrant (0.1 M HCl) was dosed by an automatic Dosimat 665 at a rate of 3 cm^3/h under constant stirring and N_2 atmosphere.

3.3. Activity Test

FTS experiments were conducted in a bench scale, fully automated facility, under a pressure of 20 bar, as described elsewhere [24]. Typically, 5 g of catalyst dispersed on inert α-SiC (in a 1:4 wt. ratio in order to minimize temperature gradients) were charged into the reactor (1 m length and 1.77 cm internal diameter, comprising a volume of 0.24 L, made of Inconel 600). The catalyst was firstly pretreated by reduction in pure H_2 (⩾99.999%) flow (1.5 NL/min) at 550 °C (5 °C/min heating rate) under atmospheric pressure for 2 h. Then, temperature was decreased to the required reaction temperature under flowing N_2 while pressure was set on 20 bars. FTS catalytic performance was then studied at 220 °C, 235 °C and 250 °C, a gas space velocity of 6000 Ncm^3/g·h and a molar ratio H_2/CO of 1.8 (considering a previous

Water Gas Shift stage). Temperature was checked (with an error of $\pm0.5\,°C$) in the bed of catalyst and along the whole tube through a series of thermocouples disposed to control temperature gradients in the fixed-bed reactor. The feed gas consisted of highly pure N_2 (10 vol %), CO (60 vol %), and H_2 (30 vol %), being N_2 the internal standard used to ensure an accurate carbon balance. Downstream, a Peltier cell was located prior to the analysis system in order to separate non-condensate water and to avoid overpressure on the input of the chromatographic unit. Light hydrocarbons and unreacted reactants were analyzed online every 90 min, using a CP-4900 Varian microgas chromatograph equipped with: (a) a Molsieve 5A capillary column (H_2, N_2, CH_4 and CO); and (b) a Pora Pack Q capillary column (CO_2, ethane and propane). C_5^+ hydrocarbons, collected for 8–12 h to ensure reaching the steady state, were extracted with n-hexane from liquid products stream. After this step, organic phase was analyzed offline in a gas chromatograph equipped with a FID detector and an ultrafast capillary column, supplied by Thermo Fisher Scientific. Hydrocarbon selectivity was estimated as reported elsewhere [23]. Three replicate experiments were randomized conducted in order to avoid systematic errors while quantifying reproducibility of each precursor system. Confidence interval was calculated from experimental data by means of the Student test (t-test, 95% confidence level), and the corresponding standard deviation measurement.

4. Conclusions

This work explored the dependence of FTS catalytic performance on cobalt precursor supported on β-SiC. The use of different parent cobalt compounds significantly modified physicochemical characteristics of Com/SiC catalyst and, therefore, their FTS catalytic performance. There was an increase in the basicity of the catalyst in the following order: CoCit/SiC < CoA/SiC < CoCl/SiC < CoN/SiC. The same trend was found for the average Co particle size, although a wide distribution was found for CoCl/SiC and CoN/SiC, affecting the extent of reduction. Consequently, FTS activity and chain growth probability were found to depend on the nature of the precursor used for the preparation of Com/SiC catalysts. The use of an organic compound such as cobalt citrate resulted in a significant increase of dispersion, exhibiting a homogeneous disposition of small cobalt crystallites on β-SiC compared with those from CoN/SiC. Consequently, citrate-based catalyst provided a lower FTS rate and an enhanced selectivity to CO_2 and C_1–C_4 fraction. Moreover, cobalt citrate also led to a C_5^+ product distribution centered on gasoline fraction ($\alpha = 0.76$). Nevertheless, cobalt nitrate, related to the highest particle size, degree of reduction and basicity, resulted in a better FTS activity with a low formation of CH_4, shifting the production of C_5^+ hydrocarbons towards kerosene-diesel fraction (0.90), with diesel yield twice that produced by CoCit/SiC.

Acknowledgments: Financial supports were provided by the Ministerio de Industria, Turismo y Comercio of Spain (CENIT-PiIBE project) and ELCOGAS S.A. Sicat catalyst is also gratefully acknowledged for providing support sample.

Author Contributions: A.R. de la Osa prepared the catalysts, performed the experiments and wrote the original manuscript. P. Sánchez helped to prepare the final manuscript and revised the final version of the paper. F. Dorado, A. Romero and J.L. Valverde revised the final version of the paper.

Conflicts of Interest: The authors declare no conflict of interest.

References

1. Borg, Ø.; Eri, S.; Blekkan, E.A.; Storsæter, S.; Wigum, H.; Rytter, E.; Holmen, A. Fischer-Tropsch synthesis over γ-alumina-supported cobalt catalysts: Effect of support variables. *J. Catal.* **2007**, *248*, 89–100.
2. Jacobs, G.; Ji, Y.; Davis, B.H.; Cronauer, D.; Kropf, A.J.; Marshall, C.L. Fischer-Tropsch synthesis: Temperature programmed EXAFS/XANES investigation of the influence of support type, cobalt loading, and noble metal promoter addition to the reduction behavior of cobalt oxide particles. *Appl. Catal. A* **2007**, *333*, 177–191.
3. Martínez, A.N.; López, C.; Márquez, F.; Díaz, I. Fischer-Tropsch synthesis of hydrocarbons over mesoporous Co/SBA-15 catalysts: The influence of metal loading, cobalt precursor, and promoters. *J. Catal.* **2003**, *220*, 486–499.
4. Iglesia, E. Design, synthesis, and use of cobalt-based Fischer-Tropsch synthesis catalysts. *Appl. Catal. A* **1997**, *161*, 59–78.
5. Iglesia, E.; Soled, S.L.; Fiato, R.A. Fischer-Tropsch synthesis on cobalt and ruthenium. Metal dispersion and support effects on reaction rate and selectivity. *J. Catal.* **1992**, *137*, 212–224.
6. Iglesia, E.; Reyes, S.C.; Madon, R.J.; Soled, S.L. Selectivity control and catalyst design in the Fischer-Tropsch synthesis: Sites, pellets, and reactors. *Adv. Catal.* **1993**, *39*, 221–302.
7. Johnson, B.G.; Bartholomew, C.H.; Goodman, D.W. The role of surface structure and dispersion in CO hydrogenation on cobalt. *J. Catal.* **1991**, *128*, 231–247.
8. Reuel, R.C.; Bartholomew, C.H. Effects of support and dispersion on the CO hydrogenation activity/selectivity properties of cobalt. *J. Catal.* **1984**, *85*, 78–88.
9. Bessell, S. Support effects in cobalt-based Fischer-Tropsch catalysis. *Appl. Catal. A* **1993**, *96*, 253–268.
10. Clause, O.; Kermarec, M.; Bonneviot, L.; Villain, F.; Che, M. Nickel(II) ion-support interactions as a function of preparation method of silica-supported nickel materials. *J. Am. Chem. Soc.* **1992**, *114*, 4709–4717.
11. Lekhal, A.; Glasser, B.J.; Khinast, J.G. Influence of pH and ionic strength on the metal profile of impregnation catalysts. *Chem. Eng. Sci.* **2004**, *59*, 1063–1077.
12. Van de Water, L.G.A.; Bezemer, G.L.; Bergwerff, J.A.; Versluijs-Helder, M.; Weckhuysen, B.M.; de Jong, K.P. Spatially resolved UV-VIS microspectroscopy on the preparation of alumina-supported Co Fischer-Tropsch catalysts: Linking activity to Co distribution and speciation. *J. Catal.* **2006**, *242*, 287–298.

13. Rosynek, M.P.; Polansky, C.A. Effect of cobalt source on the reduction properties of silica-supported cobalt catalysts. *Appl. Catal.* **1991**, *73*, 97–112.

14. Niemelä, M.K.; Krause, A.O.I.; Vaara, T.; Lahtinen, J. Preparation and characterization of Co/SiO$_2$, Co-Mg/SiO$_2$ and Mg-Co/SiO$_2$ catalysts and their activity in CO hydrogenation. *Top. Catal.* **1995**, *2*, 45–57.

15. Niemelä, M.K.; Krause, A.O.I. The long-term performance of Co/SiO$_2$ catalysts in CO hydrogenation. *Catal. Lett.* **1996**, *42*, 161–166.

16. Van de Loosdrecht, J.; Van der Haar, M.; Van der Kraan, A.M.; Van Dillen, A.J.; Geus, J.W. Preparation and properties of supported cobalt catalysts for Fischer-Tropsch synthesis. *Appl. Catal. A Gen.* **1997**, *150*, 365–376.

17. Matsuzaki, T.; Takeuchi, K.; Hanaoka, T.; Arakawa, H.; Sugi, Y. Hydrogenation of carbon monoxide over highly dispersed cobalt catalysts derived from cobalt(II) acetate. *Catal. Today* **1996**, *28*, 251–259.

18. Sun, S.; Tsubaki, N.; Fujimoto, K. The reaction performances and characterization of Fischer-Tropsch synthesis Co/SiO$_2$ catalysts prepared from mixed cobalt salts. *Appl. Catal. A* **2000**, *202*, 121–131.

19. Schanke, D.; Vada, S.; Blekkan, E.A.; Hilmen, A.M.; Hoff, A.; Holmen, A. Study of Pt-promoted cobalt CO hydrogenation catalysts. *J. Catal.* **1995**, *156*, 85–95.

20. Lacroix, M.; Dreibine, L.; De Tymowski, B.; Vigneron, F.; Edouard, D.; Bégin, D.; Nguyen, P.; Pham, C.; Savin-Poncet, S.; Luck, F.; et al. Silicon carbide foam composite containing cobalt as a highly selective and re-usable Fischer-Tropsch synthesis catalyst. *Appl. Catal. A* **2011**, *397*, 62–72.

21. De Tymowski, B.; Liu, Y.; Meny, C.; Lefèvre, C.; Begin, D.; Nguyen, P.; Pham, C.; Edouard, D.; Luck, F.; Pham-Huu, C. Co-Ru/SiC impregnated with ethanol as an effective catalyst for the Fischer-Tropsch synthesis. *Appl. Catal. A* **2012**, *419–420*, 31–40.

22. Díaz, J.A.; Calvo-Serrano, M.; De la Osa, A.R.; García-Minguillán, A.M.; Romero, A.; Giroir-Fendler, A.; Valverde, J.L. β-silicon carbide as a catalyst support in the Fischer-Tropsch synthesis: Influence of the modification of the support by a pore agent and acidic treatment. *Appl. Catal. A* **2014**, *475*, 82–89.

23. De la Osa, A.R.; De Lucas, A.; Díaz-Maroto, J.; Romero, A.; Valverde, J.L.; Sánchez, P. FTS fuels production over different Co/SiC catalysts. *Catal. Today* **2012**, *187*, 173–182.

24. De la Osa, A.R.; De Lucas, A.; Sánchez-Silva, L.; Díaz-Maroto, J.; Valverde, J.L.; Sánchez, P. Performing the best composition of supported Co/SiC catalyst for selective FTS diesel production. *Fuel* **2012**, *95*, 587–598.

25. Lee, B.; Koo, H.; Park, M.-J.; Lim, B.; Moon, D.; Yoon, K.; Bae, J. Deactivation behavior of Co/SiC Fischer-Tropsch catalysts by formation of filamentous carbon. *Catal. Lett.* **2013**, *143*, 18–22.

26. Lee, J.S.; Jung, J.S.; Moon, D.J. The effect of cobalt loading on Fischer-Tropsch synthesis over silicon carbide supported catalyst. *J. Nanosci. Nanotechnol.* **2015**, *15*, 396–399.

27. Zhu, X.; Lu, X.; Liu, X.; Hildebrandt, D.; Glasser, D. Heat transfer study with and without Fischer-Tropsch reaction in a fixed bed reactor with TiO$_2$, SiO$_2$, and SiC supported cobalt catalysts. *Chem. Eng. J.* **2014**, *247*, 75–84.

28. Torres Galvis, H.M.; Koeken, A.C.J.; Bitter, J.H.; Davidian, T.; Ruitenbeek, M.; Dugulan, A.I.; De Jong, K.P. Effect of precursor on the catalytic performance of supported iron catalysts for the Fischer-Tropsch synthesis of lower olefins. *Catal. Today* **2013**, *215*, 95–102.

29. Khodakov, A.Y. Enhancing cobalt dispersion in supported Fischer-Tropsch catalysts via controlled decomposition of cobalt precursors. *Braz. J. Phys.* **2009**, *39*, 171–175.

30. Mishra, S.K.; Kanungo, S.B. Thermal dehydration and decomposition of cobalt chloride hydrate (CoCl$_2$·xH$_2$O). *J. Therm. Anal.* **1992**, *38*, 2437–2454.

31. Yuan, X.; Lü, J.; Yan, X.; Hu, L.; Xue, Q. Preparation of ordered mesoporous silicon carbide monoliths via preceramic polymer nanocasting. *Microporous Mesoporous Mater.* **2011**, *142*, 754–758.

32. Jacobson, N.S.; Myers, D.L. Active oxidation of SiC. *Oxid. Met.* **2011**, *75*, 1–25.

33. Labuschagne, J.; Meyer, R.; Chonco, Z.H.; Botha, J.M.; Moodley, D.J. Application of water-tolerant Co/β-SiC catalysts in slurry phase Fischer-Tropsch synthesis. *Catal. Today* **2016**.

34. Kolasinski, K.W. *Surface Science Foundations of Catalysis and Nanoscience*; John Wiley and Sons: Chichester, UK, 2007.

35. Sing, K.S.W.; Everett, D.H.; Haul, R.A.W.; Moscou, L.; Pierotti, R.A.; Rouquerol, J.; Siemieniewska, T. Reporting Physisorption Data for Gas/Solid Systems with Special Reference to the Determination of Surface Area and Porosity. *Pure Appl. Chem.* **1985**, *57*, 603–619.

36. Nguyen, P.; Pham, C. Innovative porous SiC-based materials: From nanoscopic understandings to tunable carriers serving catalytic needs. *Appl. Catal. A* **2011**, *391*, 443–454.

37. Ledoux, M.J.; Pham-Huu, C. Silicon carbide a novel catalyst support for heterogeneous catalysis. *Cattech* **2001**, *5*, 226–246.

38. Lee, S.-H.; Yun, S.-M.; Kim, S.; Park, S.-J.; Lee, Y.-S. Characterization of nanoporous β-SiC fiber complex prepared by electrospinning and carbothermal reduction. *Res. Chem. Intermed.* **2010**, *36*, 731–742.

39. Cook, K.M.; Poudyal, S.; Miller, J.T.; Bartholomew, C.H.; Hecker, W.C. Reducibility of alumina-supported cobalt Fischer-Tropsch catalysts: Effects of noble metal type, distribution, retention, chemical state, bonding, and influence on cobalt crystallite size. *Appl. Catal. A* **2012**, *449*, 69–80.

40. Panpranot, J.; Kaewkun, S.; Praserthdam, P.; Goodwin, J., Jr. Effect of cobalt precursors on the dispersion of cobalt on MCM-41. *Catal. Lett.* **2003**, *91*, 95–102.

41. Jones, R.D.; Bartholomew, C.H. Improved flow technique for measurement of hydrogen chemisorption on metal catalysts. *Appl. Catal.* **1988**, *39*, 77–88.

42. Belambe, A.R.; Oukaci, R.; Goodwin, J.G., Jr. Effect of pretreatment on the activity of a Ru-promoted Co/Al$_2$O$_3$ Fischer-Tropsch catalyst. *J. Catal.* **1997**, *166*, 8–15.

43. Terörde, R.J.A.M.; Van den Brink, P.J.; Visser, L.M.; Van Dillen, A.J.; Geus, J.W. Selective oxidation of hydrogen sulfide to elemental sulfur using iron oxide catalysts on various supports. *Catal. Today* **1993**, *17*, 217–224.

44. Sun, S.; Fujimoto, K.; Yoneyama, Y.; Tsubaki, N. Fischer-Tropsch synthesis using Co/SiO$_2$ catalysts prepared from mixed precursors and addition effect of noble metals. *Fuel* **2002**, *81*, 1583–1591.

45. Zhou, W.; Chen, J.-G.; Fang, K.-G.; Sun, Y.-H. The deactivation of Co/SiO$_2$ catalyst for Fischer-Tropsch synthesis at different ratios of H$_2$ to CO. *Fuel Process. Technol.* **2006**, *87*, 609–616.

46. Solomonik, I.G.; Gryaznov, K.O.; Skok, V.F.; Mordkovich, V.Z. Formation of surface cobalt structures in SiC-supported Fischer-Tropsch catalysts. *RSC Adv.* **2015**, *5*, 78586–78597.

47. Rodrigues, E.L.; Bueno, J.M.C. Co/SiO$_2$ catalysts for selective hydrogenation of crotonaldehyde II: Influence of the Co surface structure on selectivity. *Appl. Catal. A* **2002**, *232*, 147–158.

48. Bechara, R.; Balloy, D.; Dauphin, J.-Y.; Grimblot, J. Influence of the characteristics of γ-aluminas on the dispersion and the reducibility of supported cobalt catalysts. *Chem. Mater.* **1999**, *11*, 1703–1711.

49. Ernst, B.; Bensaddik, A.; Hilaire, L.; Chaumette, P.; Kiennemann, A. Study on a cobalt silica catalyst during reduction and Fischer-Tropsch reaction: In situ EXAFS compared to XPS and XRD. *Catal. Today* **1998**, *39*, 329–341.

50. Khodakov, A.Y.; Lynch, J.; Bazin, D.; Rebours, B.; Zanier, N.; Moisson, B.; Chaumette, P. Reducibility of cobalt species in silica-supported Fischer-Tropsch catalysts. *J. Catal.* **1997**, *168*, 16–25.

51. Keyvanloo, K.; Fisher, M.J.; Hecker, W.C.; Lancee, R.J.; Jacobs, G.; Bartholomew, C.H. Kinetics of deactivation by carbon of a cobalt Fischer-Tropsch catalyst: Effects of CO and H$_2$ partial pressures. *J. Catal.* **2015**, *327*, 33–47.

52. Bao, A.; Liew, K.; Li, J. Fischer-Tropsch synthesis on CaO-promoted Co/Al$_2$O$_3$ catalysts. *J. Mole. Cataly. A* **2009**, *304*, 47–51.

53. De la Osa, A.R.; De Lucas, A.; Valverde, J.L.; Romero, A.; Monteagudo, I.; Coca, P.; Sánchez, P. Influence of alkali promoters on synthetic diesel production over Co catalyst. *Catal. Today* **2011**, *167*, 96–106.

54. Dry, M.E.; Oosthuizen, G.J. The correlation between catalyst surface basicity and hydrocarbon selectivity in the Fischer-Tropsch synthesis. *J. Catal.* **1968**, *11*, 18–24.

55. Zhang, J.; Chen, J.; Ren, J.; Sun, Y. Chemical treatment of γ-Al$_2$O$_3$ and its influence on the properties of Co-based catalysts for Fischer-Tropsch synthesis. *Appl. Catal. A* **2003**, *243*, 121–133.

56. Koo, H.-M.; Lee, B.S.; Park, M.-J.; Moon, D.J.; Roh, H.-S.; Bae, J.W. Fischer-Tropsch synthesis on Cobalt/Al$_2$O$_3$-modified SiC catalysts: Effect of cobalt-alumina interactions. *Catal. Sci. Technol.* **2014**, *4*, 343–351.

57. Zhang, J.; Chen, J.; Ren, J.; Li, Y.; Sun, Y. Support effect of Co/Al$_2$O$_3$ catalysts for Fischer-Tropsch synthesis. *Fuel* **2003**, *82*, 581–586.

58. Courty, P.; Ajot, H.; Marcilly, C.; Delmon, B. Oxydes mixtes ou en solution solide sous forme très divisée obtenus par décomposition thermique de précurseurs amorphes. *Powder Technol.* **1973**, *7*, 21–38.

59. Brunauer, S.; Emmett, P.H.; Teller, E. Adsorption of gases in multimolecular layers. *J. Am. Chem. Soc.* **1938**, *60*, 309–319.

60. Barrett, E.P.; Joyner, L.G.; Halenda, P.P. The determination of pore volume and area distributions in porous substances. I. Computations from nitrogen isotherms. *J. Am. Chem. Soc.* **1951**, *73*, 373–380.

61. Scherrer, P. Bestimmung der Größe und der inneren Struktur von Kolloidteilchen mittels Röntgenstrahlen. Nachrichten von der Gesellschaft der Wissenschaften zu Göttingen. *Mathematisch-Physikalische Klasse* **1918**, *2*, 98–100.

62. Klug, H.P.; Alexander, L.E. *X-ray Diffraction Procedures for Polycrystalline Amorphous Materials*, 2nd ed.; Wiley: New York, NY, USA, 1974; p. 992.

63. Van 't Blik, H.F.J.; Koningsberger, D.C.; Prins, R. Characterization of supported cobalt and Cobalt-Rhodium catalysts. III. Temperature-programmed reduction (TPR), oxidation (TPO), and EXAFS of CoRh SiO$_2$. *J. Catal.* **1986**, *97*, 210–218.

64. Datye, A.K.; Xu, Q.; Kharas, K.C.; McCarty, J.M. Particle size distributions in heterogeneous catalysts: What do they tell us about the sintering mechanism? *Catal. Today* **2006**, *111*, 59–67.

65. Mustard, D.G.; Bartholomew, C.H. Determination of metal crystallite size and morphology in supported nickel catalysts. *J. Catal.* **1981**, *67*, 186–206.

66. Bartholomew, C.H.; Farrauto, R.J. Chemistry of nickel-alumina catalysts. *J. Catal.* **1976**, *45*, 41–53.

67. Song, D.; Li, J. Effect of catalyst pore size on the catalytic performance of silica supported cobalt Fischer-Tropsch catalysts. *J. Mol. Catal. A* **2006**, *247*, 206–212.

68. Jacobs, G.; Das, T.K.; Zhang, Y.; Li, J.; Racoillet, G.; Davis, B.H. Fischer-Tropsch synthesis: Support, loading, and promoter effects on the reducibility of cobalt catalysts. *Appl. Catal. A* **2002**, *233*, 263–281.

Gold-Iron Oxide Catalyst for CO Oxidation: Effect of Support Structure

Hui-Zhen Cui, Yu Guo, Xu Wang, Chun-Jiang Jia and Rui Si

Abstract: Gold-iron oxide (Au/FeO_x) is one of the highly active catalysts for CO oxidation, and is also a typical system for the study of the chemistry of gold catalysis. In this work, two different types of iron oxide supports, *i.e.*, hydroxylated (Fe_OH) and dehydrated iron oxide (Fe_O), have been used for the deposition of gold via a deposition-precipitation (DP) method. The structure of iron oxide has been tuned by either selecting precipitated pH of 6.7–11.2 for Fe_OH or changing calcination temperature of from 200 to 600 °C for Fe_O. Then, 1 wt. % Au catalysts on these iron oxide supports were measured for low-temperature CO oxidation reaction. Both fresh and used samples have been characterized by multiple techniques including transmission electron microscopy (TEM) and high-resolution TEM (HRTEM), X-ray diffraction (XRD), X-ray photoelectron spectroscopy (XPS), X-ray absorption near edge structure (XANES) and temperature-programmed reduction by hydrogen (H_2-TPR). It has been demonstrated that the surface properties of the iron oxide support, as well as the metal-support interaction, plays crucial roles on the performance of Au/FeO_x catalysts in CO oxidation.

Reprinted from *Catalysts*. Cite as: Cui, H.-Z.; Guo, Y.; Wang, X.; Jia, C.-J.; Si, R. Gold-Iron Oxide Catalyst for CO Oxidation: Effect of Support Structure. *Catalysts* **2016**, *6*, 37.

1. Introduction

Since the 1990s, nanosized gold interacting with oxide supports have been reported to be active for diverse redox reactions, among those low temperature oxidation of carbon monoxide is the most studied [1–3]. Such unique catalytic properties were found to be strongly dependent on electronic structure and local coordination environment of Au atoms. During the last two decades, different reducible metal oxides such as titanium oxide (TiO_2) [4], cerium oxide (CeO_x) [5,6] and iron oxide (FeO_x) [7] have been proven to be active supports for deposition of gold. Although a full mechanism of this catalytic process still needs to be established, careful studies on strong metal-support interaction by the aid of various characterization techniques may provide further mechanistic insight [8–11]. For instance, Hutchings group reported that the delayered Au structure that was determined by high-angle annular dark-field scanning transmission electron microscopy (HAADF-STEM) characterization plays crucial roles in the CO oxidation reaction [8].

As one of the important functional materials, iron oxide (FeO_x) has been extensively investigated in heterogeneous catalysis for its potential applications in CO oxidation [12,13], water-gas shift [14–17] and preferential oxidation of CO reaction [18–20]. The general FeO_x supports include two different types: hydrated (Fe_OH), such as goethite or lepidocrocite FeOOH, and dehydrated (Fe_O), such as hematite α-Fe_2O_3, maghemite γ-Fe_2O_3 or magnetite Fe_3O_4. These complexities in composition and crystal periodicity provide rich structural models for gold deposition. On the other hand, various synthetic approaches, including so-gel method [21,22], hydrothermal method [23,24] and precipitation method [10,25], have been applied to control the size and shape of the iron oxide itself, aiming to delicately tune the interaction between metal and oxide matrix, as well as the local structure of active metals. Among them, the precipitation method is simple and easy to operate. It can easily control the synthesis parameters and achieve larger surface area. Furthermore, the precipitating conditions of Fe^{2+} or Fe^{3+} precursor has been identified as one of the key factors governing the structure and texture of the FeO_x products. Focusing on synthesis, the final pH value of the stock solution and the calcination temperature of the as-dried materials, which have not been widely studied so far [26,27], could be very important parameters for the preparation of high-quality iron oxide support.

Different techniques, including X-ray diffraction (XRD) [25], X-ray absorption fine structure (XAFS) [16], X-ray photoelectron spectroscopy (XPS) [16] and transmission electron microscopy (TEM) [28], have been used to characterize both bulk and surface structure of gold-iron oxide catalysts and further study the related active site for the low-temperature CO oxidation. Due to the complicated crystal and local structures in bulk and microdomain, the combination of multiple characterization skills, rather than a single means, is extremely important to obtain the real structural information.

Therefore, in the present work, we try to broadly explore the relationship between the nature of the oxide matrix and the catalytic reactivity of Au/FeO_x via deposition-precipitation with two series of iron oxides, namely hydrated (Fe_OH) and dehydrated (Fe_O) supports, to fully investigate the importance of preparation parameters such as precipitating pH values and calcination temperature in synthesis of FeO_x, and to deeply study the "structure-activity" relationship in Au/FeO_x system for the low-temperature CO oxidation reaction.

2. Results and Discussion

2.1. Structure and Texture of Gold-Iron Oxide Catalyst

The inductively coupled plasma atomic emission spectroscopy (ICP-AES) characterization was conducted to identify the gold loadings in both Fe_OH and

Fe_O supports. Table 1 shows that the experimental bulk Au concentrations (Au_{bulk}) in all the measured samples are in good agreement with the target value (0.54 at. %), revealing no gold loss in DP synthesis. The BET specific surface areas (S_{BET}) of the fresh Au/FeO_x catalysts are summarized in Table 1. Since the hydrated iron oxide (Fe_OH) supports were uncalcined, the corresponding S_{BET} values for Au/Fe_OH (151–212 m^2/g) are distinctly higher than those for Au/Fe_O (27–136 m^2/g), in which the dehydrated iron oxide (Fe_O) supports were calcined at different temperatures in the range of 200–600 °C. For Au/Fe_OH series, the S_{BET} number decreases with the increase of precipitating pH value of Fe_OH from 6.7 to 11.2, possibly indicating better crystallinity of iron oxide support with more hydroxyls used in preparation. For Au/Fe_O series, the S_{BET} number decreases with the increase of calcination temperature of Fe_O from 200 to 600 °C, giving hints on the elimination of surface OH groups during the heat-treatment process. Thus, the textural properties of iron oxide supports, as well as the final Au/FeO_x catalysts, are strongly dependent by the synthetic parameters used in experiments.

XRD was carried out to determine the crystal structure of the fresh Au/FeO_x catalysts. Figure 1a displays an amorphous phase for Au/Fe_OH_6.7 and Au/Fe_OH_8.2, or partially crystallized structure of Au/Fe_OH_9.7 and Au/Fe_OH_11.2 with further increasing on the applied pH value. The corresponding crystal phase is hematite α-Fe_2O_3 (JCPDS card No: 2-919), consistent with previous reports on gold-iron oxide catalysts [28]. Comparable to the above characterization results on surface area, the precipitating pH significantly affected the crystallinity of Fe_OH support, and further modified the structural properties of Au/Fe_OH catalyst. Figure 1b exhibits that the crystallinity of Au/Fe_O was clearly enhanced by applying higher calcination temperature on Fe_O support, while the crystal phase was kept as hematite α-Fe_2O_3. We also noticed that all measured samples were fully crystallized if the iron oxide support was calcined above 200 °C, which is in good agreement with the previous findings [28]. However, no Au diffraction peaks can be identified in Figure 1 due to the presence of gold in the form of amorphous or the low Au concentrations (~0.54 at. %).

TEM was conducted to identify the morphology of fresh Au/FeO_x catalysts, i.e., the size and shape for gold or iron oxide support. No Au nanoparticles have been observed in all the pictures in Figure 2, confirming that gold was still in the form of atoms or ultra-fine (<2 nm) clusters for the as-dried (60 °C) samples. This was also in good agreement with our previous results on Au/FeO_x [28]. For the iron oxide support, it can be seen from the TEM images that small-size (<5 nm) aggregates were observed for amorphous Fe_OH, Au/Fe_OH_6.7 (Figure 2d), Au/Fe_OH_8.2 (Figure 2c) and Au/Fe_O_200 (Figure 2i); big-size particles (10–50 nm) were identified for fully crystallized Au/Fe_O_600 (Figure 2e), Au/Fe_O_500 (Figure 2f), Au/Fe_O_400 (Figure 2g) and Au/Fe_O_300 (Figure 2h); and a mixture of small aggregates and big particles was confirmed for partially crystallized Au/Fe_OH_11.2 (Figure 2a) and Au/Fe_OH_9.7 (Figure 2b). These results were well consistent with the related XRD data.

Table 1. Bulk Au concentrations (Au_{bulk}), surface Au concentrations (Au_{surf}), BET (Brunner-Emmet-Teller) specific surface areas (S_{BET}) and Au 4f XPS peak positions of gold-iron oxide catalysts.

Sample	Au_{bulk} (at. %) [a]	Au_{surf} (at. %) [b]	S_{BET} (m$^2 \cdot$g^{-1}) [c]	Au 4f (eV)
Au/Fe_OH_11.2	0.53	0.67 (1.20)	151	84.55 (84.2), 88.15 (87.9)
Au/Fe_OH_9.7	0.54	-	198	-
Au/Fe_OH_8.2	0.52	-	203	-
Au/Fe_OH_6.7	0.53	0.55 (0.42)	212	84.6 (84.0), 88.15 (87.7)
Au/Fe_O_600	0.53	-	27	-
Au/Fe_O_500	0.52	-	42	-
Au/Fe_O_400	0.53	-	58	-
Au/Fe_O_300	0.52	0.68 (0.74)	97	84.7 (84.15), 88.4 (87.85)
Au/Fe_O_200	0.53	0.63 (0.54)	136	84.35 (84.2), 88.05 (87.85)

[a] Determined by ICP-AES; [b] Determined by XPS. The numbers in brackets are for the used catalysts; [c] Calculated from the adsorption branch.

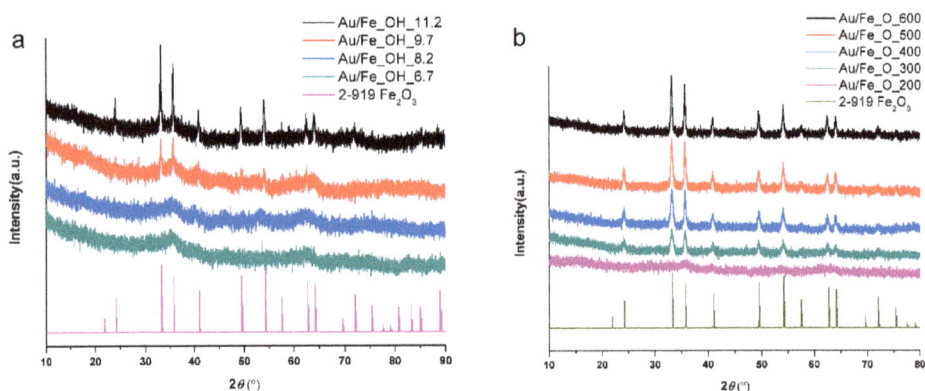

Figure 1. XRD patterns of fresh Au/FeO$_x$ catalysts: (**a**) Au/Fe_OH and (**b**) Au/Fe_O.

For the used catalysts, we found from Figure 3 that the morphologies of iron oxide supports were maintained for all the tested samples under the mild reaction conditions (up to 300 °C and less than 1.5 h). Gold particles with size of *ca.* 2 nm can be identified for gold on the fully crystallized Fe_O supports (Figure 3e–h). However, gold species cannot be distinguished clearly in Au/Fe_OH catalysts due to the amorphous state of Fe_OH particles, so the structure of gold in Au/Fe_OH is to be investigated using other techniques such as XPS and XAFS (these will be mentioned later). HRTEM was used to determine the crystallinity of iron oxide supports. Figure 4a–d exhibits the highly crystallized nature of Fe_O for Au/Fe_O_300 and Au/Fe_O_200, either fresh or used in CO oxidation.

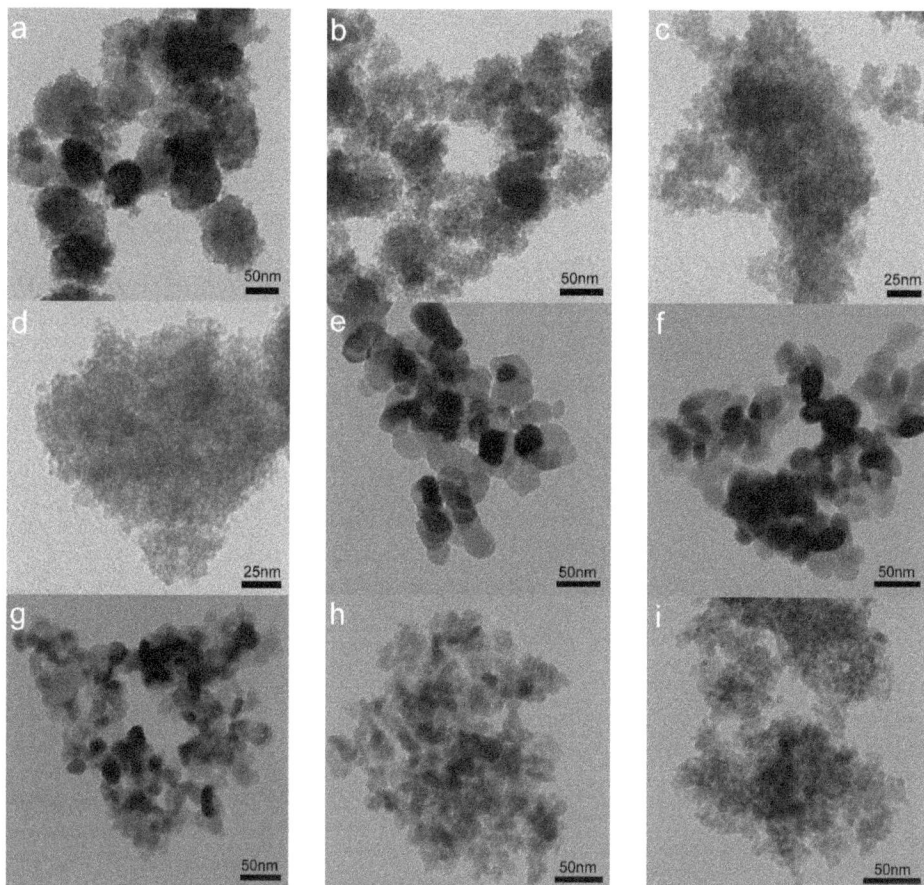

Figure 2. TEM images of fresh Au/FeO$_x$ catalysts: (**a**) Au/Fe_OH_11.2; (**b**) Au/Fe_OH_9.7; (**c**) Au/Fe_OH_8.2; (**d**) Au/Fe_OH_6.7; (**e**) Au/Fe_O_600; (**f**) Au/Fe_O_500; (**g**) Au/Fe_O_400; (**h**) Au/Fe_O_300; and (**i**) Au/Fe_OH_200.

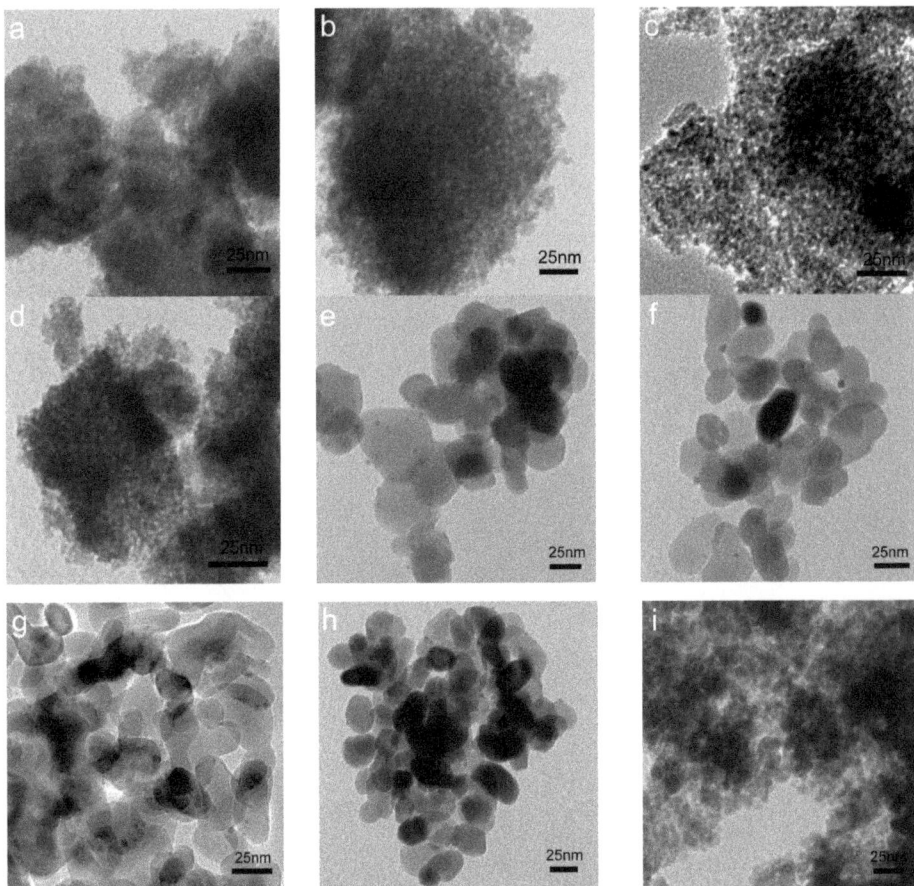

Figure 3. TEM images of used Au/FeO$_x$ catalysts: (**a**) Au/Fe_OH_11.2; (**b**) Au/Fe_OH_9.7; (**c**) Au/Fe_OH_8.2; (**d**) Au/Fe_OH_6.7; (**e**) Au/Fe_O_600; (**f**) Au/Fe_O_500; (**g**) Au/Fe_O_400; (**h**) Au/Fe_O_300; and (**i**) Au/Fe_O_200.

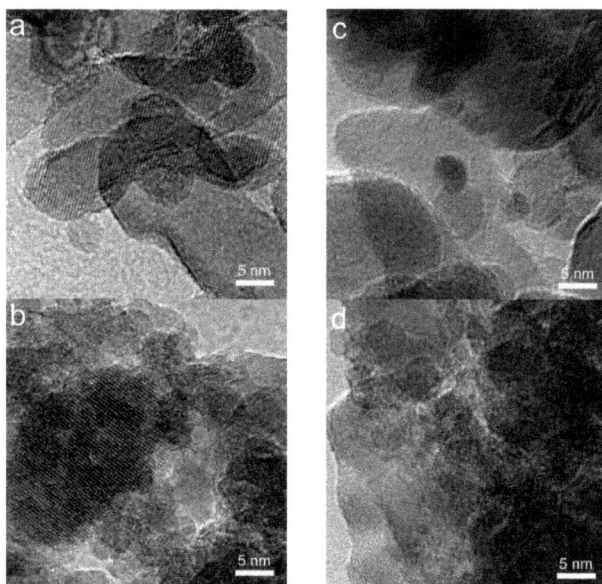

Figure 4. HRTEM images (**a–d**) of Au/FeO$_x$ catalysts: (**a**) Au/Fe_O_300, fresh; (**b**) Au/Fe_O_200, fresh; (**c**) Au/Fe_O_300, used; and (**d**) Au/Fe_O_200 used.

2.2. Catalytic Performance and Reducibility of Gold-Iron Oxide Catalyst

The catalytic performance of the Au/FeO$_x$ catalysts was evaluated for the low-temperature CO oxidation. The transient profiles in Figure 5a reveal a lower "light off" temperature or a higher activity for gold on hydrated iron oxide support synthesized with an increasing pH value. The 90% CO conversion temperature (T_{90}) was 62 °C for Au/Fe_OH_6.7, while it was 25 °C for Au/Fe_OH_11.2. The corresponding "steady-state" experiments in Figure 5c confirmed that the long-term CO conversions at 30 °C after 6 h were stabilized at 53%, 41%, 20% and 11% for the Fe_OH precipitating pH value of 11.2, 9.7, 8.2 and 6.7, respectively. All of the above demonstrate the following sequence of catalytic reactivity on Au/Fe_OH for the CO oxidation reaction: Au/Fe_OH_11.2 > Au/Fe_OH_9.7 > Au/Fe_OH_8.2 > Au/Fe_OH_6.7. In our previous work, pH = 8.2 was applied to prepare Fe_OH [28], which was based on reported synthesis [8,12]. It is noticed that the long-term stability of Au/Fe_OH_8.2 in final CO conversion was lower than that reported previously [28]. We repeated the catalytic tests at 30 °C several times and found that the deactivation behavior varied with the testing periods. Thus, we selected the catalytic data collected during the same periods between different gold-iron samples. The current results shown in Figure 5 justifies the optimized pH value is 11.2 for the hydrated iron oxide support, and thus the catalytic reactivity can be enhanced by tuning the precipitation pH values in the synthesis of Fe_OH supports.

For gold on dehydrated iron oxide support, the related CO oxidation transient profiles are shown in Figure 5b. Here, we need to control the CO conversion at the modest level and select the pH value of 8.2 for the preparation of initial Fe_OH. This can effectively avoid the too small differences on the reactivity of CO oxidation when we investigate the effect of calcination temperature towards the Fe_O support. Clearly, Au/Fe_O_300 was superior to Au/Fe_O_200, displaying distinct lower T_{90} (20 °C *vs.* 85 °C). Au/Fe_O_400 was almost identical to Au/Fe_O_300, while the reactivity of Au/Fe_O_500 and Au/Fe_O_600 dropped quickly with higher T_{90} of 35 °C and 103 °C, respectively. The long-term "steady-state" experiments in Figure 5d verify a similar trend for Au/Fe_O series. At a constant temperature of 30 °C, the stabilized CO conversions after 6–10 h on stream were 85%, 70%, 40%, 32% and 20% for the Fe_O calcination temperature of 300, 400, 500, 600 and 200 °C, respectively. All of the above demonstrate the following sequence of catalytic reactivity on Au/Fe_O for the CO oxidation reaction: Au/Fe_O_300 > Au/Fe_O_400 > Au/Fe_O_500 > Au/Fe_O_600 > Au/Fe_O_200. Previously, air-calcination at 400 °C was chosen to obtain the Fe_O support [28] which is also the optimized parameter according to our present work.

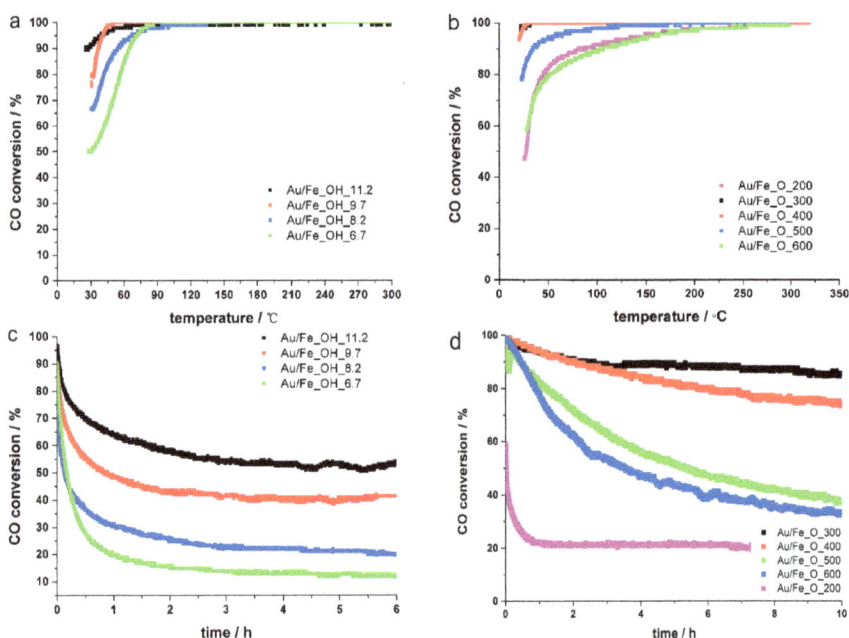

Figure 5. CO conversions of fresh Au/FeO$_x$ catalysts for the low-temperature CO oxidation reaction: (**a**) Au/Fe_OH, transient; (**b**) Au/Fe_O, transient; (**c**) Au/Fe_OH, stability at 30 °C; and (**d**) Au/Fe_O, stability at 30 °C. Reaction conditions: 1% CO/20% O_2/79% N_2, 80,000 mL·h^{-1}·g$_{cat}^{-1}$.

To correlate the catalytic reactivity of CO oxidation with the structure of gold-iron oxide, we carried out the H_2-TPR tests on both fresh Au/Fe_OH or Au/Fe_O catalysts and the corresponding Fe_OH or Fe_O supports. For the Au/Fe_OH series, the reduction of Fe_OH supports (Figure 6a) can be divided into two parts. The low-temperature reduction peaks between 180–220 °C and 350–400 °C can be attributed to the transformation of Fe_OH→Fe_3O_4 (first Fe_OH→Fe_2O_3 and then Fe_2O_3→Fe_3O_4) [29], and the hydrogen consumption amount as shown in Table 2 clearly decreased from Fe_OH_6.7 to Fe_OH_11.2, probably due to less surface hydroxyls for the high precipitating pH value. The onset of high-temperature reduction peak, assigned to the conversion of Fe_3O_4→FeO or Fe [29], was located at 350–400 °C.

By the introduction of gold, Figure 6b exhibits that the reduction of Fe_3O_4→FeO or Fe was almost maintained. However, the reduction of Fe_OH→Fe_3O_4 was significantly shifted to lower temperature range of 30–230 °C, and included a broad peak around 30–180 °C (Fe_OH→Fe_2O_3) and a sharp reduction centered at 195–214 °C (Fe_2O_3→Fe_3O_4). This shift originated from the Au activation [30], or the strong interaction between gold and iron oxide support delivered by the structure of Au–OH–Fe or Au–O–Fe [31]. It can also be seen from Figure 6b that the transformation of Fe_OH→Fe_2O_3 (Au–OH–Fe) was kept the same, while the conversion of Fe_2O_3→Fe_3O_4 (Au–O–Fe) was distinctly promoted with the increasing pH value in Fe_OH synthesis. This reveals a much stronger Au–O–Fe interaction in Au/Fe_OH_11.2 than in Au/Fe_OH_6.7, which could account for its better catalytic performance on the CO oxidation reaction.

Similarly, for the Au/Fe_O series, the reduction of Fe_O supports (Figure 6c) can be divided into two parts. The low-temperature reduction peaks between 200–300 °C and 350–400 °C can be attributed to the transformation of Fe_O (or Fe_2O_3)→Fe_3O_4 [30,32], while the onset of high-temperature reduction peak, assigned to the conversion of Fe_3O_4→FeO or Fe [32], was located at 350–400 °C. Again, by the gold doping, Figure 6d exhibits that the reduction of Fe_3O_4→FeO or Fe was kept the same, but the reduction of Fe_2O_3→Fe_3O_4 was obviously shifted to lower temperatures below 270 °C, possibly due to the presence of Au–O–Fe interaction [31]. This reduction temperature was increased with the calcination temperature of iron oxide support from 200 to 600 °C. The corresponding hydrogen consumption amount (Table 2) was raised from Au/Fe_200 to Au/Fe_400 and remained constant if the calcination temperature was further higher (400–600 °C). It hints that the strong interaction of Au–O–Fe reached a spike at Au/Fe_400. Besides, a minor peak below 170 °C can be identified for the transformation of Fe_OH→Fe_2O_3 for Au/Fe_200, Au/Fe_300 and Au/Fe_400, while it disappeared for Au/Fe_500 and Au/Fe_600 since the higher calcination temperature on Fe_O can effectively remove the surface hydroxyls.

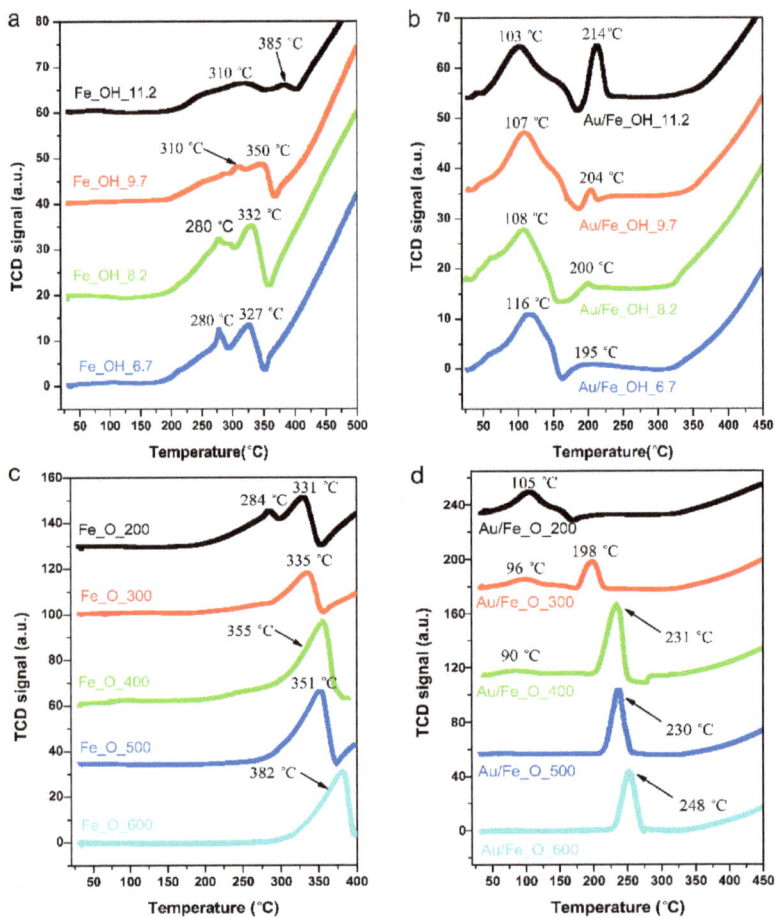

Figure 6. H$_2$-TPR profiles of fresh Au/FeO$_x$ catalysts: (**a**) Fe_OH; (**b**) Au/Fe_OH; (**c**) Fe_O; and (**d**) Au/Fe_O.

Table 2. Hydrogen consumption (H_2-consump.) of gold catalysts.

Sample	Reduction Peak (°C)	Experimental H_2-consump. (μmol·g^{-1})	Theoretical H_2-consump. [a] (μmol·g^{-1})
DP_Au/Fe_OH_11.2	102 °C, 214 °C	627	1300
DP_Au/Fe_OH_9.7	107 °C, 204 °C	549	1300
DP_Au/Fe_OH_8.2	108 °C, 200 °C	486	1300
DP_Au/Fe_OH_6.7	116 °C, 195 °C	452	1300
DP_Au/Fe_O_200	105 °C	723	1510
DP_Au/Fe_O_300	96 °C, 198 °C	778	1510
DP_Au/Fe_O_400	90 °C, 231 °C	1438	1510
DP_Au/Fe_O_500	230 °C	861	1510
DP_Au/Fe_O_600	248 °C	894	1510

[a] Calculated according to Fe_OH→Fe_3O_4 or Fe_2O_3→Fe_3O_4.

2.3. Effect of Iron Oxide Support

Previously, we utilized *in-situ* XAFS and XRD techniques to investigate the gold-iron oxide catalysts obtained by deposition-precipitation, and reported that metal–support interaction (Au–O–Fe) is key factor to govern the CO oxidation reactivity of Au/FeO$_x$ [28]. In this work, we focused on the effect of iron oxide support and optimized the synthetic parameters for the Fe_OH and Fe_O preparation. To further study the reaction mechanism in Au-Fe-O system, we selected four typical samples to reveal the "structure–activity" relationship by different characterization methods. Two of them were gold on hydrated iron oxide (Au/Fe_OH) with higher (Au/Fe_OH_11.2) and lower reactivity (Au/Fe_OH_6.7); another two were gold on dehydrated iron oxide (Au/Fe_O) with higher (Au/Fe_O_300) and lower reactivity (Au/Fe_200).

In Section 2.1, using TEM/HRTEM, we found that ~2 nm Au particles formed during the CO oxidation reaction, whether on Fe_OH (see Figure 4a,b) or on Fe_O (see Figure 4c,d). Since no such gold structure has been detected for the corresponding fresh samples, the transformation of Au species under the reaction conditions is crucial to the explanation for the different reactivity of the Au/FeO$_x$ catalysts. In Section 2.2, using H_2-TPR, the strong interaction between metal (Au) and support (Fe_OH or Fe_O) has been demonstrated for the structural evolution of Fe_OH→Fe_3O_4 and Fe_2O_3→Fe_3O_4, respectively. This could be the main reason for the origin of high activity for Au/Fe_OH_11.2 and Au/Fe_O_300. However, the information on gold chemistry, especially oxidation state and short-range local structure around Au, in the active Au/FeO$_x$ catalysts is still unknown.

XPS was conducted to determine the surface Au concentrations (Au$_{surf}$), and the related analysis results have been included in Table 1. Most of the Au$_{surf}$ values in the gold-iron oxide catalysts are close to the corresponding Au$_{bulk}$ numbers, except for the used Au/Fe_OH_11.2 (1.20 at.%) and Au/Fe_O_300 (0.74 at. %)

samples, in which their surfaces were obviously Au-richer compared with that in the corresponding fresh catalysts. Thus, the larger fraction of gold on the surface of iron oxide support during the catalytic tests can be attributed to the promotion in reactivity.

Furthermore, the XPS spectra of Fe 2p in Figure 7a,b clearly demonstrate that the oxidation state of iron was kept the same as Fe^{3+} [33,34] for all the measured samples before and after the CO oxidation reaction. This reveals that the oxidized iron species were very stable during the catalytic measurements, even with the introduction of reducing gas (CO). The XPS spectra of Au 4f in Figure 7c,d distinctly confirm that the oxidation state of gold was more ionic, exhibiting higher binding energies (also see Table 1), in all the fresh catalysts (Au/Fe_OH_11.2, Au/Fe_OH_6.7, Au/Fe_O_300, and Au/Fe_O_200) than after reaction. This indicates that CO reduced Au to form metallic species, which was in good agreement with the TEM/HRTEM results.

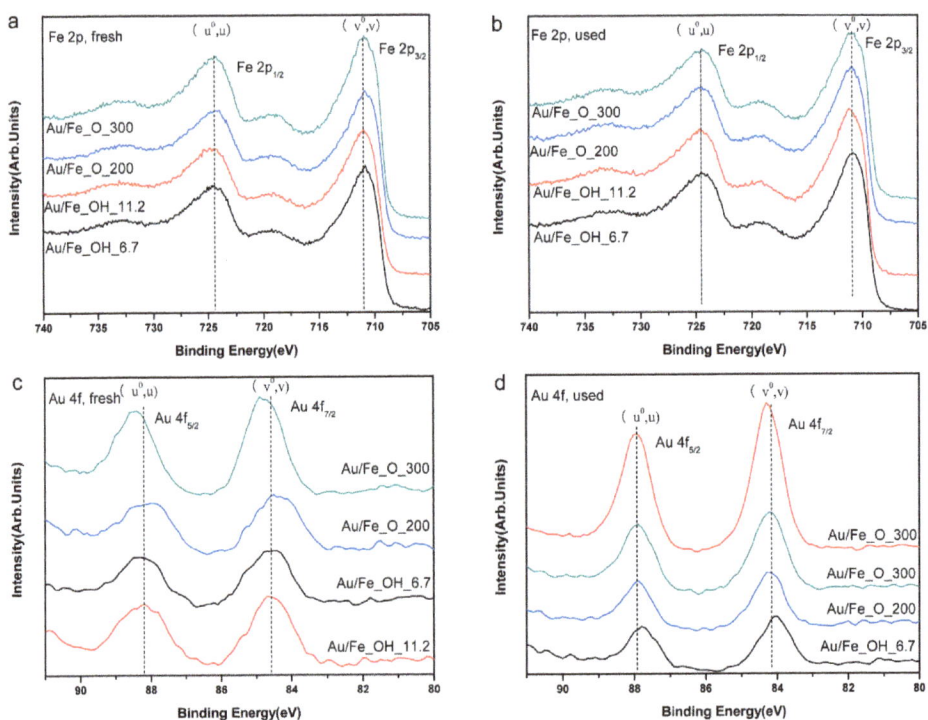

Figure 7. XPS spectra of Au/FeO$_x$ catalysts: (**a**) Fe 2p, fresh; (**b**) Fe 2p, used; (**c**) Au 4f, fresh; and (**d**) Au 4f, used.

Therefore, XAFS technique was used to investigate the electronic structure in the gold-iron oxide catalysts before and after the CO oxidation reaction. The related white line in the XANES profiles distinctly changes with the structural evolution on

Au (Au^0 and $Au^{\delta+}$) [28]. Compared to XPS, XANES test can be conducted under milder ambient conditions and exclude the effect of ultra-high vacuum level. Thus, we selected the XANES approach to identify the oxidation states of gold in this work.

The Au L3-edge XANES spectrum of Au/FeO_x catalysts is compared with that of Au foil in Figure 8. For the fresh samples (see Figure 8a), the Au L3 edges were obviously shifted to higher energies, if compared to that of Au° standard, confirming their $Au^{\delta+}$ nature. Meanwhile, strong white line was clearly observed, which verifies that the ionic gold species were dominant in the fresh catalysts [35,36]. Interestingly, according to the height of white line in Figure 8a, the oxidation state of Au in each gold-iron oxide sample follows this sequence: Au/Fe_OH_11.2 > Au/Fe_O_300 > Au/Fe_OH_6.7 > Au/Fe_O_200. This was well consistent with the catalytic reactivity order: Au/Fe_OH_11.2 > Au/Fe_OH_6.7 and Au/Fe_O_300 > Au/Fe_O_200.

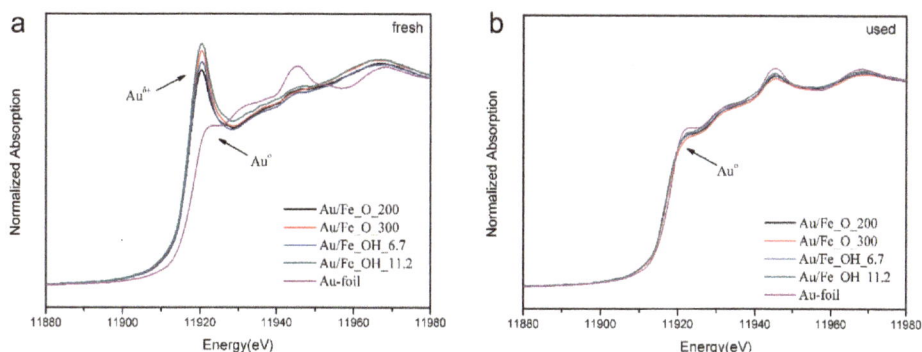

Figure 8. XANES profiles of Au L3 edge for Au/FeO_x catalysts: (**a**) fresh and (**b**) used.

However, after the reaction, gold in all the used Au/FeO_x catalysts was reduced to almost pure Au° species based on the XANES profiles in Figure 8b. This was in good agreement with the TEM/HRTEM and XPS data. Therefore, the different Au oxidation states in the fresh samples were crucial for the significant differences in reactivity between gold supported on various Fe_OH and Fe_O matrices.

3. Experimental Section

3.1. Catalyst Preparation

3.1.1. Preparation of Fe_OH with Different pH Values [8]

$Na_2CO_3 \cdot 10H_2O$ (99.9%) and $Fe(NO_3)_3 \cdot 6H_2O$ (99%) were purchased from Tianjin Kermal Chemical Reagent Factory (Tianjin, China). Typically, 0.25 mol$\cdot L^{-1}$ Na_2CO_3 aqueous solution was added drop wisely to 200 mL of 0.1 mol$\cdot L^{-1}$ $Fe(NO_3)_3$

aqueous solution under stirring at 80 °C until specific pH value (6.7, 8.2, 9.7, and 11.2) was reached. The stock solution was aged under stirring at 80 °C for another 1 h, and then the as-formed precipitate was collected by filtration. The solid product was washed with deionized (DI) water at 80 °C several times until the final pH was neutral. This powder was dried at 120 °C in still air for 12 h to generate the Fe_OH supports, marked as Fe_OH_6.7, Fe_OH_8.2, Fe_OH_9.7 and Fe_OH_11.2.

3.1.2. Preparation of Fe_O with Different Calcination Temperatures

The Fe_OH_8.2 supports were ground into fine powders, and calcined at different temperatures (200, 300, 400, 500 and 600 °C) for 2 h (rate of $5 °C \cdot min^{-1}$) to generate the Fe_O supports, marked as Fe_O_200, Fe_O_300, Fe_O_400, Fe_O_500, and Fe_O_600.

3.1.3. Deposition of Gold onto Fe_OH or Fe_O

For the sequential deposition-precipitation process, 0.5 g Fe_OH or Fe_O support powders were suspended in 23 mL Millipore water (18.25 MΩ) under stirring. Then, 2 mL of $0.0125 mol \cdot L^{-1}$ $HAuCl_4$ aqueous solution was added to the above solution at 60 °C. After 30 min, 25 mL of aqueous solution containing 0.5 g of urea was quickly added into the stock solution. Thereafter, the stock solution was heated up to 80 °C under vigorously stirring and the temperature was kept for 3 h. The pH value of this solution was raised from 4.0 to 8.6, resulting in the full decomposition of urea. The stock solution was further aged at room temperature for another 20 h. The as-prepared product was collected by filtration, and washed with Millipore water at 60 °C several times until the final pH value was neutral. After being dried at 60 °C overnight in still air, the fresh Au/Fe_OH and Au/Fe_O catalysts were obtained. The designed gold concentration in each sample was fixed at 1 wt. %, calculated as $Au/Fe(OH)_3$, or 0.54 at. %, calculated as Au/(Au + Fe). $HAuCl_4 \cdot 4H_2O$ (99.9%) were purchased from National Chemicals (Shanghai, China).

3.2. Characterization

The actual gold loadings of catalysts were determined by inductively coupled plasma atomic emission spectroscopy (ICP-AES) on an IRIS Intrepid II XSP instrument (Thermo Electron Corporation, Waitham, MA, USA).

The nitrogen adsorption-desorption measurements were performed on a NOVA 4200e instrument (Quantachrome Corporation, Boynton Beach, FL, USA) at 77 K. The samples were outgassed at 120 °C for 12 h under vacuum prior to measurements. The BET specific surface areas (S_{BET}) were calculated from the adsorption data in the relative pressure range between 0.05 and 0.30.

X-ray photoelectron spectroscopy (XPS) analysis was performed on an Axis Ultra XPS spectrometer (Kratos, UK) with 225 W of Al K_α radiation. The C 1s line at

284.8 eV was used to calibrate the binding energies. The surface gold concentrations (Au_{surf} in at. %) were determined by integrating the areas of Au 4f and Fe 2p peaks.

X-ray Diffraction (XRD) was operated on a Bruker D8 Advance diffractometer (Bruker-AXS; Karlsruhe, Germany, 40 kV, 40 mA), using Cu K_α radiation ($\lambda = 0.15406$ nm). The powder catalyst after grinding was placed inside a quartz-glass sample holder before test.

X-ray absorption near edge structure (XANES) spectra at Au K-edge ($E_0 = 11,919$ eV) were performed at BL14W1 beam line of Shanghai Synchrotron Radiation Facility (SSRF) operated at 3.5 GeV under "top-up" mode with a constant current of 240 mA. The XAFS data were recorded under fluorescence mode with 32-element Ge solid-state detector. The energy was calibrated accordingly to the absorption edge of pure Au foil. Athena codes were used to extract the data. The experimental absorption coefficients as function of energies $\mu(E)$ were processed by background subtraction and normalization procedures, and reported as "normalized absorption". Based on the normalized XANES profiles, the oxidation state of Au in each catalyst can be determined.

Transmission electron microscopy (TEM) was conducted on a field emission TEM (JEOL 2100F, Tokyo, Japan) machine equipped with a 2k × 2k CCD camera at 200 kV. High-resolution TEM (HRTEM) and the related energy dispersive X-ray analysis (EDAX) was carried out on a Philips Tecnai G^2 F20 instrument (FEI Company, Hillsboro, TX, USA) at 200 kV. All the tested sample powders were suspended in ethanol before deposition on an ultra-thin carbon film-coated copper grid.

Temperature-programmed reduction by hydrogen (H_2-TPR) was carried out in a Builder PCSA-1000 instrument (Beijing, China) equipped with a thermal conductivity detector (TCD) to detect H_2 consumption. The sieved catalysts (20–40 mesh, 30 mg) were heated (5 °C·min^{-1}) from room temperature to 400 °C in a 20% H_2/Ar (30 mL·min^{-1}) gas mixture. Before the measurements were taken, the fresh samples were pretreated in pure O_2 at 300 °C for 30 min.

3.3. Catalytic Test

CO oxidation activities of Au/FeO$_x$ catalysts were evaluated in a plug flow reactor using 50 mg of sieved (20–40 mesh) catalyst in a gas mixture of 1 vol. % CO, 20 vol. % O_2, and 79 vol. % N_2 (99.997% purity from Deyang Gas Company, Jinan, China), at a flow rate of 67 mL·min^{-1}, corresponding to a space velocity of 80,000 mL·h^{-1}·g_{cat}^{-1}. Prior to the test, the catalysts were pretreated in air at 300 °C for 30 min for activation. After the catalysts cooled down to room temperature under a flow of pure N_2 gas, reactant gases were passed through the reactor. The catalytic tests in "transient" mode were carried out in the reactant atmosphere by ramping the catalyst temperature (5 °C·min^{-1}) from room temperature to 300 °C. The outlet gas compositions of CO and CO_2 were monitored online by a non-dispersive IR spectroscopy (Gasboard-3500, Wuhan Sifang Company, Wuhan, China). A typical "steady-state" experiment (30 °C) was conducted in the same gas-mixture at 30 °C for more than 10 h.

4. Conclusions

In summary, we have prepared two series of gold-iron oxide catalysts, Au/Fe_OH and Au/Fe_O, by depositing Au onto hydrated and dehydrated supports, and further investigated their catalytic performance for the low-temperature CO oxidation reaction. Based on the related activity test results, we have demonstrated that the precipitating pH value and the calcination temperature for the iron oxide support are two key factors governing the gold catalysis. By multiple characterization techniques, including XRD, N_2 adsorption, TEM/HRTEM, XPS, XAFS and H_2-TPR, we have found that the metallic Au strongly interacting with the oxide support is the active site for CO oxidation.

Acknowledgments: Financial support from the National Science Foundation of China (NSFC) (grant Nos. 21373259 and 21301107), the Hundred Talents project of the Chinese Academy of Sciences, the Strategic Priority Research Program of the Chinese Academy of Sciences (grant No. XDA09030102), fundamental research funding of Shandong University (grant No. 2014JC005), the Taishan Scholar project of Shandong Province (China), and open funding from Beijing National Laboratory for Molecular Science and Key Laboratory of Interfacial Physics and Technology, Chinese Academy of Sciences are greatly appreciated.

Author Contributions: C.-J.J. and R.S. conceived and designed the project. H.-Z.C. and Y.G. performed the catalyst preparation, characterization and catalytic tests. X.W. conducted the XAFS and HRTEM measurements. R.S., C.-J.J. and H.-Z.C. wrote the manuscript.

Conflicts of Interest: The authors declare no conflict of interest.

References

1. Haruta, M.; Tsubota, S.; Kobayashi, T.; Kageyama, H.; Genet, M.J.; Delmon, B. Low-temperature oxidation of CO over gold supported on TiO_2, α-Fe_2O_3, and Co_3O_4. *J. Catal.* **1993**, *144*, 175–192.
2. Haruta, M.; Kobayashi, T.; Sano, H.; Yamada, N. Novel gold catalysts for the oxidation of carbon monoxide at a temperature far below 0 °C. *Chem. Lett.* **1987**, *16*, 405–408.
3. Hashmi, A.S.K.; Hutchings, G.J. Gold catalysis. *Angew. Chem. Int. Ed.* **2006**, *45*, 7896–7936.
4. Valden, M.; Lai, X.; Goodman, D.W. Onset of catalytic activity of gold clusters on titania with the appearance of nonmetallic properties. *Science* **1998**, *281*, 1647–1650.
5. Widmann, D.; Leppelt, R.; Behm, R.J. Activation of an Au/CeO_2 catalyst for the CO oxidation reaction by surface oxygen removal/oxygen vacancy formation. *J. Catal.* **2007**, *251*, 437–442.

6.   Camellone, M.F.; Fabris, S. Reaction mechanisms for the CO oxidation on Au/CeO_2 catalysts: Activity of substitutional Au^{3+}/Au^+ cations and deactivation of supported Au^+ Adatoms. *J. Am. Chem. Soc.* **2009**, *131*, 10473–10483.

7.   Haruta, M.; Yamada, N.; Kobayashi, T.; Iijima, S. Gold catalysts prepared by coprecipitation for low-temperature oxidation of hydrogen and of carbon monoxide. *J. Catal.* **1989**, *115*, 301–309.

8.   Herzing, A.A.; Kiely, C.J.; Carley, A.F.; Landon, P.; Hutchings, G.J. Identification of active gold nanoclusters on iron oxide supports for CO oxidation. *Science* **2008**, *321*, 1331–1335.

9.   Daniells, S.T.; Overweg, A.R.; Makkee, M.; Moulijn, J.A. The mechanism of low-temperature CO oxidation with Au/Fe_2O_3 catalysts: A combined Mössbauer, FT-IR, and TAP reactor study. *J. Catal.* **2005**, *230*, 52–65.

10.   Hutchings, G.J.; Hall, M.S.; Carley, A.F.; Landon, P.; Solsona, B.E.; Kiely, C.J.; Herzing, A.; Makkee, M.; Moulijn, J.A.; Overweg, A.; *et al.* Role of gold cations in the oxidation of carbon monoxide catalyzed by iron oxide-supported gold. *J. Catal.* **2006**, *242*, 71–81.

11.   Li, L.; Wang, A.-Q.; Qiao, B.-T.; Lin, J.; Huang, Y.-Q.; Wang, X.-D.; Zhang, T. Origin of the high activity of Au/FeO_x for low-temperature CO oxidation: Direct evidence for a redox mechanism. *J. Catal.* **2013**, *299*, 90–100.

12.   Liu, Y.; Jia, C.-J.; Yamasaki, J.; Terasaki, O.; Schuth, F. Highly active iron oxide supported gold catalysts for CO oxidation: How small must the gold nanoparticles be? *Angew. Chem. Int. Ed.* **2010**, *49*, 5771–5775.

13.   Zhong, Z.-Y.; Ho, J.; Teo, J.; Shen, S.-C.; Gedanken, A. Synthesis of porous α-Fe_2O_3 nanorods and deposition of very small gold particles in the pores for catalytic oxidation of CO. *Chem. Mater.* **2007**, *19*, 4776–4782.

14.   Andreeva, D. Low temperature water gas shift over gold catalysts. *Good Bull.* **2002**, *35*, 82–88.

15.   Boccuzzi, F.; Chiorino, A.; Manzoli, M.; Andreeva, D.; Tabakova, T. FTIR study of the low-temperature water-gas shift reaction on Au/Fe_2O_3 and Au/TiO_2 catalysts. *J. Catal.* **1999**, *188*, 176–185.

16.   Silberova, B.A.A.; Mul, G.; Makkee, M.; Moulijn, J.A. DRIFTS study of the water-gas shift reaction over Au/Fe_2O_3. *J. Catal.* **2006**, *243*, 171–182.

17.   Andreevaa, D.; Tabakovaa, T.; Idakieva, V.; Christova, P.; Giovanolib, R. Au/α-Fe_2O_3 catalyst for water-gas shift reaction prepared by deposition-precipitation. *Appl. Catal. A* **1998**, *169*, 9–14.

18.   Schubert, M.M.; Venugopal, A.; Kahlich, M.J.; Plzak, V.; Behm, R.J. Influence of H_2O and CO_2 on the selective CO oxidation in H_2-rich gases over Au/α-Fe_2O_3. *J. Catal.* **2004**, *222*, 32–40.

19.   Shodiya, T.; Schmidt, O.; Peng, W.; Hotz, N. Novel nano-scale Au/α-Fe_2O_3 catalyst for the preferential oxidation of CO in biofuel reformate gas. *J. Catal.* **2013**, *300*, 63–69.

20.   Landon, P.; Ferguson, J.; Solsona, B.E.; Garcia, T.; Sayari, S.A.; Carley, A.F.; Herzing, A.A.; Kiely, C.J.; Makkee, M.; Moulijn, J.A.; *et al.* Selective oxidation of CO in the presence of H_2, H_2O and CO_2 utilizing Au/α-Fe_2O_3 catalysts for use in fuel cells. *J. Mater. Chem.* **2006**, *16*, 199–208.

211

21. Ozaki, M.; Kratohvil, S.; Matijević, E. Formation of monodispersed spindle-type hematite particles. *J. Colloid Interface Sci.* **1984**, *102*, 146–151.

22. Sugimoto, T.; Sakata, K. Preparation of monodisperse pseudocubic α-Fe_2O_3 particles from condensed ferric hydroxide gel. *J. Colloid Interface Sci.* **1992**, *152*, 587–590.

23. Jia, C.-J.; Sun, L.-D.; Yan, Z.-G.; You, L.-P.; Luo, F.; Han, X.-D.; Pang, Y.-C.; Zhang, Z.; Yan, C.-H. Single-crystalline iron oxide nanotubes. *Angew. Chem. Int. Ed.* **2005**, *117*, 4402–4407.

24. Jia, C.-J.; Sun, L.-D.; Luo, F.; Han, X.-D.; Heyderman, L.J.; Yan, Z.-G.; Yan, C.-H.; Zheng, K.; Zhang, Z.; Takano, M.; *et al.* Large-scale synthesis of single-crystalline iron oxide magnetic nanorings. *J. Am. Chem. Soc.* **2008**, *130*, 16968–16977.

25. Finch, R.M.; Hodge, N.A.; Hutchings, G.J.; Meagher, A.; Pankhurst, Q.A.; Siddiqui, M.R.H.; Wagnerc, F.E.; Whyman, R. Identification of active phases in Au-Fe catalysts for low-temperature CO oxidation. *Phys. Chem. Chem. Phys.* **1999**, *1*, 485–489.

26. Qian, K.; Zhang, W.-H.; Sun, H.-X.; Fang, J.; He, B.; Ma, Y.-S.; Jiang, Z.-Q.; Wei, S.-Q.; Yang, J.-L.; Huang, W.-X. Hydroxyls-induced oxygen activation on "inert" Au nanoparticles for low-temperature CO oxidation. *J. Catal.* **2011**, *277*, 95–103.

27. Hodge, N.A.; Kiely, C.J.; Whyman, R.; Siddiqui, M.R.H.; Hutchings, G.J.; Pankhurst, Q.A.; Wagner, F.E.; Rajaram, R.R.; Golunski, S.E. Microstructural comparison of calcined and uncalcined gold/iron-oxide catalysts for low-temperature CO oxidation. *Catal. Today* **2002**, *72*, 133–144.

28. Guo, Y.; Gu, D.; Jin, Z.; Du, P.-P.; Si, R.; Tao, J.; Xu, W.-Q.; Huang, Y.-Y.; Senanayake, S.; Song, Q.-S.; *et al.* Uniform 2 nm gold nanoparticles supported on iron oxides as active catalysts for CO oxidation reaction: Structure-activity relationship. *Nanoscale* **2015**, *7*, 4920–4928.

29. Venugopal1, A.; Scurrell, M.S. Low temperature reductive pretreatment of Au/Fe_2O_3 catalysts, TPR/TPO studies and behaviour in the water-gas shift reaction. *Appl. Catal. A* **2004**, *258*, 241–249.

30. Wang, D.-H.; Hao, Z.-P.; Cheng, D.-Y.; Shi, X.-C. Influence of the calcination temperature on the $Au/FeO_x/Al_2O_3$ catalyst. *J. Chem. Technol. Biotechnol.* **2006**, *81*, 1246–1251.

31. Deng, W.-L.; Carpenter, C.; Yia, N.; Stephanopoulos, M.F. Comparison of the activity of Au/CeO_2 and Au/Fe_2O_3 catalysts for the CO oxidation and the water-gas shift reactions. *Top. Catal.* **2007**, *44*, 199–208.

32. Kadkhodayan, A.; Brenner, A. Temperature-programmed reduction and oxidation of metals supported on γ-alumina. *J. Catal.* **1989**, *117*, 311–321.

33. Huang, J.; Dai, W.-L.; Fan, K.-N. Remarkable support crystal phase effect in Au/FeO_x catalyzed oxidation of 1,4-butanediol to c-butyrolactone. *J. Catal.* **2009**, *266*, 228–235.

34. Yamashita, T.; Hayes, P. Analysis of XPS spectra of Fe^{2+} and Fe^{3+} ions in oxide materials. *Appl. Surf. Sci.* **2008**, *254*, 2441–2449.

35. Deng, W.-L.; Frenkel, A.I.; Si, R.; Flytzani-Stephanopoulos, M. Reaction-Relevant Gold Structures in the Low Temperature Water-Gas Shift Reaction on $Au-CeO_2$. *J. Phys. Chem. C* **2008**, *112*, 12834–12840.

36. Zanella, R.; Giorgio, S.; Shin, C.H.; Henry, C.R.; Louis, C. Characterization and reactivity in CO oxidation of gold nanoparticles supported on TiO_2 prepared by deposition-precipitation with NaOH and urea. *J. Catal.* **2004**, *222*, 357–367.

Competition of CO and H_2 for Active Oxygen Species during the Preferential CO Oxidation (PROX) on Au/TiO$_2$ Catalysts

Yeusy Hartadi, R. Jürgen Behm and Daniel Widmann

Abstract: Aiming at an improved mechanistic understanding of the preferential oxidation of CO on supported Au catalysts, we have investigated the competition between CO and H_2 for stable, active oxygen (O_{act}) species on a Au/TiO$_2$ catalyst during the simultaneous exposure to CO and H_2 with various CO/H_2 ratios at 80 °C and 400 °C by quantitative temporal analysis of products (TAP) reactor measurements. It is demonstrated that, at both higher and lower temperature, the maximum amount of active oxygen removal is (i) independent of the CO/H_2 ratio and (ii) identical to the amount of active oxygen removal by CO or H_2 alone. Hence, under preferential CO oxidation (PROX) reaction conditions, in the simultaneous presence of CO and H_2, CO and H_2 compete for the same active oxygen species. In addition, also the dependency of the selectivity towards CO oxidation on the CO/H_2 ratio was evaluated from these measurements. Consequences of these findings on the mechanistic understanding of the PROX reaction on Au/TiO$_2$ will be discussed.

Reprinted from *Catalysts*. Cite as: Hartadi, Y.; Behm, R.J.; Widmann, D. Competition of CO and H_2 for Active Oxygen Species during the Preferential CO Oxidation (PROX) on Au/TiO$_2$ Catalysts. *Catalysts* **2016**, *6*, 21.

1. Introduction

Gold nanoparticles supported on a variety of metal oxides are well known for their exceptional high activity and selectivity for various oxidation and reduction reactions already at very low temperatures [1–4]. One of the most prominent examples is the CO oxidation reaction on supported Au catalysts, e.g., on Au/TiO$_2$, since it also serves as a test reaction for gas phase oxidation catalysis in heterogeneous Au catalysis in general [5]. Practical applications of monometallic supported Au catalysts are rare, so far, but they are claimed to be promising candidates for the selective removal of CO from H_2-rich reformates to values below 10 ppm [6–8] for use in low temperature Polymer Electrolyte Membrane (PEM) fuel cells [6,9]. The ultra-purification of H_2-rich fuels via the selective CO oxidation (or preferential CO oxidation—PROX) reaction requires catalysts that are highly active for the CO oxidation (Equation (1)). At the same time, the catalysts should have a very low activity for the continuous oxidation of H_2 to water (Equation (2)), although H_2

reaches concentrations up to 75% (depending on the process of H_2 generation) and is, hence, present in large excess compared to CO in the reaction gas [6,9–11]. One should note, however, that an oxidation of H_2 to adsorbed hydroperoxy-like species, which have previously been proposed to represent reaction intermediates in the preferential CO oxidation in the presence of H_2, may even be beneficial for the PROX activity of supported Au catalysts [10,12].

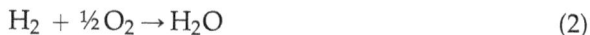

$$CO + \tfrac{1}{2}O_2 \rightarrow CO_2 \tag{1}$$

$$H_2 + \tfrac{1}{2}O_2 \rightarrow H_2O \tag{2}$$

In previous studies it has been demonstrated that metal oxide-supported Au nanoparticles (NPs) are highly selective for CO oxidation in the presence of hydrogen [9–11,13–15], where the selectivity is defined by the ratio between the oxygen consumption for CO oxidation, which equals half of the CO consumption according to the reaction stoichiometry (Equation (1)), and the overall oxygen consumption (for CO and H_2 oxidation), or the ratio between CO_2 formation and the sum of CO_2 and H_2O formation (Equation (3)):

$$S = \frac{CO\ consumption}{2 \times O_2\ consumption} = \frac{CO_2}{CO_2 + H_2O} \tag{3}$$

Despite the large excess of H_2, typically about 1% CO and up to 75% H_2 in previous studies, the selectivity of metal oxide supported Au catalysts towards CO oxidation is considerably higher than 50% under typical reaction conditions, at a reaction temperature of 80 °C [13–15], but decreases with decreasing CO content in the reaction gas atmosphere [10,13]. Considering the very low CO concentrations that can be tolerated in PEM fuel cells (<10 ppm), further improvements of existing catalysts, in particular their selectivity, is still mandatory.

Most crucial in the CO oxidation reaction on supported Au catalysts, and also most controversially debated, is the activation of molecular oxygen, which includes issues such as the active site for oxygen activation and the resulting active oxygen species for the oxidation of adsorbed CO [16]. For reaction in a CO/H_2 mixture as in the PROX reaction, this also includes the question whether CO and H_2 compete for the same adsorption sites and/or the same active oxygen species. These questions are topic of the present work, where we investigated the selective oxidation of CO in CO/H_2 mixtures on a Au/TiO$_2$ catalyst by quantitative temporal analysis of products (TAP) reactor measurements.

Before presenting the results of the present study, we will briefly summarize previous findings relevant for the present work. It is well known that CO mainly adsorbs on the surface of the Au nanoparticles under relevant reaction conditions, at room temperature and above [17–19], and also dissociatively adsorbed hydrogen is assumed to be located on these sites [20,21] or at Au atoms located at the perimeter of the Au-TiO$_2$ interface [22,23], resulting in the formation of a highly active, atomically-adsorbed hydrogen species [20,22–24]. Furthermore, there seems to be increasing agreement that CO adsorbed on the Au NPs mainly reacts with active oxygen species located at the perimeter sites of the Au-M$_x$O$_y$ interface in the dominant reaction pathway for CO oxidation [19,25,26], and that for Au catalysts supported on reducible metal oxides, e.g., for Au/TiO$_2$, and for reaction temperatures above room temperature, surface lattice oxygen of the support represents the active oxygen species [27–33]. Based on TAP reactor measurements, we recently proposed a Au-assisted Mars-van Krevelen mechanism as the dominant reaction pathway for the CO oxidation at T $\geqslant$ 80 °C on Au/TiO$_2$, where TiO$_2$ surface lattice oxygen located at the interface between Au nanoparticles and TiO$_2$ support represents the active oxygen species for CO oxidation, and is continuously removed and replenished during the reaction [27,31]. At lower temperatures, (below -120 °C), the dominant reaction pathway is expected to change and molecularly adsorbed oxygen is proposed to represent the active oxygen species [19,34]. Furthermore CO adsorption on the TiO$_2$ support becomes increasingly important for the overall reaction [19,34]. Based on similar type TAP reactor measurements, specifically from the identical amount of active oxygen species for the CO oxidation and the H$_2$ oxidation, and from their almost identical dependency on the Au particle size, we recently proposed that hydrogen oxidation at 80 °C and higher also proceeds via a Au-assisted Mars-van Krevelen mechanism, identical to CO oxidation, with surface lattice oxygen located at the perimeter of the Au-TiO$_2$ interface acting as active oxygen species [35]. This conclusion was supported recently by density functional theory (DFT) based calculations, where the lowest H$_2$ dissociation barrier was obtained for H$_2$ interacting with O^{2-} surface lattice oxygen species in TiO$_2$ close to the Au/TiO$_2$ perimeter [36]. We had speculated already from those TAP results that CO and H$_2$ compete for this O$_{act}$ species when present simultaneously, and that this competition is a crucial factor for the catalysts selectivity towards CO oxidation [35]. Experimental proof for this mechanism, in the simultaneous presence of CO and H$_2$, is still lacking. Here one has to consider that under PROX reaction conditions the formation of hydroxyl groups and adsorbed water on the catalysts surface upon interaction with H$_2$ may significantly influence its CO oxidation activity [14,37–40].

In the following we report results of TAP reactor measurements, where the absolute amount of active oxygen removal by CO and H$_2$ and its replenishment by O$_2$ was evaluated upon exposure of a Au/TiO$_2$ catalyst to alternate sequences of CO/H$_2$/Ar and O$_2$/Ar pulses, using different CO/H$_2$ mixtures. From these data we will derive whether CO and H$_2$ compete for the same active oxygen species also in the simultaneous presence of CO and H$_2$. Moreover, the selectivity for CO oxidation during mixed CO/H$_2$ pulses can be determined. Finally we will discuss the consequences of the present finding for the mechanism of the PROX reaction on Au/TiO$_2$ catalysts.

2. Results and Discussions

2.1. Active Oxygen Removal by CO and H$_2$

Before focusing on the reaction with mixed CO/H$_2$/Ar pulses, we determined the removal of stable, active oxygen from the Au/TiO$_2$ catalyst surface by CO/Ar and H$_2$/Ar pulses only for calibration. The absolute amounts of active oxygen determined in these measurements will then be used as references for the active oxygen removal in the simultaneous presence of CO and H$_2$. Similar to our previous study [35], the catalyst was alternately exposed to several sequences of (i) CO/Ar and O$_2$/Ar or (ii) H$_2$/Ar and O$_2$/Ar, respectively, in order to reactively remove and replenish the available active oxygen from the catalyst surface. These measurements were performed after *in situ* calcination (O400) of the Au/TiO$_2$ catalyst and subsequent saturation of its surface by adsorbed water by exposure to 5000 H$_2$/Ar pulses at 80 °C. Additional experiments including the exposure of the catalyst to various amounts of H$_2$/Ar pulses and subsequent temperature programmed desorption (TPD) measurements revealed that (i) there is formation of stable adsorbed water upon exposure to H$_2$/Ar pulses at 80 °C and (ii) treatment with 5000 H$_2$/Ar pulses results in a surface saturation and, hence, equal surface concentrations of adsorbed water during all subsequent experiments. Note that, at 400 °C, in contrast, water formed during the H$_2$ oxidation readily desorbs, and does not need to be considered in the following measurements.

The multi-pulse experiments following this treatment (O400 and H$_2$ pulses) revealed an oxygen consumption of the fully-reduced catalyst during O$_2$/Ar pulses at 80 °C of 3.3×10^{18} O atoms·$\mathrm{g_{cat}}^{-1}$. This is almost identical to the CO consumption/CO$_2$ formation in the subsequent sequence of CO/Ar pulses (3.4×10^{18} molecules·$\mathrm{g_{cat}}^{-1}$), as well as to the oxygen consumption during subsequent O$_2$/Ar pulses (after 250 CO/Ar pulses). Hence, this (partial) reduction and re-oxidation of the catalyst surface is fully reversible under present conditions. Furthermore, in agreement with previous findings, the amount of CO$_2$ formation in each CO/Ar sequence always equals the amount of CO consumed, indicating that there is no significant build-up of stable adsorbed, carbon containing surface species (surface carbonates, *etc.*) under present reaction conditions [25,41]. Finally, all CO$_2$ is formed during CO pulses (and not during O$_2$ pulses), as expected for CO weakly adsorbed on the Au NPs at 80 °C, and completely desorbed before O$_2$ is introduced [31].

The oxygen storage capacity (OSC) at 80 °C is slightly lower in the present study (3.3×10^{18} O atoms·$\mathrm{g_{cat}}^{-1}$) than determined previously for a comparable

Au/TiO$_2$ catalyst, which was also saturated with adsorbed water by H$_2$ pulses prior to the pulse experiments (4.6 × 10^{18} O atoms· g$_{cat}$$^{-1}$) [35], but of a similar order of magnitude. More important for the present study, however, is the comparison between the absolute amounts of active oxygen on (fully oxidized) Au/TiO$_2$ for the CO oxidation and the H$_2$ oxidation. From the oxygen uptake during O$_2$/Ar pulses after 1500 H$_2$/Ar pulses on the Au/TiO$_2$ catalyst at 80 °C we obtained a value of 3.3 × 10^{18} O atoms· g$_{cat}$$^{-1}$ for the total amount of stable oxygen which is active for the H$_2$ oxidation (see Table 1). Note that compared to CO/Ar a significantly higher number of H$_2$/Ar pulses is necessary to completely remove the active surface oxygen. While for a complete removal of all availabe O$_{act}$ species via the CO oxidation about 200 CO/Ar pulses were needed, almost 1500 H$_2$/Ar pulses are necessary for its removal via the H$_2$ oxidation. This reflects the much lower efficiency of hydrogen for active oxygen removal compared to CO, as was already demonstrated previously [31]. The total amounts of active oxygen for CO and H$_2$ oxidation, however, are almost identical also for the present Au/TiO$_2$ catalyst [31].

Table 1. Absolute amounts of active stable oxygen (O$_{act}$) removal and CO$_2$ formation during CO, H$_2$, and CO/H$_2$ pulses with different CO/H$_2$ ratios on the fully oxidized Au/TiO$_2$ catalyst at 80 °C, and corresponding selectivities for CO oxidation.

Ratio CO/H$_2$	H$_2$/(CO + H$_2$)/%	O$_{act}$ Removal */ 10^{18} O atoms· g$_{cat}$$^{-1}$	CO$_2$ Formation */ 10^{18} molecules· g$_{cat}$$^{-1}$	Selectivity/%
CO only	0	3.3 ± 0.2	3.4 ± 0.2	100
1/0.5	33	2.9 ± 0.2	2.6 ± 0.2	92
1/1	50	2.6 ± 0.2	2.3 ± 0.2	90
1/2	67	3.1 ± 0.2	2.2 ± 0.1	71
1/4	80	3.3 ± 0.2	2.0 ± 0.1	60
1/8	89	3.1 ± 0.2	1.5 ± 0.1	49
H$_2$ only	100	3.3 ± 0.2	0	0

* Error estimations are based on the repeated reproduction of individual pulse sequences.

Additionally, for reaction at 400 °C, the total amounts of active oxygen for CO oxidation and for H$_2$ oxidation are also almost identical (12.7 × 10^{18} O atoms· g$_{cat}$$^{-1}$ and 12.6 × 10^{18} O atoms· g$_{cat}$$^{-1}$, respectively; see Table 2). The higher amount of O$_{act}$ species available and, accordingly, a higher amount of TiO$_2$ surface lattice oxygen removal at 400 °C is in agreement with findings in previous studies, where this was explained by the migration of TiO$_2$ surface lattice oxygen to Au-TiO$_2$ perimeter sites at elevated temperatures [27]. This enables the removal also of O$_{act}$ species which were originally located further away from the Au-TiO$_2$ interface perimeter. At 80 °C, in contrast, the mobility of surface lattice oxygen is negligible, and only surface lattice oxygen species located directly at Au-TiO$_2$ interface perimeter sites can be removed by reaction with CO$_{ad}$ or H$_{ad}$ [25,35]. Considering the Au loading and the Au particle size distribution (from TEM imaging), and assuming hemispherical Au nanoparticles, the total amount of these species can be estimated to be 6.2×10^{18} O atoms$\cdot g_{cat}^{-1}$.

Hence, the data are in full agreement with our previous proposal that CO and hydrogen are both oxidized by the same active oxygen species at both 80 °C and 400 °C, by TiO_2 surface lattice oxygen close to Au-TiO_2 perimeter sites [27]. For more details see Ref. [35].

Table 2. Absolute amounts of active stable oxygen (O_{act}) removal and CO_2 formation during CO, H_2, and CO/H_2 pulses with different CO/H_2 ratios on the fully oxidized Au/TiO_2 catalyst at 400 °C.

Ratio CO/H_2	H_2/(CO+H_2)/%	O_{act} Removal */ 10^{18} O atoms$\cdot g_{cat}^{-1}$	CO_2 Formation */ 10^{18} molecules$\cdot g_{cat}^{-1}$
CO only	0	12.7 ± 0.6	10.0 ± 0.5
1/0.5	33	11.5 ± 0.6	9.3 ± 0.5
1/1	50	11.5 ± 0.6	8.8 ± 0.4
1/2	67	12.9 ± 0.6	8.7 ± 0.4
1/4	80	12.9 ± 0.6	7.7 ± 0.3
1/8	89	11.8 ± 0.6	6.1 ± 0.3
H_2 only	100	12.6 ± 0.6	0

* Error estimations are based on the repeated reproduction of individual pulse sequences.

2.2. Active Oxygen Removal by CO/H_2 Mixtures

Similar multi-pulse experiments were performed with alternate sequences of O_2/Ar and CO/H_2/Ar pulses with various CO/H_2 ratios. At both temperatures (80 °C and 400 °C) five different CO/H_2 mixtures were investigated, with CO/H_2 ratios ranging from 1/0.5 to 1/8 (see also Tables 1 and 2). The trends in the pulse signals of CO, CO_2, H_2, H_2O, and O_2 during reduction and re-oxidation with CO/H_2/Ar pulses (CO/H_2 = 1/2) and subsequent O_2/Ar pulses, respectively, are illustrated in Figure 1 for reaction at 80 °C on Au/TiO_2. Only the first 50 pulses of each sequence are shown, where differences in the measured intensities due to consumption of educts (CO, H_2, and O_2)/formation of products (CO_2, H_2O) are visible in the raw data, while the actual number of pulses per sequence are significantly higher (see Section 3.2). For CO/H_2/Ar and O_2/Ar at 80 °C the number of pulses in each sequence was 250 and 150, respectively.

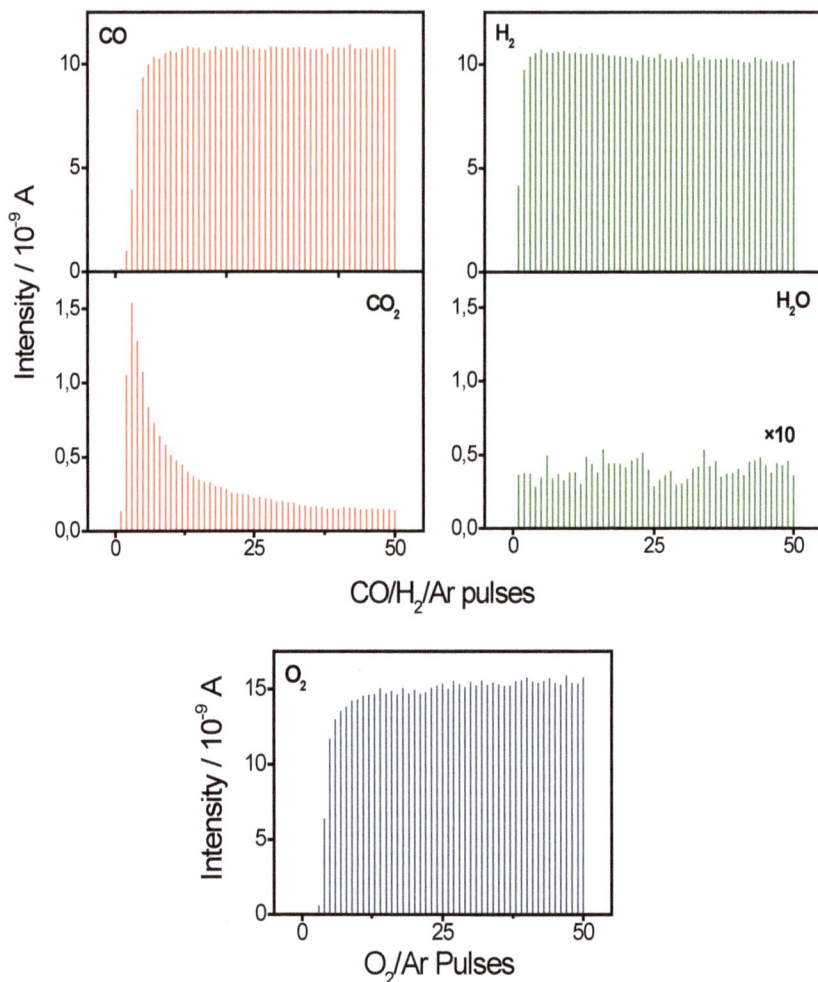

Figure 1. Mass spectrometer signals of CO, CO_2, H_2, and H_2O measured during a sequence of $CO/H_2/Ar$ pulses (CO/H_2 ratio = 1/2) on Au/TiO_2 at 80 °C after oxidation by O_2/Ar pulses, as well as the O_2 signal measured during subsequent O_2/Ar pulses at 80 °C.

During the first 10 $CO/H_2/Ar$ pulses on the fully oxidized Au/TiO_2 catalyst there is a significant consumption of CO, indicated by the much lower CO intensity compared to that after about 50 or more pulses, and in parallel the formation of CO_2 (Figure 1). Hence, also under these conditions CO oxidation proceeds by reaction with stable surface oxygen of the catalyst. With increasing number of pulses the CO consumption/CO_2 formation during each pulse decreases, until there is no visible discrepancy to the steady-state situation after about 10 pulses. The absolute amount of CO consumed/CO_2 formed during the 250 CO/H_2/Ar pulses in this sequence at 80 °C was 2.2×10^{18} molecules· g_{cat}^{-1}. The significantly lower amount of CO_2 formation during the exposure of the fully oxidized Au/TiO_2 catalyst to CO/H_2 pulses as compared to CO pulsing (3.3×10^{18} molecules· g_{cat}^{-1}, see above) indicates already that stable adsorbed active oxygen is not only consumed by reaction with CO, but also by reaction with hydrogen. From the H_2 signal shown in Figure 1, however, no (or only very little) H_2 consumption can be detected, in contrast to the CO signal. A somewhat lower H_2 signal is observed only during the first 1–2 pulses, with the lower intensity measured during the first pulse originating from a generally lower overall pulse size for the first pulse in every sequence (an artifact of our TAP reactor system). Hence, from the raw data there is no indication for reduction of Au/TiO_2 by H_2 during these pulses, and also the quantitative evaluation of all CO/H_2/Ar pulses could not resolve significant H_2 consumption. From previous TAP reactor studies on Au/TiO_2 as well as on Au/CeO_2 it is already known that the efficiency of H_2 for active oxygen removal is considerably lower compared to CO [35,42]. The H_2 consumption during a single pulse is, accordingly, too low to be resolved in these measurements, and the same is true for the formation of water during exposure to H_2 or CO/H_2 pulses at 80 °C [35]. As described in Section 3.2 (see below), the consumption of active oxygen by H_2 can, nevertheless, be calculated from the total oxygen uptake during subsequent O_2/Ar pulses, from the more than stoichiometric consumption of O_2 compared to the preceding CO_2 formation during two consecutive runs at 80 °C.

The O_2 signal measured during re-oxidation of Au/TiO_2 by O_2/Ar pulses, after reduction of the catalyst by CO and H_2, is also shown in Figure 1 (lower panel). Similar to CO consumption during CO/H_2/Ar pulses, O_2 consumption during the first 10–15 O_2/Ar pulses is obvious already from the raw data by the lower intensity compared to that after saturation. Quantitative evaluation of the O_2 consumption during the whole sequence (150 O_2/Ar pulses) revealed that 3.1×10^{18} O atoms· g_{cat}^{-1} are needed to fully re-oxidize the catalyst at 80 °C (see also Table 1). This value is almost identical to the one determined after reduction by CO or H_2 only (3.3×10^{18} O atoms· g_{cat}^{-1} in both cases), but significantly higher than the CO_2 formation during the preceding CO/H_2/Ar pulses (2.2×10^{18} molecules· g_{cat}^{-1}). Accordingly, the amount of active oxygen removed by reaction with hydrogen amounts to 0.9×10^{18} O atoms· g_{cat}^{-1}. These results clearly demonstrate that under present reaction conditions, with CO and hydrogen being simultaneously present on the surface, these species compete for the same active oxygen species. The higher amount of active oxygen removal by CO, in spite of the considerable excess of H_2 during the CO/H_2/Ar pulses (CO/H_2 ratio of 1/2) illustrates the high selectivity of Au/TiO_2 for the CO oxidation (71% in this case) under present reaction conditions.

The results of additional measurements with different CO/H_2 ratios during the $CO/H_2/Ar$ pulses, where the amount of CO molecules per pulse was kept constant at 3×10^{15} molecules CO per pulse, while the amount of H_2 was varied between 1.5×10^{15} and 24×10^{15} molecules H_2 per pulse, are plotted in Figure 2a and listed in Table 1, together with the absolute amounts of active oxygen consumption/replenishment for CO only (0% H_2) and H_2 only (100% H_2).

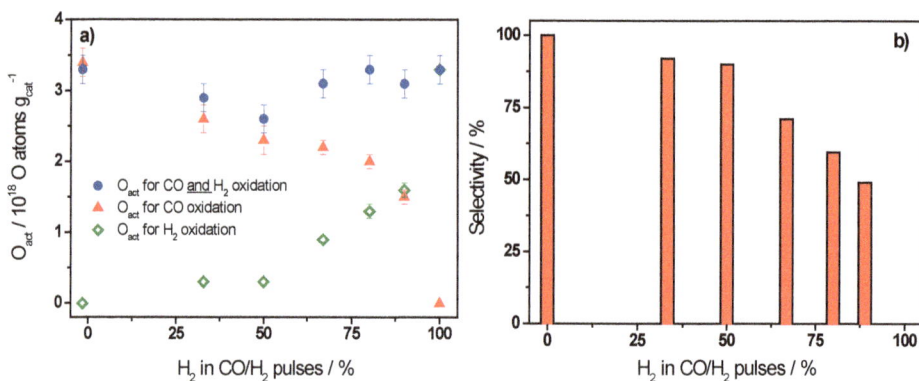

Figure 2. (a) Absolute amounts of active oxygen (O_{act}) removal from a fully oxidized Au/TiO_2 catalyst (after O_2/Ar pulses) by CO and/or H_2 during CO, H_2, and CO/H_2 pulses with varying ratios of H_2 (33%–89% H_2) at 80 °C, and (b) corresponding selectivities for CO oxidation/CO_2 formation during the $CO/H_2/Ar$ pulses at 80 °C.

These data clearly demonstrate that the total amount of oxygen that can be removed by CO/H_2 is almost identical for all mixtures and, moreover, also for "CO only" and "H_2 only" experiments. Obviously, under present reaction conditions, in the simultaneous presence of CO and H_2, these reactants are oxidized by the same active oxygen species as we have already concluded for the individual reactants before (see Section 2.1 and [35]). The existence of other active oxygen species, which may be active only for H_2 oxidation when CO and H_2 are present simultaneously, can be ruled out, since this would result in a higher amount of active oxygen species for the simultaneous CO and H_2 oxidation as compared to the CO oxidation only.

On a first glance these findings seem to contradict recent conclusions from Saavedra et al., who studied, in detail, the role of water in the catalytic CO oxidation on Au/TiO_2 [40]. Based on experimental results and DFT calculations they concluded (i) that weakly adsorbed water on TiO_2 determines the catalysts activity at room temperature by changing the effective number of active sites; and (ii) that O_2 activation in the dominant reaction pathway under these conditions requires the presence of both water and CO [40]. Considering that our experiments were performed at 80 °C and under vacuum conditions (not during pulses, but between individual pulses and pulse sequences), the presence of weakly-adsorbed water on the catalyst surface can be excluded. Temperature programmed desorption (TPD) measurements performed after saturation of the catalyst with water by 1500 H_2/Ar pulses had shown that stable adsorbed water is formed upon H_2 pulsing, which desorbs with a maximum in the TPD spectra at about 230 °C. This is the only water species present during O_2 pulsing at 80 °C. Hence, our results clearly demonstrate that as long as only stable adsorbed water is present on the catalyst surface, coexisting adsorbed CO and hydrogen both are oxidized by TiO_2 surface lattice oxygen and compete for this O_{act} species. Weakly-adsorbed water species, whose presence during continuous PROX reaction cannot be ruled out from our data, may also enable another oxygen activation mechanism as described in the recent study by Saavedra et al. [40]. In that case, however, the dominant reaction mechanisms of CO and H_2 oxidation under PROX reaction conditions are expected to sensitively depend on the reaction temperature, due to the facile desorption of these species at higher temperatures. Similar kinds of temperature effects have been proposed earlier for the CO oxidation reaction under dry reaction conditions [31]. Another possibility that has to be considered under continuous PROX conditions, in the simultaneous presence of CO, O_2, and H_2 in the catalyst bed, is the H-assisted O_2 activation. Rousset and coworkers previously demonstrated that the presence of H_2 promotes the CO oxidation activity over various Au catalysts supported on different (reducible and non-reducible) metal oxides [10,12]. According to them, this beneficial effect of H_2 results most probably from the reaction of adsorbed hydrogen atoms with molecular oxygen to active hydroperoxy-like species that represent additional reaction intermediates in the CO oxidation. In the present approach there is an inherent difficulty in studying this effect, since CO/H_2 and O_2 are never present simultaneously in the catalyst bed during the alternate pulse sequences. Here, further work is planned to more closely investigate the influence of H_2 on the O_2 activation in pulse experiments, in particular by exposing the catalyst to mixed $CO/H_2/O_2/Ar$ pulses. In that approach it will also be highly interesting to compare the oxygen activation in the presence of H_2 on different Au catalysts supported on reducible and non-reducible metal oxides (Au/TiO_2, Au/ZnO, Au/Al_2O_3, and Au/SiO_2, for example). Numerous previous studies have demonstrated that the activity for CO oxidation largely depends on the reducibility of the metal oxide support, with Au catalysts supported on reducible oxides being much more active [43,44]. Applying TAP reactor pulse experiments we could even demonstrate a quantitative relation between the catalysts reducibility (measured by its oxygen storage capacity) and its activity for CO oxidation [41]. In the presence of H_2, however, the activity of Au catalysts supported on reducible and non-reducible metal oxides becomes very similar [10,12], reflecting a more pronounced promotional effect of H_2 on the CO oxidation activity (see above) for non-reducible than for reducible metal oxide supports.

In addition to the absolute amount of active oxygen on Au/TiO_2 we also calculated the selectivity for active oxygen removal by CO at 80 °C for different CO/H_2 ratios, where the selectivity was given by the ratio between the absolute amount of CO_2 formation and the overall oxygen removal during $CO/H_2/Ar$ pulses (see Section 3.2). Figure 2a illustrates already that the amount of CO consumption/CO_2 formation continuously decreases with increasing H_2 content in the CO/H_2 pulses at 80 °C, despite the almost constant amount of O_{act} removal (for combined CO and hydrogen oxidation). This results in a decrease of the selectivity for CO_2 formation from 92% in the presence of 33% H_2 to slightly below 50% in the presence of 89% H_2 (see Figure 2b and Table 1). A similar trend of a decreasing selectivity with decreasing CO/H_2 ratio was also observed in kinetic measurements of the PROX reaction on a Au/TiO_2 catalyst under continuous flow, but otherwise similar reaction conditions (80 °C) [14,45]. The high selectivities fully agree with the much lower efficiency for O_{act} removal by H_2 (including H_2 adsorption, dissociation, and reaction with O_{act}) compared to reaction with CO. This was already demonstrated by the much higher number of H_2 pulses needed for the complete removal of the available active oxygen species compared to reaction with CO in multi-pulse TAP reactor experiments (see Section 2.1 and Refs. [35,42]). Considering the selectivity of about 50% in the presence of 89% H_2 (CO/H_2 ratio of 1/8), this difference in the efficiency for O_{act} removal can be estimated to be about 8. This value is also in close agreement to the approximately eight times higher amount of H_2 molecules compared to CO molecules that are needed for a complete removal of all available O_{act} species (see Section 2.1). One has to keep in mind, however, that the latter averages over the entire pulse sequence, and does not necessarily reflect the difference in efficiency of a single CO or H_2 molecule. The selectivity largely originates from the competition between adsorbed CO and hydrogen for the same active oxygen species when simultaneously present on the catalyst surface. Additional effects arising, e.g., from the competition of the two adsorbing species (CO, H_2) for adsorption sites and from (de-)stabilizing interaction between the coadsorbed species, which had been suggested previously [13,14], can neither be excluded nor supported from our TAP reactor experiments measurements. It should be noted, however, that the steady-state coverages under typical reaction conditions ($\leqslant$0.5% CO, 60%–70% H_2, 80 °C, atmospheric pressure) are well below saturation. Accordingly, such effects are expected to be small, and similar conclusions were put forward also in the above studies [13,14].

Finally, we performed similar multi-pulse experiments with alternate sequences of O_2/Ar pulses and $CO/H_2/Ar$ pulses at 400 °C, using the same reaction gas mixtures for O_{act} removal from Au/TiO_2 as described above (CO, H_2, and

$CO/H_2 = 1/0.5 - CO/H_2 = 1/8$). The main characteristics during reduction and re-oxidation were identical to those obtained in the experiments at 80 °C: For all measurements, CO_2 and H_2O are only formed during CO and/or H_2 pulses, respectively, and the processes of catalyst reduction and re-oxidation are fully reversible, as evidenced by at least three cycles of reduction and re-oxidation with identical absolute amounts of O_{act} removal/replenishment for each gas mixture. The absolute amounts of active oxygen removal upon exposure to CO only, H_2 only, and CO/H_2 mixtures are presented in Figure 3 and listed in Table 2.

Similar to the results at 80 °C also in this case the absolute amount of active oxygen (12×10^{18} O atoms· g_{cat}^{-1}) is essentially independent of the composition of the reaction gas pulses used for reduction. Hence, also at 400 °C CO and H_2 seem to compete for TiO_2 surface lattice oxygen close to the Au nanoparticles when present simultaneously. The only difference to the measurements at lower reaction temperature is that at 400 °C the increasing thermal mobility of surface lattice oxygen and surface oxygen vacancies enables the migration and, therefore, the removal and replenishment of oxygen and oxygen vacancies which were originally located further away from the perimeter sites. At 80 °C, in contrast, only oxygen species directly adjacent to the Au nanoparticles participate in the oxidation reactions [27]. Nevertheless, also at 400 °C removal of O_{act} species upon reaction with CO or H_2 is expected to occur only at $Au-TiO_2$ interface perimeter sites.

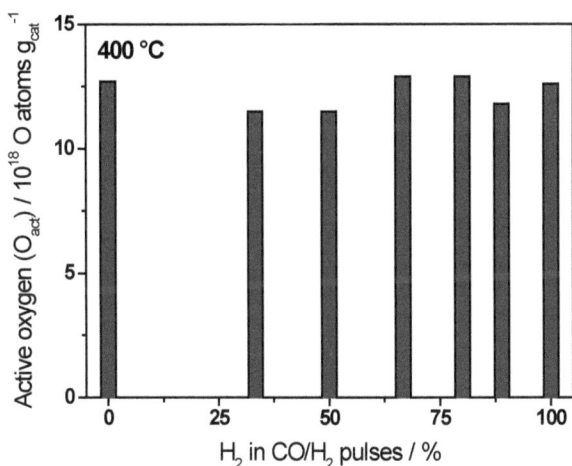

Figure 3. Absolute amounts of active oxygen (O_{act}) removal from a fully oxidized Au/TiO_2 catalyst (after O_2/Ar pulses) by CO and/or H_2 during CO, H_2, and CO/H_2 pulses with varying ratios of H_2 (33%–89% H_2) at 400 °C.

As described below (Section 3.3) it is not possible to calculate the selectivity for reaction at 400 °C, since even in the absence of H_2 there is no stoichiometric

225

consumption of O_2 and CO (with respect to the CO oxidation). Hence, the differences between O_2 consumption and CO consumption in consecutive sequences of O_2/Ar and CO/H_2/Ar pulses cannot be assigned to active oxygen removal via the H_2 oxidation only. Nevertheless, from the continuously-decreasing amount of CO_2 formation with increasing H_2 ratio in CO/H_2/Ar pulses, from 9.3×10^{18} CO_2 molecules·g_{cat}^{-1} to 6.1×10^{18} CO_2 molecules·g_{cat}^{-1} at CO/H_2 = 1/0.5 and CO/H_2 = 1/8 (see Table 2), it is evident that also at 400 °C the reaction is dominated by the same trend of decreasing selectivity towards CO_2 formation with increasing H_2 content. Moreover, considering that even for the highest amount of H_2 during CO/H_2 pulses (CO/H_2 = 1/8) and, hence, in an eight-fold excess of H_2 compared to CO, still 6.1×10^{18} CO_2 molecules·g_{cat}^{-1} are formed, which is about 60% of the amount of CO_2 formed upon reduction by CO only (see Table 2), it can be concluded that also at 400 °C the catalyst still selectively catalyses the CO oxidation under present reaction conditions.

In total, these results, obtained at different temperatures as well as for reaction in different atmospheres, point to a competition of CO and H_2 for the same active oxygen species when present simultaneously and, hence, under PROX reaction conditions. This competition will mainly determine the catalyst selectivity towards CO oxidation, and strategies for even more selective Au catalysts under these conditions (80 °C and higher) should accordingly focus on the efficiency of H_2 and CO for active oxygen removal, which may be achieved by a systematic modification of the support material and/or the Au-support interface perimeter sites, which both significantly influence the activity for CO and H_2 oxidation via a Au-assisted Mars-van Krevelen mechanism [31,41].

3. Materials and Methods

3.1. Preparation and Characterization of Au/TiO₂

For all measurements we used a home-made Au/TiO_2 catalyst with a Au loading of 2.2 wt. %, which was prepared via a deposition-precipitation method using commercial, non-porous TiO_2 as support material (P5, surface area 56 $m^2 \cdot g^{-1}$). Prior to all measurements the catalyst was pre-treated *in situ* by calcination at 400 °C in a continuous flow of 10% O_2/N_2 (20 Nml·min^{-1}) at atmospheric pressure for 30 min (hereafter denoted as O400), resulting in a surface-volume mean diameter of 2.4 ± 0.5 nm. The mean Au particle size was determined by transmission electron microscopy (TEM), evaluating the diameter of more than 500 Au nanoparticles after pre-treatment (O400).

3.2. TAP Reactor Measurements

Pulse experiments were performed in a TAP reactor system, which was built in our laboratory and which is described in detail in Ref. [46]. This system consists of a gas mixing unit, two piezo-electrically driven pulse valves, and a quartz glass micro-reactor (length: 90 mm, inner diameter: 4 mm), which is connected to an analysis chamber (under ultrahigh vacuum conditions), which houses the mass spectrometer for gas phase analysis. For *in situ* catalyst pre-treatment at atmospheric pressure, the micro-reactor with the catalyst inside can be separated from the analysis chamber via a differentially-pumped gate valve. This latter gate valve enables fast transitions from vacuum to ambient conditions in the micro-reactor and *vice versa* [46].

Gas pulses containing between 8×10^{15} and 3×10^{16} molecules per pulse were generated by the piezo-electrically driven pulse valves and directed to the micro-reactor. At the centre of this reactor the catalyst bed was fixed by stainless steel sieves. For all measurements the catalyst bed consisted of two outer layers of SiO_2 powder (inert under present reaction conditions) and a central catalyst zone with 10 mg of the Au/TiO_2 catalyst diluted with 10 mg SiO_2 (dilution 1:1) in between the SiO_2 layers.

Multi-pulse experiments were performed with a separation time of 5 s between two consecutive pulses. During these measurements the catalyst was alternately exposed to sequences of (i) CO/Ar; (ii) H_2/Ar; or (iii) CO/H_2/Ar and O_2/Ar pulses for removal and replenishment of stable, active oxygen species. The number of pulses in each sequence was chosen such that there is no measurable reactive removal or oxygen uptake at the end of each sequence, *i.e.*, at the end of CO/Ar, H_2/Ar, and CO/H_2/Ar pulse sequences or after the O_2/Ar pulse sequences. The exact numbers of pulses in each sequence are stated together with the results obtained (see Sections 2.1 and 2.2). Moreover, these cycles with alternate reduction and re-oxidation of the catalyst surface were always repeated at least three times in order to demonstrate the reversibility of the observed processes and, hence, to exclude permanent changes of the catalysts state during pulse experiments. The absolute amount of reversible oxygen removal and replenishment is defined as the catalysts oxygen storage capacity (OSC). Note that in each pulse Ar was included in order to enable the quantitative evaluation of the amount of CO, H_2, or O_2 consumed during every single pulse on an absolute scale. This is based on the comparison of the ratio of the intensities of the respective species and Ar during reaction (uptake/consumption of CO, H_2 and O_2) to that after saturation, where no more consumption of the corresponding species was detected. CO/Ar and O_2/Ar pulses always consisted of 50% CO or 50% O_2 and 50% Ar at a pulse size of 8×10^{15} molecules per pulse. Hence, the catalyst was exposed to about 4×10^{15} molecules CO or O_2 during each of the pulses. For pulses with a CO/H_2/Ar mixture, in contrast, the overall pulse size was between 1.2×10^{16} and 3.0×10^{16} molecules per pulse. Here we kept the amount of CO constant at 3×10^{15} molecules CO per pulse, while varying the amount of H_2 between 1.5×10^{15} and 24×10^{15} molecules H_2 per pulse. This way, CO/H_2 ratios between $1/0.5$ and $1/8$ were realized for the $CO/H_2/Ar$ pulses. Note that in the discussions the H_2 ratio during $CO/H_2/Ar$ pulses is always stated with respect to CO and H_2 only (not considering Ar), which are accordingly between 33% H_2 ($CO/H_2 = 1/0.5$) and 89% H_2 ($CO/H_2 = 1/8$).

In order to avoid complications in the direct comparison of active oxygen removal by CO and H_2 at 80 °C due to adsorption of water, which is always formed on the catalyst surface during H_2 oxidation and has a distinct impact on the CO oxidation activity (see above), we saturated the Au/TiO_2 catalyst surface with adsorbed water by exposing it to 5000 H_2/Ar pulses before starting the measurements. Note that this treatment additionally resulted in a complete removal of stable reactive oxygen (O_{act}) species for H_2 oxidation. The starting point of subsequent measurements is, hence, a fully reduced Au/TiO_2 catalyst.

3.3. Calculation of the Selectivity

Usually the selectivity for CO oxidation in CO/H_2 mixtures is defined by the ratio between the CO_2 formation rate and the overall oxygen consumption rate (for CO oxidation and H_2 oxidation, see also Equation (3)), which are determined in kinetic measurements. This is not possible in the present study, where we focused on the absolute amounts of oxygen removal and replenishment during alternate sequences of $CO/H_2/Ar$ and O_2/Ar pulses rather than on their kinetics. In this case we calculated the selectivity for CO oxidation at 80 °C by the ratio between the absolute amount of CO consumption/CO_2 formation (N_{CO} during a sequence of $CO/H_2/Ar$ pulses) and the absolute amount of atomic oxygen consumption ($2 \cdot N_{O2}$) during a subsequent sequence of O_2/Ar pulses (Equation (4)). Note that it was not possible to determine the absolute amount of H_2O formation during $CO/H_2/Ar$ pulses at 80°C, since water desorption is too slow to be resolved.

$$S = \frac{N_{CO}}{2 \cdot N_{O_2}} \tag{4}$$

For reaction at 400 °C it was not possible to calculate the selectivity in that way. From previous studies it is known that at this temperature, even in the absence of H_2, the O_2 uptake is always slightly higher than the oxygen consumption during subsequent exposure to CO during alternate sequences of O_2/Ar and CO/Ar pulses [27,35]. Therefore, the difference between CO consumption/CO_2 formation during exposure to $CO/H_2/Ar$ pulses to that of O_2 consumption during subsequent exposure to O_2/Ar pulses is not solely due to active oxygen consumption by H_2 oxidation/H_2O formation, and can not be used for determination of the selectivity. For this reason we can only provide values for O_{act} removal and CO_2 formation at 400 °C.

4. Conclusions

From the comparison of the active oxygen removal from a Au/TiO_2 catalyst by CO, H_2, and CO/H_2 pulses with varying CO/H_2 ratios in quantitative temporal analysis of products (TAP) reactor measurements at 80 °C and 400 °C, we arrived at the following conclusions on the PROX reaction mechanism:

(1) CO and hydrogen are oxidized by the same stable active oxygen species under present reaction conditions, also in the simultaneous presence of CO and H_2 in the reaction atmosphere, as evidenced by the similar amounts of active oxygen removal from a O_2/Ar pulse oxidized Au/TiO_2 catalyst upon exposure to multi-pulse sequences of CO, H_2, and CO/H_2 mixtures. This is independent of the CO/H_2 ratio. Hence, also under PROX reaction conditions CO and H_2 compete for TiO_2 surface lattice oxygen close to the Au nanoparticles/at the perimeter of the $Au-TiO_2$ interface as active oxygen species.

(2) The selectivity of Au/TiO_2 catalysts for CO oxidation in CO/H_2 containing gas mixtures observed in kinetic measurements is proposed to mainly result from the much higher efficiency of CO for active oxygen removal compared to hydrogen under present reaction conditions, in the simultaneous presence of H_2 and CO in the reaction atmosphere. The latter is illustrated also by the much higher amount of H_2 pulses required for complete removal of active oxygen from the catalyst compared to CO pulsing.

(3) Since we can rule out the presence of weakly-bound adsorbed water under present reaction conditions, in multi-pulse sequences, the much higher efficiency of CO for reaction with active oxygen species or the preference for this reaction pathway cannot result from effects caused by weakly-bound water species. On the other hand, stable adsorbed (strongly bound) water species, which are known to be present on the surface upon H_2 pulsing, leave the dominant reaction pathway for CO oxidation under present reaction conditions, namely reaction of CO_{ad} with surface lattice oxygen at the perimeter of the Au NP-support interface, unchanged, although the reaction rate may change.

Since the selectivity of pure Au/TiO_2 catalysts originates mainly from the intrinsic difference in the reactivity of TiO_2 surface lattice oxygen located at $Au-TiO_2$ perimeter sites for CO oxidation and H_2 oxidation under present reaction conditions, further improvements in the selectivity of TiO_2-based Au catalysts should include a modification of these perimeter sites, for example by doping of the TiO_2 support.

We cannot rule out from the present data, however, that other reaction pathways contribute as well at lower reaction temperature and/or during reaction under continuous flow conditions at atmospheric pressure. Under those conditions, weakly-adsorbed water, as well as adsorbed H_{ad}, are present and may also be involved in the reaction, as recently proposed by Saavedra *et al.* [40] and Rousset *et al.* [12].

Overall, the present findings confirm previous ideas on the CO oxidation mechanism and the physical origin of the high selectivity for CO oxidation in the PROX reaction on supported Au catalysts, specifically on Au/TiO_2 catalysts, which were based on plausibility arguments extending mechanistic insights from CO oxidation and H_2 oxidation to the present case of reaction in CO/H_2 mixtures.

Acknowledgments: We gratefully acknowledge financial support by the Deutsche Forschungsgemeinschaft (DFG) and Ulm University for covering the costs to publish in open access.

Author Contributions: Daniel Widmann conceived and designed the experiments; Yeusy Hartadi performed the experiments; Yeusy Hartadi, R. Jürgen Behm, and Daniel Widmann wrote the paper.

Conflicts of Interest: The authors declare no conflict of interest.

Abbreviations

The following abbreviations are used in this manuscript:

O_{act}	Active Oxygen
PROX	Preferential CO Oxidation
TAP	Temporal Analysis of Products
PEM	Polymer Electrolyte Membrane
TEM	Transmission Electron Microscopy
TPD	Temperature Programmed Desorption

References

1. Hashmi, A.S.K.; Hutchings, G.J. Gold Catalysis. *Angew. Chem. Int. Ed.* **2006**, *45*, 7896–7936.
2. Min, B.K.; Friend, C.M. Heterogeneous Gold-Based Catalysis for Green Chemistry: Low-Temperature CO Oxidation and Propene Oxidation. *Chem. Rev.* **2007**, *107*, 2709–2724.
3. Freakley, S.J.; He, Q.; Kiely, C.J.; Hutchings, G.J. Gold Catalysis: A Reflection on Where We are Now. *Catal. Lett.* **2015**, *145*, 71–79.
4. Haruta, M. Role of perimeter interfaces in catalysis by gold nanoparticles. *Faraday Discuss.* **2011**, *152*, 11–32.
5. Freund, H.J.; Meijer, G.; Scheffler, M.; Schlögl, R.; Wolf, M. CO Oxidation as a Prototypical Reaction for Heterogeneous Processes. *Angew. Chem. Int. Ed.* **2011**, *50*, 10064–10094.

6. Trimm, D.L.; Önsan, Z.I. On-board fuel conversion for hydrogen-fuel-cell-driven vehicles. *Catal. Rev.* **2001**, *43*, 31–84.

7. Ghenciu, A.F. Review of fuel processing catalysts for hydrogen production in PEM fuel cell systems. *Curr. Opin. Solid State Mater. Sci.* **2002**, *6*, 389–399.

8. Corti, C.W.; Holliday, R.J.; Thompson, D.T. Progress towards the commercial application of gold catalysts. *Top. Catal.* **2007**, *44*, 331–343.

9. Park, E.D.; Lee, D.; Lee, H.C. Recent progress in selective CO removal in a H_2-rich stream. *Catal. Today* **2009**, *139*, 280–290.

10. Rossignol, C.; Arrii, S.; Morfin, F.; Piccolo, L.; Caps, V.; Rousset, J.-L. Selective oxidation of CO over model gold-based catalysts in the presence of H_2. *J. Catal.* **2005**, *230*, 476–483.

11. Bion, N.; Epron, F.; Moreno, M.; Marino, F.; Duprez, D. Preferential Oxidation of Carbon Monoxide in the Presence of Hydrogen (PROX) over Noble Metals and Transition Metal Oxides: Advantages and Drawbacks. *Top. Catal.* **2008**, *51*, 76–88.

12. Quinet, E.; Piccolo, L.; Morfin, F.; Avenier, P.; Diehl, F.; Caps, V.; Rousset, J.L. On the mechanism of hydrogen-promoted gold-catalyzed CO oxidation. *J. Catal.* **2009**, *268*, 384–389.

13. Kahlich, M.J.; Gasteiger, H.A.; Behm, R.J. Kinetics of the Selective Low-Temperature Oxidation of CO in H_2-rich Gas over Au/Fe_2O_3. *J. Catal.* **1999**, *182*, 430–440.

14. Schumacher, B.; Denkwitz, Y.; Plzak, V.; Kinne, M.; Behm, R.J. Kinetics, mechanism and the influence of H_2 on the CO oxidation reaction on a Au/TiO_2 catalyst. *J. Catal.* **2004**, *224*, 449–462.

15. Denkwitz, Y.; Schumacher, B.; Kucěrová, G.; Behm, R.J. Activity, stability and deactivation behavior of supported Au/TiO_2 catalysts in the CO oxidation and preferential CO oxidation reaction at elevated temperatures. *J. Catal.* **2009**, *267*, 78–88.

16. Grzybowska, B. Nano-Au/oxide support catalysts in oxidation reactions: Provenance of active oxygen species. *Catal. Today* **2006**, *112*, 3–7.

17. Bollinger, M.A.; Vannice, M.A. A kinetic and DRIFTS study of low-temperature carbon monoxide oxidation over Au-TiO_2 catalysts. *Appl. Catal. B* **1996**, *8*, 417–443.

18. Boccuzzi, F.; Chiorino, A. FTIR study on Au/TiO_2 at 90K and room temperature. An insight into the nature of the reaction centers. *J. Phys. Chem. B* **2000**, *104*, 5414–5416.

19. Green, I.X.; Tang, W.; Neurock, M.; Yates, J.T. Spectroscopic Observation of Dual Catalytic Sites During Oxidation of CO on a Au/TiO_2 Catalyst. *Science* **2011**, *333*, 736–739.

20. Boronat, M.; Francesc, I.; Corma, A. Active Sites for H_2 Adsorption and Activation in Au/TiO_2 and the Role of the Support. *J. Phys. Chem. A* **2009**, *113*, 3750–3757.

21. Panayotov, D.A.; Burrows, V.A.; Yates, J.T., Jr.; Morris, L. Mechanistic Studies of Hydrogen Dissociation and Spillover on Au/TiO_2: IR Spectroscopy of Coadsorbed CO and H-Donated Electrons. *J. Phys. Chem. C* **2011**, *115*, 22400–22408.

22. Fujitani, T.; Nakamura, I.; Akita, T.; Okumura, M.; Haruta, M. Hydrogen Dissociation by Gold Clusters. *Angew. Chem.* **2009**, *121*, 9679–9682.

23. Nakamura, I.; Mantoku, H.; Furukawa, T.; Fujitani, T. Active Sites for Hydrogen Dissociation over $TiO_x/Au(111)$ Surfaces. *J. Phys. Chem. C* **2011**, *115*, 16074–16080.

24. Bus, E.; Miller, J.T.; van Bokhoven, J.A. Hydrogen Chemisorption on Al_2O_3-Supported Gold Catalysts. *J. Phys. Chem. B* **2005**, *109*, 14581–14587.
25. Kotobuki, M.; Leppelt, R.; Hansgen, D.; Widmann, D.; Behm, R.J. Reactive oxygen on a Au/TiO_2 supported catalyst. *J. Catal.* **2009**, *264*, 67–76.
26. Vilhelmsen, L.B.; Hammer, B. Indentification of the Catalytic Site at the Interface Perimeter of Au clusters on Rutile TiO_2(110). *ACS Catal.* **2014**, *4*, 1626–1631.
27. Widmann, D.; Behm, R.J. Active oxygen on a Au/TiO_2 catalyst: Formation, stability and CO oxidation activity. *Angew. Chem. Int. Ed.* **2011**, *50*, 10241–10245.
28. Maeda, Y.; Iizuka, Y.; Kohyama, M. Generation of Oxygen Vacancies at a Au/TiO_2 Perimeter Interface during CO Oxidation Detected by *in Situ* Electrical Conductance Measurement. *J. Am. Chem. Soc.* **2013**, *135*, 906–909.
29. Li, L.; Wang, A.; Qiao, B.; Lin, J.; Huang, Y.; Wang, X.; Zhang, T. Origin of the high activity of Au/FeO_x for low-temperature CO oxidation: Direct evidence for a redox mechanism. *J. Catal.* **2013**, *299*, 90–100.
30. Kim, H.Y.; Henkelman, G. CO Oxidation at the Interface of Au Nanoclusters and the Stepped-CeO_2(111) Surface by the Mars-van Krevelen Mechanism. *J. Phys. Chem. Lett.* **2013**, *4*, 216–221.
31. Widmann, D.; Behm, R.J. Activation of Molecular Oxygen and the Nature of the Active Oxygen Species for CO Oxidation on Oxide Supported Au Catalysts. *Acc. Chem. Res.* **2014**, *47*, 740–749.
32. Duan, Z.; Henkelman, G. CO Oxidation at the Au/TiO_2 Boundary: The Role of the Au/Ti5c Site. *ACS Catal.* **2015**, *5*, 1589–1595.
33. Saqlain, M.A.; Hussain, A.; Siddiq, M.; Ferreira, A.R.; Leitao, A.A. Thermally activated surface oxygen defects at the perimeter of Au/TiO_2: A DFT + U study. *Phys. Chem. Chem. Phys.* **2015**, *17*, 25403–25410.
34. Green, I.X.; Tang, W.; Neurock, M.; Yates, J.T. Insights into Catalytic Oxidation at the Au/TiO_2 Dual Perimeter Sites. *Acc. Chem. Res.* **2013**, *47*, 805–815.
35. Widmann, D.; Hocking, E.; Behm, R.J. On the origin of the selectivity in the preferential CO oxidation on Au/TiO_2—Nature of the active oxygen species for H_2 oxidation. *J. Catal.* **2014**, *317*, 272–276.
36. Sun, K.; Kohyama, M.; Tanaka, S.; Takeda, S. A Study on the Mechanism for H_2 Dissociation on Au/TiO_2 Catalysts. *J. Phys. Chem. C* **2014**, *118*, 1611–1617.
37. Daté, M.; Haruta, M. Moisture effect on CO oxidation over Au/TiO_2 catalyst. *J. Catal.* **2001**, *201*, 221–224.
38. Diemant, T.; Bansmann, J.; Behm, R.J. CO oxidation on planar Au/TiO_2 model catalysts: Deactivation and the influence of water. *Vacuum* **2009**, *84*, 193–196.
39. Gao, F.; Wood, T.E.; Goodman, D.W. The Effects of Water on CO Oxidation over TiO_2 Supported Au Catalysts. *Catal. Lett.* **2010**, *134*, 9–12.
40. Saavedra, J.; Doan, H.A.; Pursell, C.J.; Grabow, L.C.; Chandler, B.D. The critical role of water at the gold-titania interface in catalytic CO oxidation. *Science* **2014**, *345*, 1599–1602.

41. Widmann, D.; Liu, Y.; Schüth, F.; Behm, R.J. Support effects in the Au catalyzed CO oxidation—Correlation between activity, oxygen storage capacity and support reducibility. *J. Catal.* **2010**, *276*, 292–305.

42. Wang, L.-C.; Widmann, D.; Behm, R.J. Reactive removal of surface oxygen by H_2, CO and CO/H_2 on a Au/CeO_2 catalyst and its relevance to the preferential CO oxidation (PROX) and reverse water gas shift (RWGS) reaction. *Catal. Sci. Technol.* **2015**, *5*, 925–941.

43. Schubert, M.M.; Plzak, V.; Garche, J.; Behm, R.J. Activity, Selectivity, and Long-Term Stability of Different Metal Oxide Supported Gold Catalysts for the Preferential CO Oxidation in H_2-rich Gas. *Catal. Lett.* **2001**, *76*, 143.

44. Arrii, S.; Morfin, F.; Renouprez, A.J.; Rousset, J.L. Oxidation of CO on Gold Supported Catalysts Prepared by Laser Vaporization: Direct Evidence of Support Contribution. *J. Am. Chem. Soc.* **2004**, *126*, 1199–1205.

45. Lakshmanan, P.; Park, J.E.; Park, E.D. Recent Advances in Preferential Oxidation of CO in H_2 over Gold Catalysts. *Catal. Surv. Asia* **2014**, *18*, 75–88.

46. Leppelt, R.; Hansgen, D.; Widmann, D.; Häring, T.; Bräth, G.; Behm, R.J. Design and characterization of a temporal analysis of products reactor. *Rev. Sci. Instrum.* **2007**, *78*, 104103.

Characterization and Catalytic Activity of Mn-Co/TiO$_2$ Catalysts for NO Oxidation to NO$_2$ at Low Temperature

Lu Qiu, Yun Wang, Dandan Pang, Feng Ouyang, Changliang Zhang and Gang Cao

Abstract: A series of Mn-Co/TiO$_2$ catalysts were prepared by wet impregnation method and evaluated for the oxidation of NO to NO$_2$. The effects of Co amounts and calcination temperature on NO oxidation were investigated in detail. The catalytic oxidation ability in the temperature range of 403–473 K was obviously improved by doping cobalt into Mn/TiO$_2$. These samples were characterized by nitrogen adsorption-desorption, X-ray diffraction (XRD), X-ray photoelectron spectroscopy (XPS), transmission electron microscope (TEM) and hydrogen temperature programmed reduction (H$_2$-TPR). The results indicated that the formation of dispersed Co$_3$O$_4 \cdot$ CoMnO$_3$ mixed oxides through synergistic interaction between Mn-O and Co-O was directly responsible for the enhanced activities towards NO oxidation at low temperatures. Doping of Co enhanced Mn^{4+} formation and increased chemical adsorbed oxygen amounts, which also accelerated NO oxidation.

Reprinted from *Catalysts*. Cite as: Qiu, L.; Wang, Y.; Pang, D.; Ouyang, F.; Zhang, C.; Cao, G. Characterization and Catalytic Activity of Mn-Co/TiO$_2$ Catalysts for NO Oxidation to NO$_2$ at Low Temperature. *Catalysts* **2016**, *6*, 9.

1. Introduction

Nitrogen oxides (NO$_x$) in the exhaust from stationary and mobile sources are toxic to human's health and have brought environmental problems, such as photochemical smog, acid rain and ozone depletion. NO$_x$ storage-reduction (NSR) and selective catalytic reduction (SCR) are considered to be promising technologies to achieve high NO$_x$ reduction efficiency. It is known that most nitrogen oxides from exhaust emissions exist in the form of NO (>90%). NO oxidation is considered to be a key step for NO$_x$ reduction on both the NSR and SCR reactions. In the process of NSR, NO is first oxidized to NO$_2$ and then stored on the basic components of catalysts as nitrates [1]. For SCR, NO$_2$ is favored for NO$_x$ conversion according to the so-called Fast SCR reaction, which is thought to be faster by one order of magnitude than the Standard SCR reaction under oxidizing conditions [2]. The light-off temperature in the NO oxidation to NO$_2$ is significant for the low-temperature SCR of NO$_x$ reduction. Therefore the research into NO oxidation catalysts is important for the

NO_x removal and various catalysts have been researched, including the supported noble metals [3–6] and other metal oxides [7–9].

Among these catalysts, there has been some attention focused on cobalt oxides [10] and manganese oxides [11]. It was reported that cobalt oxides possess better oxidation ability [12–14]. Co^{2+}/Co^{3+} oxidation states were found to interconvert readily in oxidizing and reducing conditions [15]. The presence of Co_3O_4 means high activity for oxidation reactions [10,16]. In addition, Mn-based catalysts have been proven to be highly active for the low-temperature SCR reaction, which would avoid the disadvantages associated with the commercial high-temperature catalysts. Several manganese based catalysts, such as the Mn/TiO_2 [17,18] and the Mn catalysts doped with other metal (Fe, Co, Ni, Cu, Ce, *etc.*) oxides [19–21], have been reported for the low-temperature SCR reaction. As reported in the literature [22,23], Mn oxides not only acted as highly efficient SCR catalysts, but also showed certain catalytic activity for NO oxidation. The $Mn-Co/TiO_2$ catalysts have been evaluated for the low temperature SCR reaction [19]. However, the activities of $Mn-Co/TiO_2$ for NO oxidation to NO_2 have been reported rarely.

The present work aims to investigate the catalysts of Mn-Co composite oxides loaded on TiO_2 for NO oxidation at low temperatures. The catalysts with different Co contents and calcined at different temperatures were investigated to determine the interactions between cobalt oxides and manganese oxides. Special attention was paid to the Mn-O-Co mixed oxides which can change the properties of the catalysts and enhance the oxidation ability. The $Mn-Co/TiO_2$ catalysts were characterized by means of N_2 adsorption-desorption, X-ray diffraction (XRD), transmission electron microscopy (TEM), X-ray photoelectron spectroscopy (XPS) and hydrogen temperature-programmed reduction (H_2-TPR).

2. Results and Discussion

2.1. Activity of NO Oxidation to NO_2

Figure 1 shows the oxidation efficiency of NO to NO_2 over a series of $Mn-Co/TiO_2$ catalysts with different cobalt contents and a constant loading of 10 wt. % manganese. These catalysts were calcined at 773 K. As a reference, the activities of $10Co/TiO_2$, $10Mn/TiO_2$ and $20Mn/TiO_2$ were also tested. As shown in Figure 1, the NO conversion efficiencies of $10Mn/TiO_2$ and $20Mn/TiO_2$ were higher than that of $10Co/TiO_2$ below 513 K. Dramatic increase in the NO oxidation ratio within the temperature range of 403–473 K was observed upon addition of cobalt oxides to Mn/TiO_2. However, the NO conversion increased slowly with increasing Co content above 473 K. The $10Mn-10Co/TiO_2$ exhibited a maximum oxidation efficiency of 61% at 543 K, which was nearly equal to that of $10Co/TiO_2$. The activity results implied that strong interactions might exist between Co and Mn oxides, which could promote the NO oxidation to NO_2 at low temperatures.

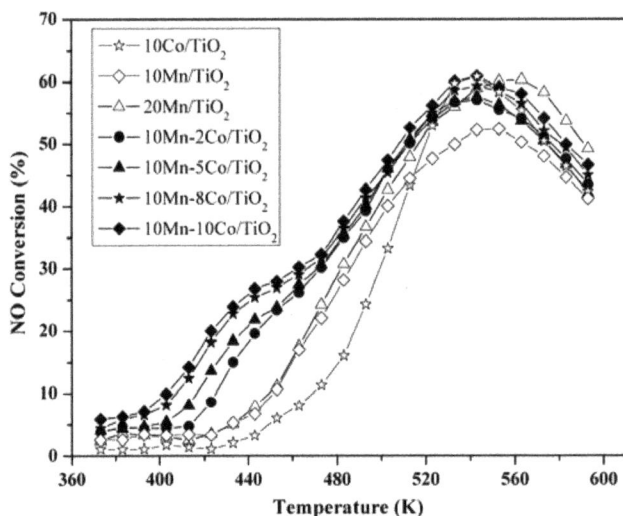

Figure 1. NO oxidation efficiency of Mn-Co/TiO$_2$ catalysts with different Co contents. Reaction conditions: [NO] = 400 ppm, [O$_2$] = 5%, Ar balance, GHSV = 42,000 h^{-1}.

Figure 2 shows the oxidation efficiency of NO to NO_2 over a series of 10Mn-5Co/TiO$_2$ catalysts calcined at various temperatures. It can be seen that the catalytic activity increased with the calcination temperature increase from 573 to 773 K, and then decreased. The catalyst Mn-Co/TiO$_2$ (773 K) had the highest oxidation efficiency at 543 K among these samples. However, the Mn-Co/TiO$_2$ calcined at 573 and 673 K exhibited excellent catalytic activities in the temperature range of 403–483 K.

2.2. Nitrogen Adsorption-Desorption Characterization

The specific surface areas and pore volumes for Mn-Co/TiO$_2$ catalysts were determined by nitrogen adsorption-desorption method. The results are listed in Table 1. A consistently decreasing trend of surface areas with increasing calcination temperature was noted for 10Mn-5Co/TiO$_2$ samples, and a little decrease with the increased Co contents was also showed.

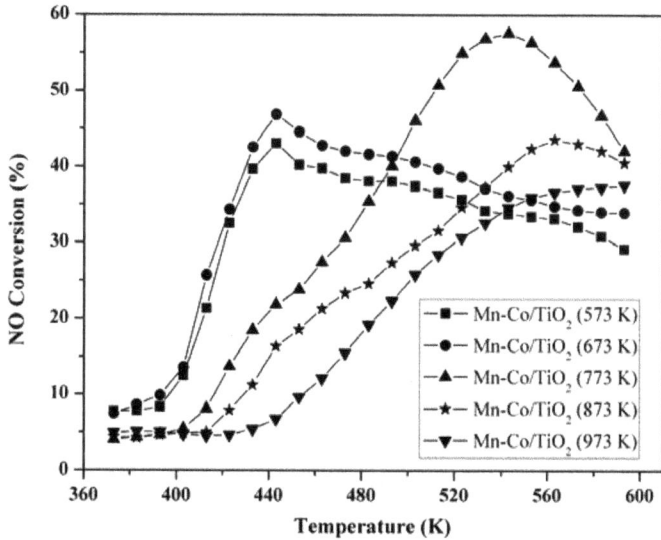

Figure 2. NO oxidation efficiency of 10Mn-5Co/TiO$_2$ catalysts calcined at different temperatures. Reaction conditions: [NO] = 400 ppm, [O$_2$] = 5%, Ar balance, GHSV = 42,000 h^{-1}.

Table 1. BET surface areas and average pore diameters of Mn-Co/TiO$_2$ catalysts with various Co contents and calcined at different temperatures.

Catalyst	S_{BET} (cm$^2 \cdot$ g^{-1})	Average Pore Diameter (nm)
10Mn/TiO$_2$ (773 K)	48.35	8.67
10Mn-5Co/TiO$_2$ (573 K)	50.57	7.21
10Mn-5Co/TiO$_2$ (673 K)	50.38	6.86
10Mn-5Co/TiO$_2$ (773 K)	45.79	8.52
10Mn-5Co/TiO$_2$ (873 K)	30.51	8.38
10Mn-5Co/TiO$_2$ (973 K)	7.01	6.27
10Mn-10Co/TiO$_2$ (773 K)	42.13	8.02

2.3. XRD Characterization

Figure 3 shows the X-ray powder diffraction patterns of the Mn-Co/TiO$_2$ catalysts calcined at 773 K, with different Co contents ranged from 0 to 10 wt. %. It can be seen that the relative strong patterns showed diffraction peaks corresponding to anatase TiO$_2$ phase (PDF#21-1272) and rutile phase (PDF#21-1276), while anatase was the main form in these catalysts. In the pattern of 10Co/TiO$_2$ (Figure 3a), several weak diffraction peaks were detected at 2θ values of 18.8°, 31.2°, 44.7°, 59.3° and 65.2°, corresponding to Co$_3$O$_4$ phase (PDF#43-1003). In the pattern of 10Mn/TiO$_2$ (Figure 3b), these weak peaks at 2θ values of 37.3° and 42.7° can be

237

attributed to MnO_2 (PDF#50-0866). However, the crystalline MnO_2 was poor since its corresponding diffraction peaks were very weak.

For the Mn-Co/TiO_2 samples with different Co contents, several weak peaks appeared at 18.4°, 30.8°, 33.4°, 44.1°, 58.9° and 65.5° (Figure 3d–g). The intensities of these peaks were enhanced slightly by the increase of Co contents. Compared with the pattern of 10Co/TiO_2, these peaks had a slight shift toward lower or higher 2θ value. These peaks at 30.8°, 33.4°, 58.9° and 65.5° can be assigned to $CoMnO_3$, while the peaks at 18.4° and 44.1° were ascribed to $CoMn_2O_4$. These peaks were also weak, inferring that most of the binary metal oxides existed in highly dispersed state. However, the intensities of these peaks were enhanced when the content of Mn increased to 20 wt. % and the Co content was 10 wt. % (Figure 3g). The existence of $CoMnO_3$ and $CoMn_2O_4$ was also demonstrated by Meng et al. [24] when Co oxides and Mn oxides were mixed without the support TiO_2. It can be deduced that $CoMnO_3$ and $CoMn_2O_4$ were two types of Mn-Co composite oxides formed on surface of TiO_2, in which Mn ions existed mainly in valence states of Mn^{4+} and Mn^{3+}, respectively.

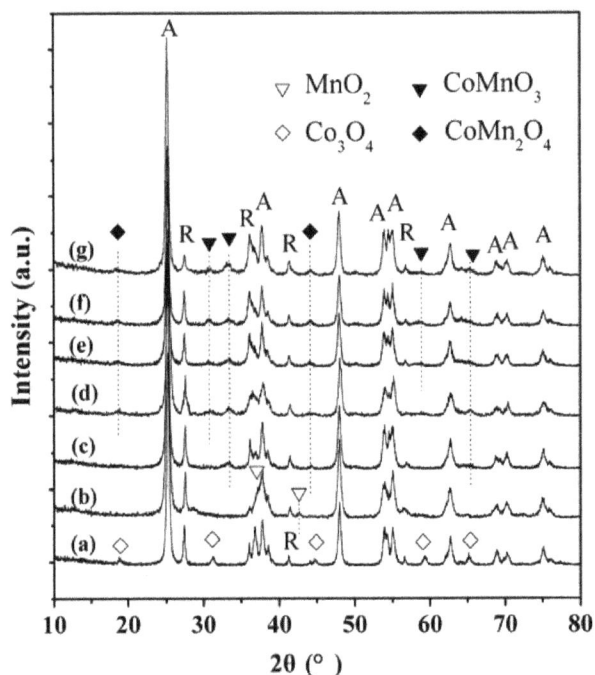

Figure 3. XRD patterns of the catalysts Mn-Co/TiO_2 (773 K) with different Co contents: (a) 10Co/TiO_2; (b) 10Mn/TiO_2; (c) 10Mn-2Co/TiO_2; (d) 10Mn-5Co/TiO_2; (e) 10Mn-8Co/TiO_2; (f) 10Mn-10Co/TiO_2; and (g) 20Mn-10Co/TiO_2. (A = anatase phase of TiO_2, R = rutile phase of TiO_2).

Figure 4 shows the XRD patterns of 10Mn-5Co/TiO_2 catalysts calcined at various temperatures. It can be seen that the intensities of rutile TiO_2 peaks were enhanced with calcination temperature increasing while the intensities of anatase peaks were decreased, indicating the transformation of anatase to rutile. The samples calcined at 573 and 673 K were low crystallized with the Mn-O-Co mixed oxides (Figure 4a,b). The weak crystal structure of $CoMnO_3$ and $CoMn_2O_4$ were only detected until the calcination temperature reached 773 K, represented by the peaks at 18.4°, 30.8°, 33.4°, 44.1° and 65.5°. These peaks decreased or disappeared above 873 K. Evidently, NO oxidation efficiency is related to the dispersion of $CoMnO_3$ and the crystal types of TiO_2.

As the calcination temperature increased to 873 and 973 K, the peaks at 18.4°, 30.8° and 33.4° shifted slightly toward lower 2θ value. We considered that was due to the transformation of Mn^{4+} to Mn^{3+} and Mn^{3+} to Mn^{2+} in the process of sintering. The peak at 32.9° can be attributed to $CoMn_2O_4$ phase and the peaks appeared at 18.1°, 29.2° and 60.6° can be attributed to a more complex $(Co,Mn)(Co,Mn)_2O_4$ phase, which had a higher crystallinity.

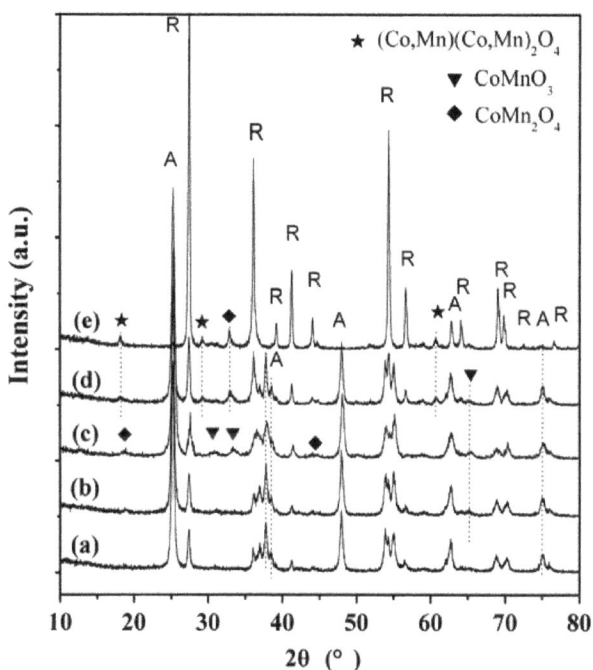

Figure 4. XRD patterns of the 10Mn-5Co/TiO_2 catalysts calcined at various temperatures: (**a**) 573 K; (**b**) 673 K; (**c**) 773 K; (**d**) 873 K; and (**e**) 973 K. (A = anatase phase of TiO_2, R = rutile phase of TiO_2).

2.4. TEM Characterization

Figure 5 shows the TEM images of 10Mn-5Co/TiO$_2$ calcined at 673 and 773 K. For Mn-Co/TiO$_2$ (673 K) in Figure 5A, the lattice fringes a and b were determined to be 0.36 and 0.45 nm, matched CoMnO$_3$ (012) (standard value is 0.362 nm) and (003) (standard value is 0.457 nm) crystal faces, respectively. For Mn-Co/TiO$_2$ (773 K) in Figure 5B, the lattice fringe c was determined to be 0.48 nm, corresponding to CoMn$_2$O$_4$ (standard value is 0.485 nm). The lattice fringes d and e were determined to be 0.35 and 0.52 nm, corresponding to anatase (101) of TiO$_2$ and MnO$_2$ (200) (standard value is 0.515 nm) crystal faces, respectively. The lattice fringes f and g were determined to be 0.36 and 0.41 nm, matched CoMnO$_3$ (012) and (101) (standard value is 0.408 nm) crystal faces, respectively. These results demonstrated the existence of CoMnO$_3$ and CoMn$_2$O$_4$ phases in Mn-Co/TiO$_2$ samples.

2.5. XPS Characterization

The surface oxidation states and atomic concentrations of 10Mn/TiO$_2$ (773 K) and 10Mn-5Co/TiO$_2$ calcined at 673, 773 and 973 K were determined by XPS analysis. The photoelectron profiles of Mn 2p and O 1s are shown in Figure 6. The atomic concentrations of Mn, Co and O are listed in Table 2. To evaluate Mn valence states, a peak-fitting deconvolution was performed for these profiles fitted with a mixing of Gaussian and Lorentzian peaks after removal of a Shirley background.

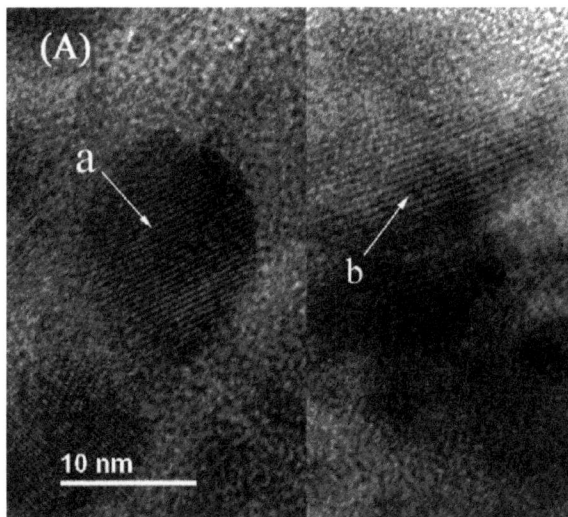

Figure 5. *Cont.*

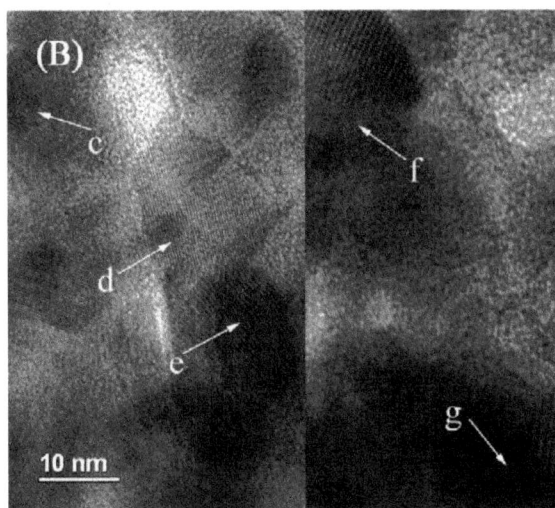

Figure 5. TEM images of the catalysts: (**A**) 10Mn-5Co/TiO$_2$ (673 K); and (**B**) 10Mn-5Co/TiO$_2$ (773 K). (a. CoMnO$_3$ (012); b. CoMnO$_3$ (003); c. CoMn$_2$O$_4$ (111); d. anatase (101); e. MnO$_2$ (200); f. CoMnO$_3$ (012); g. CoMnO$_3$ (101)).

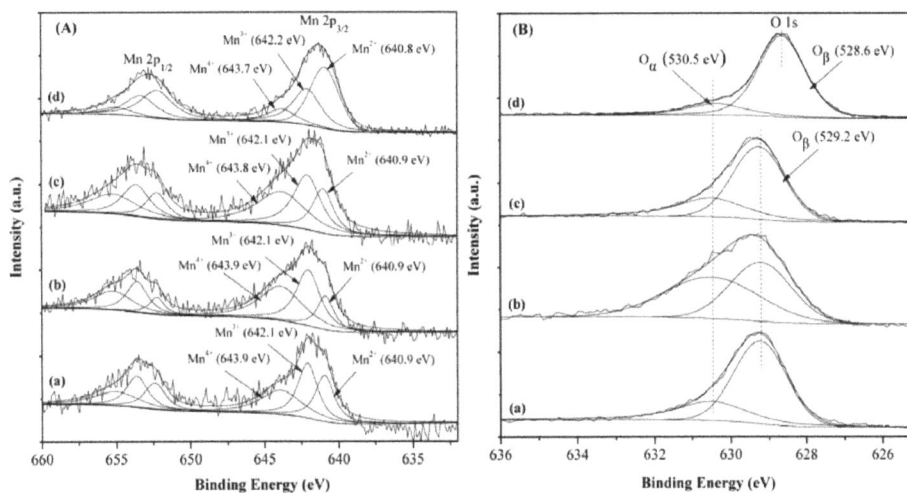

Figure 6. XPS spectra for (**A**) Mn 2p and (**B**) O 1s of the catalysts: (**a**) 10Mn/TiO$_2$ (773 K); (**b**) 10Mn-5Co/TiO$_2$ (673 K); (**c**) 10Mn-5Co/TiO$_2$ (773 K); and (**d**) 10Mn-5Co/TiO$_2$ (973 K).

In Figure 6A, two peaks were detected at 641.2–642.1 eV and 652.8–653.7 eV, belonging to Mn 2p$_{3/2}$ and Mn 2p$_{1/2}$, respectively. The spectra of Mn 2p$_{3/2}$ could be separated into three peaks. The peaks corresponding to the higher, middle and

241

lower binding energy were attributed to Mn^{4+}, Mn^{3+} and Mn^{2+}, respectively [25,26]. Their relative atomic ratios in the surface layer are listed in Table 2. The increase of Mn^{4+} ratio after Co doping indicated the transformation of Mn^{2+} to Mn^{3+} and Mn^{3+} to Mn^{4+} by the addition of cobalt oxides. The reduction was resulted from the strong interaction between Mn and Co oxides. In addition, as the calcination temperature increased, the relative ratio of Mn^{4+} decreased gradually while the relative ratio of Mn^{2+} increased. The main valence state of manganese in Mn-Co/TiO_2 (673 K) was Mn^{4+}, whereas Mn^{2+} was the dominant valence state in the catalyst calcined at 973 K. It can be concluded that the Mn^{4+} oxidation state was responsible for the activity of NO oxidation at low temperatures.

Figure 6B shows the O 1s XPS spectra for these catalysts. The O 1s spectra can be separated into two peaks at around 528.6–529.2 eV and 530.5 eV. The higher binding energy one with less intensity was ascribed to adsorbed oxygen or surface hydroxyl species [27,28], referred to as O_α, whereas the lower binding energy one was due to lattice oxygen O^{2-} [28,29], denoted as O_β. The value of binding energy of O_β decreased from 529.2 to 528.6 eV when the calcination temperature increased from 773 to 973 K. The binding energies of 529.2 and 528.6 eV were attributed to the lattice oxygen of anatase TiO_2 and rutile TiO_2, respectively. The ratios of O_α:O_β in Table 2 showed an increase in chemical adsorbed oxygen by doping of Co oxides. The reason is that as a nonstoichiometric compound-like, Co oxides can adsorb and exchange oxygen easily in several surface layers [15,30], and then promote oxygen adsorption. In addition, it was found out that the relative ratio of O_α reduced from 54% to 17% when calcination temperature increased from 673 K to 973 K, indicating the inhibition for adsorption of oxygen after high-temperature calcination. The XPS results indicated that the decline of NO oxidation efficiency was related to the decrease of chemical adsorbed oxygen. Wu *et al.* [31] have suggested that chemisorbed oxygen was helpful to the oxidation of NO to NO_2.

Table 2. Surface atomic concentrations and ratios for the catalysts determined by XPS spectra.

Catalyst	Mn^{4+}:Mn^{3+}:Mn^{2+}	O_α:O_β	Mn at. %	Co at. %
10Mn/TiO_2 (773 K)	34:36:30	30:70	6.48	-
10Mn-5Co/TiO_2 (673 K)	43:39:17	54:46	5.47	2.32
10Mn-5Co/TiO_2 (773 K)	40:37:23	32:68	6.19	3.73
10Mn-5Co/TiO_2 (973 K)	14:36:50	17:83	6.73	1.54

2.6. H_2-TPR Characterization

Figure 7 presents the H_2-TPR profiles of the Mn-Co/TiO_2 catalysts with various Co contents and calcined at 773 K. The reduction profile of 5Co/TiO_2 was also provided to identify the peaks of Co oxides (Figure 7a). The profile of $5Co/TiO_2$ was characterized by two reduction peaks at 693 and 824 K, due to reduction of Co_3O_4 to CoO and CoO to Co°, respectively [32]. In terms of $10Mn/TiO_2$ (Figure 7b), two separate reduction peaks were observed. According to the literature [33], the lower temperature one at 670 K (T_1) could be ascribed to the reduction of MnO_2 to Mn_2O_3. The higher temperature peak could be separated to two peaks at 763 and 796 K, corresponding to the reduction of Mn_2O_3 to Mn_3O_4 and Mn_3O_4 to MnO, respectively [33]. The prominent peak at 670 K indicated that MnO_2 was present in Mn/TiO_2.

For the $Mn-Co/TiO_2$ catalysts with various Co contents (Figure 7c–f), it can be seen that after the addition of Co oxides into Mn/TiO_2, the profiles of the first and second peaks were similar to the reduction peaks of Mn/TiO_2. These two peaks became more and more narrow and their intensities increased with increasing Co content. Meanwhile, the summits shifted toward low temperature with the increase of Co content. As can be seen, after 10 wt. % Co was added into the catalyst, the first peak at 625 K was lower for about 45 K than the first reduction peak of Mn/TiO_2, and lower for about 68 K than that of Co/TiO_2. Consequently, it can be deduced that an interaction existed between Mn and Co oxides, leading to the down-shift of the reduction temperature and related to the higher NO oxidation efficiency at low temperatures for $Mn-Co/TiO_2$. Combining with the XPS results, the interaction between Mn and Co ions can be represented as reaction (1):

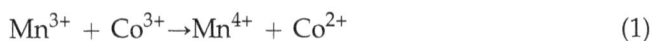

$$Mn^{3+} + Co^{3+} \rightarrow Mn^{4+} + Co^{2+} \tag{1}$$

It is known that the Co^{3+} probably corresponds to $[Co^{2+}Co^{3+}_2O_4]$ with a normal spinel structure [30]. Based on XRD and TEM results, $CoMnO_3$ is the main physical phase on $Mn-Co/TiO_2$ (773 K). From the perspective of standard electrode potential, the standard electrode potential of $Co^{3+} \rightarrow Co^{2+}$ is higher than that of $Mn^{4+} \rightarrow Mn^{3+}$ [34], and the lower the standard electrode potential, the more easily the coordinated oxygen is obtained. Apparently, if the oxygen vacancy exists in the mixed oxide of Co_3O_4 and $CoMnO_3$, the oxygen ion will preferentially coordinate to Mn^{4+} ion rather than Co^{3+} ion, which will lead to oxygen vacancy of Co_3O_4. Meanwhile, Co_3O_4 adsorbs oxygen easily on Co-contained surface and exchange with the oxygen on several surface atom layers to form $Co_3O_{4+y}\cdot CoMnO_3$ [15,30]. These oxygen ions are highly active and reduced easily, leading to the shift of the reduction peak toward lower temperatures compared with the Co/TiO_2. Consequently, Co doping into Mn/TiO_2 can lower the reduction temperature. When the y is small, we write $Co_3O_{4+y}\cdot CoMnO_3$ as $Co_3O_4\cdot CoMnO_3$. Therefore, the first peak of $Mn-Co/TiO_2$ could be ascribed to the reduction of the compound $Co_3O_4\cdot CoMnO_3$. The second and third reduction peak temperatures were also lower than those of Mn/TiO_2 in the similar way. Combining with XRD and XPS results, the second peak could be ascribed to the reduction of the composite oxide $CoMn_2O_4$ and the third peak could be ascribed to the reduction that the final product was $CoO \cdot MnO$. These two peaks trended to combination with increasing Co content, probably due to the strong oxygen adsorption ability of Co oxides. Consequently, the Mn^{3+} ions were easy to be directly reduced to Mn^{2+}. The fourth peak was due to the reduction of Co^{2+} to Co°. It shifted toward higher temperature after co-doping of Mn oxides. The Mn^{3+} ions are much reducible than the Co^{2+} ions and request lower activation energy for reduction. Accordingly, the Co^{2+} can be reduced only until the Mn^{3+} ions were reduced completely.

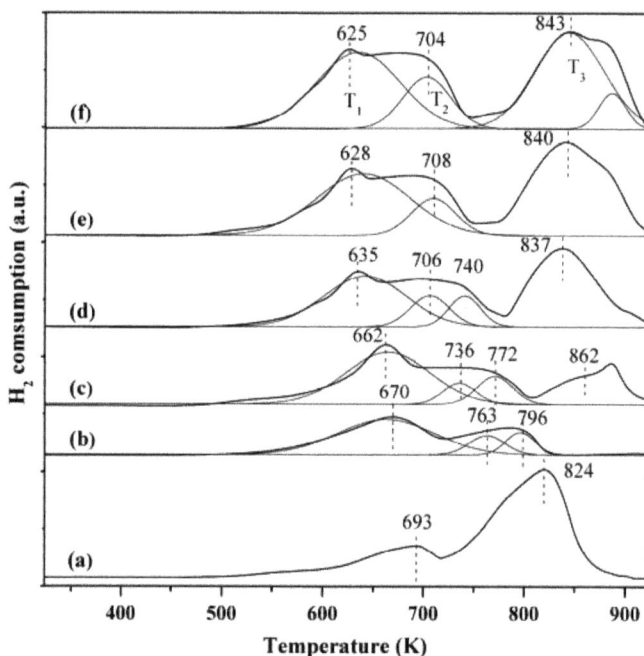

Figure 7. H_2-TPR profiles of the catalysts Mn-Co/TiO_2 (773 K) with various Co contents: (**a**) 5Co/TiO_2; (**b**) 10Mn/TiO_2; (**c**) 10Mn-2Co/TiO_2; (**d**) 10Mn-5Co/TiO_2; (**e**) 10Mn-8Co/TiO_2; and (**f**) 10Mn-10Co/TiO_2.

Figure 8 shows the H_2-TPR profiles of 10Mn-5Co/TiO_2 catalysts calcined at various temperatures. The catalysts calcined at 573 and 673 K clearly showed similar reduction peaks. The T_1 and T_2 peaks were mainly due to the reduction of $Co_3O_4 \cdot CoMnO_3$ and $CoMn_2O_4$, respectively. The intensities of the two peaks decreased gradually and the peaks broadened with the increase of calcination temperature. It indicated that the Mn-O-Co mixed oxides formation and the oxygen mobility of Mn-O-Co were affected by the calcination temperature. The

TPR results were associated with the NO oxidation ability in Figure 2. For the Mn-Co/TiO$_2$ calcined at 573 and 673 K, the H$_2$ reduction peaks of T$_1$ corresponded to their NO oxidation peaks at 443 K. This also demonstrated that the formation of highly dispersed Co$_3$O$_4$·CoMnO$_3$ through synergistic interaction between Mn-O and Co-O was responsible for the enhanced activities towards NO oxidation. The high calcination temperature leads to catalyst sintering, difficulty in oxygen mobility of Mn-O-Co, the reduction of Mn^{4+} to Mn^{3+} and Mn^{2+}, and the transformation of CoMnO$_3$ to (Co,Mn)(Co,Mn)$_2$O$_4$. These factors are responsible for the decrease of NO oxidation efficiency.

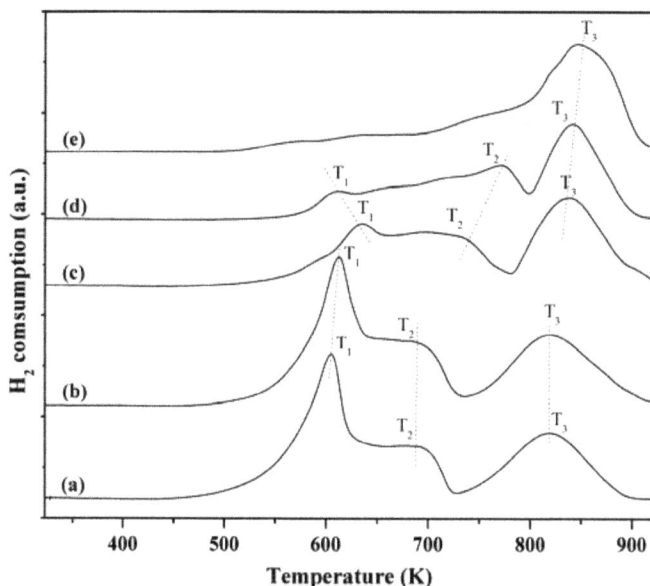

Figure 8. H$_2$-TPR profiles of 10Mn-5Co/TiO$_2$ catalysts calcined at various temperatures: (a) 573 K; (b) 673 K; (c) 773 K; (d) 873 K; and (e) 973 K.

3. Experimental Section

3.1. Catalyst Preparation

The catalysts were prepared by wet impregnating method using TiO$_2$ P25 (99.5%, Evonic Degussa, Germany) as support materials. Mn(NO$_3$)$_2$ solution (50.0%, Damao, Tianjing, China) and Co(NO$_3$)$_3$· 6H$_2$O (99.0%, Damao, Tianjing, China) were used as the precursors. The required amounts of precursors were dissolved into 5 mL deionized water and then 6.0 g of support was added into the solution. The mixed solution was stirred for 1 h and left at room temperature for 24 h. Subsequently, the samples were dried at 383 K for 12 h, followed by calcination in air for 3 h at 573–973 K. The manganese loadings were selected as 10 wt. %, and the cobalt contents were varied ranging from 0 to 10 wt. %. The catalysts were simply denoted as xMn-yCo/TiO$_2$ (T), where x and y represented the weight percentages of Mn and Co to TiO$_2$, respectively, and T represented the calcination temperature.

3.2. Catalyst Evaluation

The catalytic activity tests were carried out in a fixed bed quartz reactor (i.d. 8 mm) containing 0.15 g of the catalyst (60–100 mesh). The simulated gas mixture (contained 400 ppm NO, 5 vol. % O$_2$ and Ar balance) was fed to the reactor with a gas hourly space velocity (GHSV) of 42,000 h^{-1}. The reaction temperature was increased from 373 to 593 K at a heating rate of 1 K/min. The reactor temperature was controlled by a thermocouple and a PID-regulation system (CKW-2200, Bachy, Beijing, China). The outlet gas was monitored using an online nitrogen oxides analyzer (EC9841B, Ecotech, Ferntree Gully, Australia) to test the concentrations of NO, NO$_2$ and NO$_x$.

The percentage of NO oxidation to NO$_2$ was calculated as Equation (2):

$$[\text{Conv.}]_{NO} = ([NO]_{in} - [NO]_{out})/[NO]_{in} \times 100\% \tag{2}$$

3.3. Catalyst Characterization

Specific surface area and pore size distribution of the samples were measured using a BELSORP-mini II instrument (Ankersmid, Holland), through nitrogen adsorption at liquid nitrogen temperature (77 K) after degassing samples in vacuum at 573 K for 3 h.

The crystalline phases of the catalysts were determined by X-ray diffractometer (RIGAKU, D/Max 2500PC, Tokyo, Japan) in the 2θ angle range of 10–80° using Cu Kα radiation combined with nickel filter.

Detailed physical structural characteristics were observed with a transmission electron microscope (Tecnai G^2 F30, FEI, Hillsboro, OR, USA). Samples were prepared by ultrasonic dispersion in ethanol. The suspension was deposited on a Lacey-carbon film (Beijing, China), which was supported on a copper grid.

The surface oxidation states and atomic concentrations of samples were analyzed by a X-ray photoelectron spectrometer (ULVAC-PHI 1800, Tokyo, Japan) using Al Kα as a radiation source. The binding energy of the C 1s peak at 284.6 eV was taken for correcting the obtained spectra.

The H$_2$-Temperature programmed reduction experiments were carried out with 0.05 g catalysts under a total flow rate of 40 mL/min. Before the TPR measurements, the catalysts were pretreated in a flow of N$_2$ at 573 K for 1 h and subsequently cooled to 323 K. Then the TPR runs were carried out from 323 to 923 K with a flow of 5% H$_2$/N$_2$ at a heating rate of 10 K/min. The consumption of H$_2$ was continuously monitored using an online gas chromatograph (GC 6890, Qingdao, China) with a thermal conductivity detector (TCD).

4. Conclusions

A series of Mn-Co/TiO$_2$ catalysts were prepared by wet impregnation method and developed for the catalytic oxidation of NO to NO$_2$ below 593 K. The bimetallic catalysts Mn-Co/TiO$_2$ showed higher catalytic abilities than the Mn/TiO$_2$ and Co/TiO$_2$ within the temperature range of 403–473 K. The Mn-Co/TiO$_2$ calcined at 573 and 673 K showed excellent activities for NO oxidation at low temperatures. The correlations among the catalytic performances and the redox properties of the catalysts were investigated. It was found from the XRD, TEM and H$_2$-TPR results that the CoMnO$_3$ mixed oxides were formed, which led to the shift of the reduction peak toward low temperatures. The XPS results revealed the increase of chemical adsorbed oxygen and Mn^{4+} ratio by doping Co into Mn/TiO$_2$. Increasing calcination temperature led to the decrease of chemical adsorbed oxygen amounts and the reduction of Mn^{4+} to Mn^{3+} and Mn^{2+}. In conclusion, the higher the amount of chemically adsorbed oxygen, the higher the amount of Mn^{4+} ions, and the formation of dispersed Co$_3$O$_4$·CoMnO$_3$ mixed oxides were helpful to the oxidation of NO to NO$_2$ at low temperatures.

Acknowledgments: This project was financially supported by Foundation Science and Technology innovation Committee of Shenzhen, China (No. JCYJ20140417172417138 and No. ZDSYS201405081616222508).

Author Contributions: Feng Ouyang, Lu Qiu, and Yun Wang contributed to the experimental design. Lu Qiu, Yun Wang and Changliang Zhang contributed to all the experimental data collection. Lu Qiu wrote the first draft of the manuscript, which was then extensively improved by Feng Ouyang, Dandan Pang and Gang Cao.

Conflicts of Interest: The authors declare no conflict of interest.

References

1. Epling, W.S.; Campbell, L.E.; Yezerets, A.; Currier, N.W.; Parks, J.E. Overview of the Fundamental Reactions and Degradation Mechanisms of NO$_x$ Storage/Reduction Catalysts. *Catal. Rev. Sci. Eng.* **2004**, *46*, 163–245.
2. Nova, I.; Ciardelli, C.; Tronconi, E.; Chatterjee, D.; Bandl-Konrad, B. NH$_3$-NO/NO$_2$ chemistry over V-based catalysts and its role in the mechanism of the Fast SCR reaction. *Catal. Today* **2006**, *114*, 3–12.
3. Amberntsson, A.; Fridell, E.; Skoglundh, M. Influence of platinum and rhodium composition on the NO$_x$ storage and sulphur tolerance of a barium based NO$_x$ storage catalyst. *Appl. Catal. B* **2003**, *46*, 429–439.
4. Li, L.; Qu, L.; Cheng, J.; Li, J.; Hao, Z. Oxidation of nitric oxide to nitrogen dioxide over Ru catalysts. *Appl. Catal. B* **2009**, *88*, 224–231.

5. Salasc, S.; Skoglundh, M.; Fridell, E. A comparison between Pt and Pd in NO_x storage catalysts. *Appl. Catal. B* **2002**, *36*, 145–160.

6. Weiss, B.M.; Iglesia, E. Mechanism and site requirements for NO oxidation on Pd catalysts. *J. Catal.* **2010**, *272*, 74–81.

7. Guillén-Hurtado, N.; Atribak, I.; Bueno-López, A.; García-García, A. Influence of the cerium precursor on the physico-chemical features and NO to NO_2 oxidation activity of ceria and ceria-zirconia catalysts. *J. Mol. Catal. A* **2010**, *323*, 52–58.

8. Metkar, P.S.; Balakotaiah, V.; Harold, M.P. Experimental and kinetic modeling study of NO oxidation: Comparison of Fe and Cu-zeolite catalysts. *Catal. Today* **2012**, *184*, 115–128.

9. Vijay, R.; Hendershot, R.J.; Rivera-Jiménez, S.M.; Rogers, W.B.; Feist, B.J.; Snively, C.M.; Lauterbach, J. Noble metal free NO_x storage catalysts using cobalt discovered via high-throughput experimentation. *Catal. Commun.* **2005**, *6*, 167–171.

10. Wang, H.; Wang, J.; Wu, Z.; Liu, Y. NO Catalytic Oxidation Behaviors over CoO_x/TiO_2 Catalysts Synthesized by Sol-Gel Method. *Catal. Lett.* **2010**, *134*, 295–302.

11. Wu, Z.; Tang, N.; Xiao, L.; Liu, Y.; Wang, H. MnO_x/TiO_2 composite nanoxides synthesized by deposition-precipitation method as a superior catalyst for NO oxidation. *J. Colloid Interface Sci.* **2010**, *352*, 143–148.

12. Brik, Y.; Kacimi, M.; Ziyad, M.; Bozon-Verduraz, F. Titania-Supported Cobalt and Cobalt-Phosphorus Catalysts: Characterization and Performances in Ethane Oxidative Dehydrogenation. *J. Catal.* **2001**, *202*, 118–128.

13. Yang, W.-H.; Kim, M.H.; Ham, S.-W. Effect of calcination temperature on the low-temperature oxidation of CO over CoO_x/TiO_2 catalysts. *Catal. Today* **2007**, *123*, 94–103.

14. Zafeiratos, S.; Dintzer, T.; Teschner, D.; Blume, R.; Hävecker, M.; Knop-Gericke, A.; Schlögl, R. Methanol oxidation over model cobalt catalysts: Influence of the cobalt oxidation state on the reactivity. *J. Catal.* **2010**, *269*, 309–317.

15. Petitto, S.C.; Marsh, E.M.; Carson, G.A.; Langell, M.A. Cobalt oxide surface chemistry: The interaction of CoO(100), Co_3O_4(110) and Co_3O_4(111) with oxygen and water. *J. Mol. Catal. A* **2008**, *281*, 49–58.

16. Irfan, M.F.; Goo, J.H.; Kim, S.D. Co_3O_4 based catalysts for NO oxidation and NO_x reduction in fast SCR process. *Appl. Catal. B* **2008**, *78*, 267–274.

17. Ettireddy, P.R.; Ettireddy, N.; Boningari, T.; Pardemann, R.; Smirniotis, P.G. Investigation of the selective catalytic reduction of nitric oxide with ammonia over Mn/TiO_2 catalysts through transient isotopic labeling and *in situ* FT-IR studies. *J. Catal.* **2012**, *292*, 53–63.

18. Boningari, T.; Pappas, D.K.; Ettireddy, P.R.; Kotrba, A.; Smirniotis, P.G. Influence of SiO_2 on M/TiO_2 (M = Cu, Mn, and Ce) formulations for low-temperature selective catalytic reduction of NO_x with NH_3: Surface properties and key components in relation to the activity of NO_x reduction. *Ind. Eng. Chem. Res.* **2015**, *54*, 2261–2273.

19. Thirupathi, B.; Smirniotis, P.G. Co-doping a metal (Cr, Fe, Co, Ni, Cu, Zn, Ce, and Zr) on Mn/TiO_2 catalyst and its effect on the selective reduction of NO with NH_3 at low-temperatures. *Appl. Catal. B* **2011**, *110*, 195–206.

248

20. Liu, Z.; Yi, Y.; Zhang, S.; Zhu, T.; Zhu, J.; Wang, J. Selective catalytic reduction of NO_x with NH_3 over Mn-Ce mixed oxide catalyst at low temperatures. *Catal. Today* **2013**, *216*, 76–81.

21. Yu, J.; Guo, F.; Wang, Y.; Zhu, J.; Liu, Y.; Su, F.; Gao, S.; Xu, G. Sulfur poisoning resistant mesoporous Mn-base catalyst for low-temperature SCR of NO with NH_3. *Appl. Catal. B* **2010**, *95*, 160–168.

22. Lee, S.M.; Park, K.H.; Kim, S.S.; Kwon, D.W.; Hong, S.C. Effect of the Mn oxidation state and lattice oxygen in Mn-based TiO_2 catalysts on the low-temperature selective catalytic reduction of NO by NH_3. *J. Air Waste Manag.* **2012**, *62*, 1085–1092.

23. Qi, G.; Yang, R.T. Low-temperature selective catalytic reduction of NO with NH_3 over iron and manganese oxides supported on titania. *Appl. Catal. B* **2003**, *44*, 217–225.

24. Meng, B.; Zhao, Z.; Chen, Y.; Wang, X.; Li, Y.; Qiu, J. Low-temperature synthesis of Mn-based mixed metal oxides with novel fluffy structures as efficient catalysts for selective reduction of nitrogen oxides by ammonia. *Chem. Commun.* **2014**, *50*, 12396–12399.

25. Venkataswamy, P.; Jampaiah, D.; Lin, F.; Alxneit, I.; Reddy, B.M. Structural properties of alumina supported Ce-Mn solid solutions and their markedly enhanced catalytic activity for CO oxidation. *Appl. Surf. Sci.* **2015**, *349*, 299–309.

26. Yao, Y.; Cai, Y.; Wu, G.; Wei, F.; Li, X.; Chen, H.; Wang, S. Sulfate radicals induced from peroxymonosulfate by cobalt manganese oxides $Co_xMn_{3-x}O_4$ for Fenton-Like reaction in water. *J. Hazard. Mater.* **2015**, *296*, 128–137.

27. Rida, K.; Benabbas, A.; Bouremmad, F.; Peña, M.A.; Martínez-Arias, A. Surface properties and catalytic performance of $La_{1-x}Sr_xCrO_3$ perovskite-type oxides for CO and C_3H_6 combustion. *Catal. Commun.* **2006**, *7*, 963–968.

28. Sutthiumporn, K.; Kawi, S. Promotional effect of alkaline earth over Ni-La_2O_3 catalyst for CO_2 reforming of CH_4: Role of surface oxygen species on H_2 production and carbon suppression. *Int. J. Hydrog. Energ.* **2011**, *36*, 14435–14446.

29. Merino, N.A.; Barbero, B.P.; Eloy, P.; Cadús, L.E. $La_{1-x}Ca_xCoO_3$ perovskite-type oxides: Identification of the surface oxygen species by XPS. *Appl. Surf. Sci.* **2006**, *253*, 1489–1493.

30. Petitto, S.C.; Langell, M.A. Surface composition and structure of Co_3O_4(110) and the effect of impurity segregation. *J. Vac. Sci. Technol. A* **2004**, *22*, 1690–1696.

31. Wu, Z.; Jin, R.; Liu, Y.; Wang, H. Ceria modified MnO_x/TiO_2 as a superior catalyst for NO reduction with NH_3 at low-temperature. *Catal. Commun.* **2008**, *9*, 2217–2220.

32. Wang, Z.L.; Yin, J.S.; Mo, W.D.; Zhang, Z.J. *In-Situ* Analysis of Valence Conversion in Transition Metal Oxides Using Electron Energy-Loss Spectroscopy. *J. Phys. Chem. B* **1997**, *101*, 6794–6798.

33. Sreekanth, P.M.; Pena, D.A.; Smirniotis, P.G. Titania Supported Bimetallic Transition Metal Oxides for low-temperature SCR of NO with NH_3. *Ind. Eng. Chem. Res.* **2006**, *45*, 6444–6449.

34. Standard Electrode Potential (Data Page). Available online: https://en.wikipedia.org/wiki/Standard_electrode_potential_(data_page) (accessed on 19 October 2015).

Aerobic Catalytic Oxidation of Cyclohexene over TiZrCo Catalysts

Tong Liu, Haiyang Cheng, Weiwei Lin, Chao Zhang, Yancun Yu and Fengyu Zhao

Abstract: The aerobic oxidation of hydrocarbon is of great significance from the viewpoints of both fundamental and industry studies as it can transfer the petrochemical feedstock into valuable chemicals. In this work, we investigated the aerobic oxidation of cyclohexene over TiZrCo catalysts, in which 2-cyclohexen-1-one was produced with a high selectivity of 57.6% at a conversion of 92.2%, which are comparable to the best results reported for the aerobic oxidation of cyclohexene over heterogeneous catalysts. The influences of kinds of solvent, substrate concentration and reaction temperature were evaluated. Moreover, the catalytic performance of the TiZrCo catalyst and the main catalytic active species were also discussed. The results of SEM, XRD and XPS suggested that the surface CoO and Co_3O_4 species are the catalytic active species and contribute to the high activity and selectivity in the present cyclohexene oxidation. The present catalytic system should have wide applications in the aerobic oxidation of hydrocarbons.

Reprinted from *Catalysts*. Cite as: Liu, T.; Cheng, H.; Lin, W.; Zhang, C.; Yu, Y.; Zhao, F. Aerobic Catalytic Oxidation of Cyclohexene over TiZrCo Catalysts. *Catalysts* **2016**, *6*, 24.

1. Introduction

The selective oxidation of hydrocarbon is of great importance in the chemical industry, and the oxidation of alkenes to value-added chemicals has been paid more attention [1–4]. For example, 2-cyclohexen-1-one, which can be produced from oxidation of the C–H bond at the allylic site of cyclohexene, is one of the important fine chemicals because it is widely used in the manufacture of perfumes, pharmaceuticals, dyestuff and agrochemicals [5–7]. However, the yield of 2-cyclohexen-1-one produced from cyclohexene oxidation is quite low due to the existence of two active sites (C–H and C=C bond) on the cyclohexene molecule [8]. Thus, it is still a great challenge to enhance the selectivity of 2-cyclohexen-1-one by designing an efficient catalyst. For oxidation of cyclohexene, the traditional oxidants are iodosylbenzene, sodium hypochlorite, chromium trioxide, t-butyl hydroperoxide, H_2O_2, *etc.* These strong oxidants can improve the oxidation rate and increase the yield of product, but they are expensive and harmful to the environment. Therefore, developing a green oxidation process is of importance in view of the academic research and industrial application. Recently, the oxidation with molecular oxygen as an oxidant has received much attention as it is a cheap, abundant and environmentally benign process and some achievements have been obtained [9–13]. For example, the ionic liquids with metal chelate anion $[C_{10}mim][Co(F_6\text{-}acac)_3]$ exhibited high catalytic activity for the allylic oxidation of cyclohexene in the absence of solvent with a high selectivity of 81% to 2-cyclohexen-1-one at a conversion of 100% [10], and the Cu-[cationic salphen][Br$^-$]$_2$ complex presented a selectivity of 64.1% to 2-cyclohexen-1-one at a 100% conversion for the aerobic oxidation of cyclohexene in acetonitrile [11]. Although these homogeneous catalysts are efficient for the selective allylic oxidation of cyclohexene, their industrial application is limited due to the difficult separation, and the residual metal ions will affect the quality of the product. Hence, heterogeneous catalysts have been developed by immobilizing metal complexes on solid supports, such as resinate-immobilized Co(II), which exhibited high catalytic activity with a 44.4% selectivity to 2-cyclohexen-1-one at a 94.5% conversion of cyclohexene [12]. Core-shell type Fe_3O_4@chitosan-Schiff base-immobilized Co(II), Cu(II) and Mn(II) complexes were also reported active for the cyclohexene oxidation, and a selectivity of 77.2% to 2-cyclohexen-1-one was obtained at a conversion of 46.8% [13]. On the other hand, the supported transition metal or oxides were also employed as heterogeneous catalysts for the allylic oxidation of cyclohexene [6,14–20]. Recently, Au nanoparticles supported on modified bentonite and silica gave a high conversion (92%) and an excellent selectivity (97%) to 2-cyclohexen-1-one in the aerobic oxidation of cyclohexene without solvent [21]. It was also reported that PdO/SBA-15 was an active catalyst for the oxidation of cyclohexene in acetonitrile, and a conversion of 56% and a selectivity of 82% to 2-cyclohexen-1-one were obtained [6]. In addition, nitrogen-doped carbon nanotubes, and graphitic carbon nitride-supported FeO and CoO were also effective catalysts for the oxidation of cyclohexene [14,22]. Until now, it was still a hot topic to design a heterogeneous catalyst for the aerobic allylic oxidation of cyclohexene.

In our previous work, we found that Ti-Zr-Co metallic catalyst was effective for the oxidation of cyclohexane and ethylbenzene [23–25], in which cyclohexanol and cyclohexanone were produced with a high selectivity of 90% at a conversion around 7%, and acetophenone was produced with a 69.2% selectivity at a high conversion of 61.9%. The Ti-Zr-Co metallic catalyst is simple and cheap in production and sturdy to wearing in the utilization, as compared to those reported catalysts such as metal-organic complex, metal nanoparticles and nanocarbon materials. Therefore, it inspires us to study its efficiency in catalyzing other hydrocarbon oxidations. In the present work, the catalytic performances of the Ti-Zr-Co catalysts were discussed for the aerobic oxidation of cyclohexene. We found that 2-cyclohexen-1-one was produced with a high selectivity of 57.6% at a conversion of 92.2%, which are comparable to the best results reported for the aerobic oxidation of cyclohexene over heterogeneous catalysts. It was confirmed the surface CoO and Co_3O_4 acted as the catalytic active sites and contributed to the excellent conversion and selectivity.

2. Results and Discussion

For the oxidation of cyclohexene, it is very difficult to control product selectivity due to the existence of the two active groups of the C-H bond at the allylic site and the C=C bond, as when the C–H bond is oxidized, 2-cyclohexene-1-ol, 2-cyclohexene-1-one or cyclohexene hydroperoxide will be generated; as the C=C bond is oxidized, cyclohexene oxide, cyclohexanol, cyclohexanone, cyclohexanediol and dialdehyde will be produced (Scheme 1) [26].

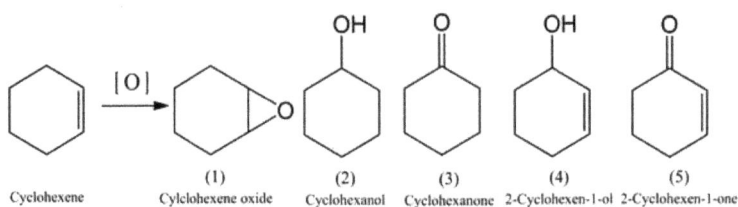

(1)	(2)	(3)	(4)	(5)
Cyclohexene	Cylclohexene oxide	Cyclohexanol	Cyclohexanone	2-Cyclohexen-1-ol 2-Cyclohexen-1-one

Scheme 1. Reaction route for the aerobic oxidation of cyclohexene.

Firstly, we examined several organic solvents for the oxidation of cyclohexene and the results are listed in Table 1. Among the examined solvents, acetone and acetonitrile are more effective, and the higher conversion and selectivity of 2-cyclohexen-1-one were obtained; in contrast, ethanol and cyclohexane are less efficient. However, it was found that acetone could be oxidized (to 2,2-diethoxypropane) during the reaction. As a result, acetonitrile is a suitable solvent which led to a conversion of cyclohexene of 38.0% and a selectivity to 2-cyclohexen-1-one of 60.6%. Therefore, the acetonitrile was selected as solvent in the following studies.

Next, the reaction conditions were evaluated for the oxidation of cyclohexene in acetonitrile. The results for the effect of the concentration of cyclohexene are shown in Table 2. The conversion of cyclohexene increased significantly with the concentration of cyclohexene; it increased from 33.8% to 97.5% when the cyclohexene concentration was raised from 4.8% to 13.0%. However, the selectivity to 2-cyclohexen-1-one decreased linearly due to deep oxidation of 2-cyclohexen-1-one to undesired byproducts at a higher conversion. The aerobic oxidation of cyclohexene is a radical reaction; it contains the chain-initiation, -propagation and -termination steps [1]. The produced radical in the initial step could promote the following steps and was more efficient at higher cyclohexene concentrations as it enhanced the impact probability of radicals, resulting in an increase of the conversion of cyclohexene. It is notable that when the concentration of cyclohexene increased to 16.7%, the reaction became very violent with a sharp increase of pressure (up to 10 MPa at 3.3 h); for safety, the concentration of cyclohexene was controlled below 16.7% under the present reaction conditions.

Table 1. Effect of solvent on the oxidation of cyclohexene over $Ti_{60}Zr_{10}Co_{30}$ catalyst.

Solvent	Conversion (%)	Selectivity (%) [a]					
		(1)	(2)	(3)	(4)	(5)	Others [b]
Acetone	43.9	7.3	0	1.5	13.3	71.8	6.1
Acetonitrile	38.0	3.0	0.4	1.4	13.6	60.6	21.0
Ethanol	22.0	1.4	0	0.3	4.7	25.0	68.6 [c]
Cyclohexane	6.2	-	0.2	7.4	9.9	37.8	44.7 [d]

Reaction conditions: cyclohexene 1 mL (at a concentration of 4.8%), solvent 20 mL, $Ti_{60}Zr_{10}Co_{30}$ 20 mg, O_2 2 MPa, 100 °C, 12 h. [a] (1) cyclohexene oxide; (2) cyclohexanol; (3) cyclohexanone; (4) 2-cyclohexen-1-ol; (5) 2-cyclohexen-1-one; [b] Others may consist of reaction intermediate such as cyclohexene hydrogen peroxide, deeply oxidized products such as some ring-opening acids or the byproduct from solvent reacting with substrate. [c] The ethanol was oxidized into acetic acid and ethyl acetate. [d] A certain amount of cyclohexane was found to be oxidized to cyclohexanol and cyclohexanone. The result could not exact calculated.

Table 2. Effect of the cyclohexene concentration on the oxidation reaction.

Entry	Cyclohexene (%) [a]	Time (h)	Conversion (%)	Selectivity (%)					
				(1)	(2)	(3)	(4)	(5)	Others [b]
1	4.8	5	33.8	3.1	-	1.7	19.1	71.3	4.9
2	9.1	5	67.0	5.5	-	0.7	11.6	62.5	19.8
3	13.0	5	97.5	-	-	0.6	3.7	49.5	46.1
4 [c]	16.7	3.3	92.8	1.9	0.9	1.4	6.2	30.2	59.4

Reaction conditions: acetonitrile 20 mL, $Ti_{60}Zr_{10}Co_{30}$ 20 mg, O_2 2 MPa, 120 °C. [a] Concentration of cyclohexene = $V_{cyclohexene}/(V_{cyclohexene} + V_{acetonitrile}) \times 100\%$; [b] Others may consist of reaction intermediate such as cyclohexene hydrogen peroxide, deeply oxidized products such as some ring-opening acids; [c] When the concentration of cyclohexene reached 16.7%, the reaction proceeded very quickly with a suddenly pressure rising, and the reaction was stopped at 3.3 h for safety.

It is well known that temperature is one of the most important factors for oxidation reactions. Generally, high temperature is in favor for the oxidation of hydrocarbons. As the results show in Figure 1, the conversion increased from 17.8% to 98.2% while the selectivity of 2-cyclohexen-1-one decreased from 70.1% to 43.7% with the temperature rising from 80 to 140 °C. Deep oxidation is serious at higher temperatures and undesirable byproducts such as ring-opening acids are produced. The selectivity of cyclohexene oxide, cyclohexanol and cyclohexanone changed slightly, within 5%. The optimal temperature was 120 °C, at which a 57.6% selectivity of 2-cyclohexene-1-one was obtained at a high conversion of 92.2%.

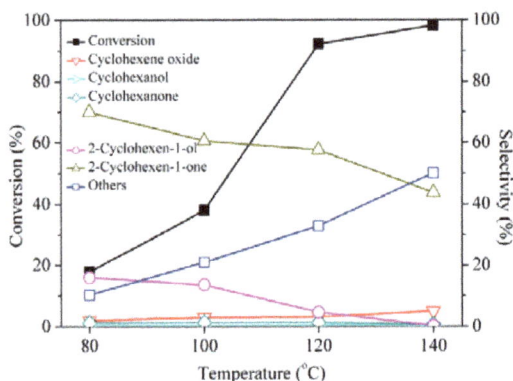

Figure 1. Effect of temperature on the oxidation of cyclohexene. Reaction conditions: acetonitrile 20 mL, cyclohexene 1 mL (at a concentration of 4.8%), $Ti_{60}Zr_{10}Co_{30}$ 20 mg, 2 MPa O_2, 12 h.

In addition, the influence of reaction time on the conversion and product selectivity were examined at conditions of 120 °C, a cyclohexene concentration of 9.1% and an oxygen pressure of 2 MPa. As shown in Figure 2, the conversion increased to 96.1% with extending the reaction time to 8 h. The selectivity of 2-cyclohexen-1-ol and 2-cyclohexen-1-one decreased from 20.3% to 4.0% and 68.8% to 55.3% due to the formation and accumulation of deep oxidation products with the reaction proceeding. The other products derived from deep oxidation such as the ring-opening acids increased with a selectivity up to 40% (8 h) from 6% (1 h), and the intermediates such as cyclohexene oxide, cyclohexanol and cyclohexanone changed very little (<5.5%).

Generally, the aerobic radical oxidation of hydrocarbons can automatically occur in the absence of catalyst [27], but it is very slow and the selectivity of the desired product is poor. The oxidation of cyclohexene can also automatically occur somewhat at certain conditions without catalyst; however, the conversion and selectivity to 2-cyclohexen-1-one are quite low [22]. Herein, we also found that the cyclohexene can convert with a conversion of 14.4% under the aerobic oxidation conditions without catalyst. As expected, the presence of TiZrCo catalysts can promote the reaction rate significantly as shown in Figure 3. The catalytic performances of TiZrCo catalysts were discussed according to the composition and the surface active species. It is clear that the catalytic activity depends on the Co content in the TiZrCo metallic catalysts, as the conversion of cyclohexene increased to 38.5% over $Ti_{50}Zr_{10}Co_{40}$ from 32% over $Ti_{70}Zr_{10}Co_{20}$, while the selectivity to 2-cyclohexene-1-one was around 66%–69%, and changed very little. These results indicate that the ternary TiZrCo metallic catalysts are effective for the present cyclohexene oxidation, and the content of Co affects the catalytic activity significantly. In order to check the catalytic efficiency of Co species in the present oxidation, CoTi$_2$ and Co$_3$O$_4$/TiO$_2$ with the higher content of Co species were also examined. Unfortunately, both of them gave lower activity compared to the ternary TiZrCo metallic catalysts, as seen in Figure 3. These results suggested not only the Co species but the surface and bulk structure of the TiZrCo catalysts play an important role in the present oxidation. It is assumed that the addition of Zr may produce surface defects and induce the formation of many more CoO and Co$_3$O$_4$ active species.

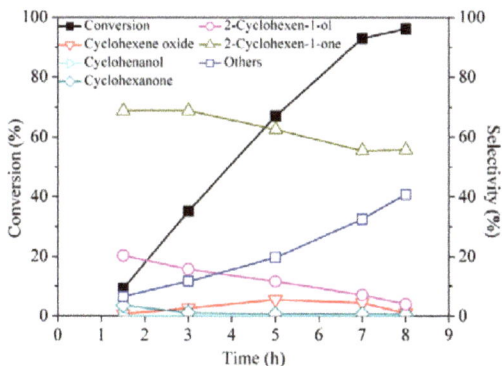

Figure 2. The variation of the conversion and selectivity with extending reaction time. Reaction conditions: cyclohexene 2 mL (at a concentration of 9.1%), acetonitrile 20 mL, Ti$_{60}$Zr$_{10}$Co$_{30}$ 20 mg, 2 MPa O$_2$, 120 °C.

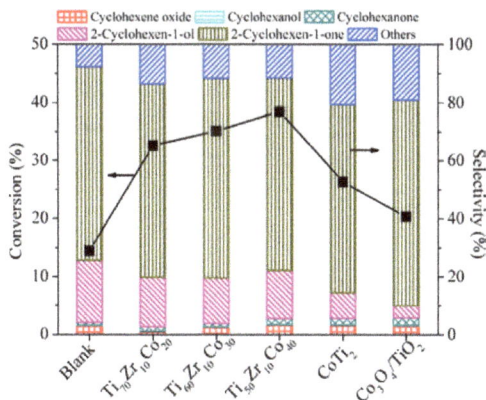

Figure 3. Comparison of the catalytic performances of different catalysts in the oxidation of cyclohexene. Reaction conditions: cyclohexene 2 mL, acetonitrile 20 mL, catalyst 20 mg, O$_2$ 2 MPa, 120 °C, 3 h.

Therefore, the structure of the TiZrCo catalysts and their relationship to the catalytic performances were discussed. The SEM image of a representative sample of $Ti_{60}Zr_{10}Co_{30}$ is shown in Figure 4a. The size of the particles is at a range of 50–100 μm, and the other two samples with different compositions should have a similar morphology as all the samples were crushed and screen-separated by 140 meshes before being examined. The bulk structure of these samples was characterized with XRD as shown in Figure 4b. For $Ti_{60}Zr_{10}Co_{30}$, with a molar ratio of Co/Ti of 1/2, the diffraction patterns are in accordance with the $CoTi_2$ phase without other crystal phases. For $Ti_{70}Zr_{10}Co_{20}$, an I-phase was found beside the $CoTi_2$ phase [28]. However, for $Ti_{50}Zr_{10}Co_{40}$, a new CoTi phase was detected beside the $CoTi_2$ phase.

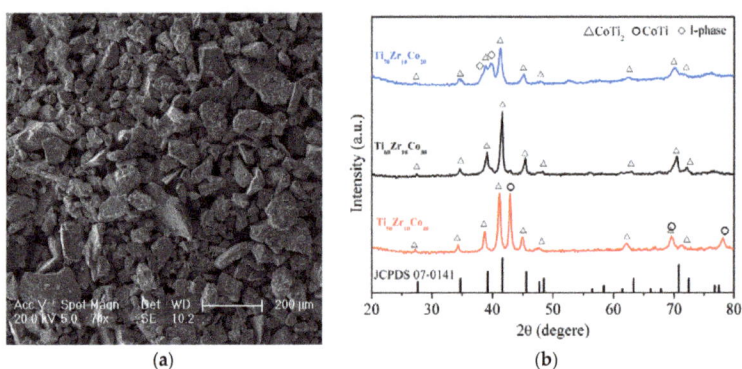

(a)   (b)

Figure 4. The SEM image of $Ti_{60}Zr_{10}Co_{30}$ (**a**); and the XRD patterns of Ti-Zr-Co catalysts and the JCPDS 07-0141 file identifying $CoTi_2$ (**b**).

Therefore, the variation of Co content will impact the bulk structure of TiZrCo significantly, which may have an effect on the catalytic activity. By comparison, the surface species will play a more important role in the catalysis due to the catalytic reaction always occurring on the surface of the heterogeneous catalyst. The surface composition of TiZrCo was studied by XPS. As shown in Figure 5, the Ti and Zr existed on the surface with oxidation states of TiO_2 and ZrO_2, and Co mainly existed on the surface with metallic Co, oxides of CoO and Co_3O_4, as confirmed by the peak at the binding energy of 777.6 eV and intense shake-up satellites Co 2p XPS spectra around 6 eV above the primary spin-orbit Bes [29–31]; Co_3O_4 was confirmed by Co^{3+}/Co^{2+} (2/1) [32]. The surface compositions calculated according to the results of XPS (Figure 5) are listed in Table 3. To consider the main active species of the Co element, it is clear that all the catalysts contained the same species of Co, CoO and Co_3O_4 on their surfaces, including $CoTi_2$. The most active $Ti_{50}Zr_{10}Co_{40}$ catalyst contains a higher ratio of CoO on the surface and different bulk phases, which suggested CoO may be more effective than the Co_3O_4 species, and the bulk phase may be also involved in the present catalysis based on the reaction results in Figure 3. In addition, the surface area of TiZrCo catalysts was also examined, as shown in Table 3; it should have less effect on the present oxidation compared to the surface active species, as the $Ti_{70}Zr_{10}Co_{20}$ with the largest surface area did show a lower conversion, which contains a lower ratio of CoO and Co_3O_4 on the surface.

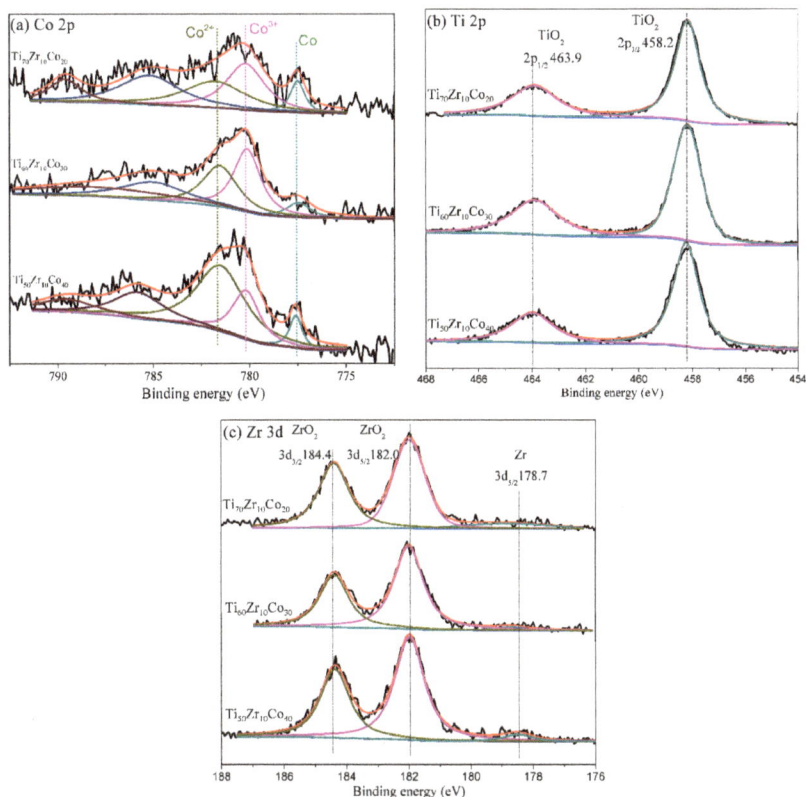

Figure 5. The XPS spectra of (**a**) Co 2p; (**b**) Ti 2p and (**c**) Zr 3d for Ti-Zr-Co catalysts.

Table 3. Textural and surface properties.

Entry	Catalyst	Bulk Phase [a]	Surface Content of Co (%) [b]			S_{BET} (m²/g) [c]
			Co	CoO	Co_3O_4	
1	$CoTi_2$	$CoTi_2$, I-phase	31.5	37.8	30.7	-
2	$Ti_{70}Zr_{10}Co_{20}$	$CoTi_2$, I-phase	26.9	31.2	40.9	189
3	$Ti_{60}Zr_{10}Co_{30}$	$CoTi_2$	20.1	32.8	47.1	70
4	$Ti_{50}Zr_{10}Co_{40}$	$CoTi_2$, CoTi	11.5	70.3	18.2	76

[a] The bulk phase of the alloy catalysts was obtained based on the XRD patterns; [b] The surface composition of the alloy catalysts was calculated from XPS results; [c] The surface areas were calculated using the BET equation.

3. Experimental Section

3.1. Ti-Zr-Co Alloy Preparation

The series of Ti-Zr-Co alloys, as reported in our previous works [25], were prepared by arc-melting of Ti (99 wt. %), Zr (97 wt. %) and Co (99 wt. %) metals with a certain mole ratio on a water-cooled cuprum hearth in a high-purity argon atmosphere at 250 A. To make the chemical compositions homogenous, the ingot of alloy was turned over and remelted at least three times. After that, the surface of the cast ingot was burnished in order to eliminate the oxide layer. Then the alloy ingot was crushed by repeated manual beating with a steel pestle and mortar, and the alloy powders were screen separated by 140 meshes.

3.2. Catalyst Characterization

The phase composition and microstructure of the alloys were examined by X-ray diffraction (XRD) on a Bruker-AXS D8 ADVANCE (Bruker AXS, Karlsruhe, Germany) with Kα. The leaching of Ti, Zr or Co in the filtrate was not detected by ICP-OES measurement (iCAP6300, Thermo Waltham, MA, USA). XPS measurements were performed by using a VG Microtech 3000 Multilab. The electronic states of Co 2p, Zr 3d and Ti 2p were determined. All XPS spectra were corrected to the C 1s peak at 284.6 eV. Scanning electron microscopy (SEM) image was performed on a Hitachi S-4800 field emission scanning electron microscope (HITACHI, Tokyo, Japan) at an accelerating voltage of 10 kV, and the size of particles was in a range of 50–100 μm. Nitrogen porosimetry measurement was performed on a Micromeritics ASAP 2020M instrument (Micromeritics, Norcross, GA, USA). The surface areas were calculated using the BET equation.

3.3. Catalytic Tests

The catalytic performance was tested in a stainless steel autoclave with a Teflon inner liner (50 mL). Typically, the Ti-Zr-Co catalyst (20 mg), cyclohexene (2 mL) and acetonitrile (20 mL) were added. The reactor was then sealed, and placed into an oil bath preset to 120 °C for 5 min. O_2 (2 MPa) was introduced and the reaction was started with a continuously stirring at 1200 rpm. When the reaction finished, the reactor was cooled to room temperature and then depressurized carefully. Then the reaction solution was diluted with ethanol to 50 mL, the composition of reaction products was confirmed by gas chromatography/mass spectrometry (Agilent 5890, Santa Clara, CA, USA) and analyzed with a gas chromatograph (Shimadzu, Kyoto, Japan, GC-2010) equipped with a capillary column (RTX-50, Bellefonte, PA, USA, 30 m × 0.25 mm × 0.25 μm, carrier: N_2) and a flame ionization detector (FID).

Safety warning: The use of compressed O_2 in the presence of organic substrates requires appropriate safety precautions and must be carried out in suitable equipment. When the pressure decreased far more quickly, the reaction must be stopped immediately. Note, the reaction should be diluted by solvent, otherwise, the blast occur as the reaction is very violent.

4. Conclusions

In summary, TiZrCo catalysts were studied for the first time for the oxidation of cyclohexene. High selectivity (57.6%) to 2-cyclohexen-1-one was obtained at a high conversion (92.2%) of cyclohexene. CoO and Co_3O_4 on the surface are the main active species and contribute to the high activity and selectivity in the present cyclohexene oxidation. These results indicate that TiZrCo metallic catalysts are effective for the aerobic oxidation of cyclohexene. It is important and significant to extend the studies of the metallic alloy catalyst in catalyzing the aerobic oxidation of hydrocarbons. It is expected that the TiZrCo catalysts may have a broad prospect in industrial applications.

Acknowledgments: The authors gratefully acknowledge the financial support from the One Hundred Talent Program of CAS and Youth Innovation Promotion Association CAS.

Author Contributions: Tong Liu did the experiment and wrote the paper, Haiyang Cheng discussed the results and revised the paper, Chao Zhang, Weiwei Lin, Yancun Yu, Fengyu Zhao supervised this work.

Conflicts of Interest: The authors declare no conflict of interest.

References

1. Suresh, A.K.; Sharma, M.M.; Sridhar, T. Engineering aspects of industrial liquid-phase air oxidation of hydrocarbons. *Ind. Eng. Chem. Res.* **2000**, *39*, 3959–3997.
2. Thomas, J.M.; Raja, R.; Sankar, G.; Bell, R.G. Molecular sieve catalysts for the regioselective and shapeselective oxyfunctionalization of alkanes in air. *Acc. Chem. Res.* **2001**, *34*, 191–200.
3. Patil, N.S.; Uphade, B.S.; Jana, P.; Bharagava, S.K.; Choudhary, V.R. Epoxidation of styrene by anhydrous *t*-butyl hydroperoxide over reusable gold supported on MgO and other alkaline earth oxides. *J. Catal.* **2004**, *223*, 236–239.
4. Kuznetsov, M.L.; Rocha, B.G.M.; Pombeiro, A.J.L.; Shul'pin, G.B. Oxidation of olefins with hydrogen peroxide catalyzed by bismuth salts: A mechanistic study. *ACS Catal.* **2015**, *5*, 3823–3835.
5. Wu, H.Y.; Zhang, X.L.; Yang, C.Y.; Chen, X.; Zheng, X.C. Alkali-hydrothermal synthesis and characterization of W-MCM-41 mesoporous materials with various Si/W molar ratios. *Appl. Surf. Sci.* **2013**, *270*, 590–595.
6. Ganji, S.; Bukya, P.; Vakati, V.; Rao, K.S.R.; Burri, D.R. Highly efficient and expeditious PdO/SBA-15 catalysts for allylic oxidation of cyclohexene to cyclohexenone. *Catal. Sci. Technol.* **2013**, *3*, 409–414.

7. Cai, Z.Y.; Zhu, M.Q.; Chen, J.; Shen, Y.Y.; Zhao, J.; Tang, Y.; Chen, X.Z. Solvent-free oxidation of cyclohexene over catalysts with molecular oxygen. *Catal. Commun.* **2010**, *12*, 197–201.

8. Weiner, H.; Trovarelli, A.; Finke, R.G. Expanded product, plus kinetic and mechanistic, studies of polyoxoanion-based cyclohexene oxidation catalysis: The detection of similar to 70 products at higher conversion leading to a simple, product-based test for the presence of olefin autoxidation. *J. Mol. Catal. A* **2003**, *191*, 217–252.

9. Chen, K.X.; Zhang, P.F.; Wang, Y.; Li, H.R. Metal-free allylic/benzylic oxidation strategies with molecular oxygen: Recent advances and future prospects. *Green Chem.* **2014**, *16*, 2344–2374.

10. Zhang, P.F.; Gong, Y.T.; Lv, Y.Q.; Guo, Y.; Wang, Y.; Wang, C.M.; Li, H.R. Ionic liquids with metal chelate anions. *Chem. Commun.* **2012**, *48*, 2334–2336.

11. Abdolmaleki, A.; Adariani, S.R. Copper-cationic salphen catalysts for the oxidation of cyclohexene by oxygen. *Catal. Commun.* **2015**, *59*, 97–100.

12. Yin, C.X.; Yang, Z.H.; Li, B.; Zhang, F.M.; Wang, J.Q.; Ou, E.C. Allylic oxidation of cyclohexene with molecular oxygen using cobalt resinate as catalyst. *Catal. Lett.* **2009**, *131*, 440–443.

13. Cai, X.; Wang, H.; Zhang, Q.; Tong, J.; Lei, Z. Magnetically recyclable core-shell Fe_3O_4@chitosan-schiff base complexes as efficient catalysts for aerobic oxidation of cyclohexene under mild conditions. *J. Mol. Catal. A* **2014**, *383*, 217–224.

14. Yang, D.X.; Jiang, T.; Wu, T.B.; Zhang, P.; Han, H.L.; Han, B.X. Highly selective oxidation of cyclohexene to 2-cyclohexene-1-one in water using molecular oxygen over Fe-Co-g-C_3N_4. *Catal. Sci. Technol.* **2016**, *6*, 193–200.

15. Silva, F.P.; Jacinto, M.J.; Landers, R.; Rossi, L.M. Selective allylic oxidation of cyclohexene by a magnetically recoverable cobalt oxide catalyst. *Catal. Lett.* **2011**, *141*, 432–437.

16. Wang, L.; Wang, H.; Hapala, P.; Zhu, L.; Ren, L.; Meng, X.; Lewis, J.P.; Xiao, F.S. Superior catalytic properties in aerobic oxidation of olefins over Au nanoparticles on pyrrolidone-modified SBA-15. *J. Catal.* **2011**, *281*, 30–39.

17. Donoeva, B.G.; Ovoshchnikov, D.S.; Golovko, V.B. Establishing a au nanoparticle size effect in the oxidation of cyclohexene using gradually changing Au catalysts. *ACS Catal.* **2013**, *3*, 2986–2991.

18. Ghiaci, M.; Dorostkar, N.; Victoria Martinez-Huerta, M.; Fierro, J.L.G.; Moshiri, P. Synthesis and characterization of gold nanoparticles supported on thiol functionalized chitosan for solvent-free oxidation of cyclohexene with molecular oxygen. *J. Mol. Catal. A* **2013**, *379*, 340–349.

19. Zhou, L.; Lu, T.; Xu, J.; Chen, M.; Zhang, C.; Chen, C.; Yang, X.; Xu, J. Synthesis of hierarchical MeAPO-5 molecular sieves-catalysts for the oxidation of hydrocarbons with efficient mass transport. *Microporous Mesoporous Mater.* **2012**, *161*, 76–83.

20. Sang, X.X.; Zhang, J.L.; Wu, T.B.; Zhang, B.X.; Ma, X.; Peng, L.; Han, B.X.; Kang, X.C.; Liu, C.C.; Yang, G.Y. Room-temperature synthesis of mesoporous CuO and its catalytic activity for cyclohexene oxidation. *RSC Adv.* **2015**, *5*, 67168–67174.

21. Nejad, M.S.; Ghasemi, G.; Martinez-Huerta, M.V.; Ghiaci, M. Synthesis and characterization of Au nanocatalyst on modifed bentonite and silica and their applications for solvent free oxidation of cyclohexene with molecular oxygen. *J. Mol. Catal. A* **2015**, *406*, 118–126.

22. Cao, Y.; Yu, H.; Peng, F.; Wang, H. Selective allylic oxidation of cyclohexene catalyzed by nitrogen-doped carbon nanotubes. *ACS Catal.* **2014**, *4*, 1617–1625.

23. Hao, J.M.; Wang, J.Y.; Wang, Q.; Yu, Y.C.; Cai, S.X.; Zhao, F.Y. Catalytic oxidation of cyclohexane over Ti-Zr-Co catalysts. *Appl. Catal. A* **2009**, *368*, 29–34.

24. Hao, J.M.; Liu, B.Z.; Cheng, H.Y.; Wang, Q.; Wang, J.Y.; Cai, S.X.; Zhao, F.Y. Cyclohexane oxidation on a novel $Ti_{70}Zr_{10}Co_{20}$ catalyst containing quasicrystal. *Chem. Commun.* **2009**, 3460–3462.

25. Liu, T.; Cheng, H.Y.; Sun, L.S.; Liang, F.; Zhang, C.; Ying, Z.; Lin, W.W.; Zhao, F.Y. Synthesis of acetophenone from aerobic catalytic oxidation of ethylbenzene over Ti-Zr-Co alloy catalyst: Influence of annealing conditions. *Appl. Catal. A* **2016**, *512*, 9–14.

26. Dapurkar, S.E.; Kawanami, H.; Komura, K.; Yokoyama, T.; Ikushima, Y. Solvent-free allylic oxidation of cycloolefins over mesoporous CrMCM-41 molecular sieve catalyst at 1 atm dioxygen. *Appl. Catal. A* **2008**, *346*, 112–116.

27. Hermans, I.; Jacobs, P.A.; Peeters, J. Understanding the autoxidation of hydrocarbons at the molecular level and consequences for catalysis. *J. Mol. Catal. A* **2006**, *251*, 221–228.

28. Kim, W.J.; Kelton, K.F. Icosahedral-phase formation and stability in Ti Zr Co alloys. *Philos. Mag. Lett.* **1996**, *74*, 439–448.

29. Kim, D.H.; Lee, S.Y.; Jin, J.E.; Kim, G.T.; Lee, D.J. Electrical conductivity enhancement of metallic single-walled carbon nanotube networks by CoO decoration. *Phys. Chem. Chem. Phys.* **2014**, *16*, 6980–6985.

30. Barreca, D.; Massignan, C.; Daolio, S.; Fabrizio, M.; Piccirillo, C.; Armelao, L.; Tondello, E. Composition and microstructure of cobalt oxide thin films obtained from a novel cobalt(ii) precursor by chemical vapor deposition. *Chem. Mater.* **2001**, *12*, 588–593.

31. Dupin, J.C.; Gonbeau, D.; Vinatier, P.; Levasseur, A. Systematic XPS studies of metal oxides, hydroxides and peroxides. *Phys. Chem. Chem. Phys.* **2000**, *2*, 1319–1324.

32. Todorova, S.; Kolev, H.; Holgado, J.P.; Kadinov, G.; Bonev, C.; Pereñíguez, R.; Caballero, A. Complete *n*-hexane oxidation over supported Mn-Co catalysts. *Appl. Catal. B* **2010**, *94*, 46–54.

Phosphotungstate-Based Ionic Silica Nanoparticles Network for Alkenes Epoxidation

Xiaoting Li, Pingping Jiang, Zhuangqing Wang and Yuandan Huang

Abstract: An inorganic-organic porous silica network catalyst was prepared by linking silica nanoparticles using ionic liquid and followed by anion-exchange with phosphotungstate. Characterization methods of FT-IR, TG, SEM, TEM, BET, *etc.*, were carried out to have a comprehensive insight into the catalyst. The catalyst was used for catalyzing cyclooctene epoxidation with high surface area, high catalytic activity, and convenient recovery. The conversion and selectivity of epoxy-cyclooctene could both reach over 99% at 70 °C for 8 h using hydrogen peroxide (H_2O_2) as an oxidant, and acetonitrile as a solvent when the catalyst was 10 wt. % of cyclooctene.

Reprinted from *Catalysts*. Cite as: Li, X.; Jiang, P.; Wang, Z.; Huang, Y. Phosphotungstate-Based Ionic Silica Nanoparticles Network for Alkenes Epoxidation. *Catalysts* **2016**, *6*, 2.

1. Introduction

Inorganic-organic hybrid [1–8] materials have attracted considerable interest as the combination of the features of different parts can generate high performance. As the development of nanoscale materials, SiO_2 has indeed stimulated remarkable scientific interest because of its excellent performance and promising applications in scientific and technological fields. In contrast to other inorganic materials, the preparation of SiO_2 nanoparticles has been very mature with a wide source of raw materials, and they possess the merits of high specific surfaces, as well as excellent thermal stabilities. Furthermore, the presence of a large number of silanol (Si–OH) groups on the surface makes it easily for surface organic functionalization [9,10]. Ionic liquids, a class of new type of green environmental protection organic compounds with outstanding properties, were introduced as the organic part, recently. The resulting materials can be applied into numerous fields, like catalysts, anion exchange, selective gas trapping, drug delivery, or electrochemistry [11].

The modification of silica nanoparticles with ionic liquids has been already reported by some researchers [12–14]. Ionic liquids are usually just grafted or absorbed onto the surface for next use. However, this is no longer a hot spot and novel method. Meanwhile, it remains separation problems if the nanoparticles are smaller than a certain size. For a few years now, the newly-arising challenge in the field of nanoparticle research is focusing on the development of specific materials based on assemblies of nanoparticles, which approaches to make use of the nanoparticle collective properties [15–20]. Thus, a silica network is being prepared by connecting silica nanoparticles with ionic liquid to take the advantage of covalent link of ionic liquid by two different parts.

Epoxides are important raw materials in chemical industrial production [21]. Polyoxometalates (POMs), as is known to all, have been widely used as epoxidation catalysts with H_2O_2 for their high efficient active centers [22–24]. However, POMs, themselves, are low efficient and soluble in some epoxidation systems which results in the difficulty in separation of the catalysts. Therefore, ionic liquid-based POM hybrid catalysts come into being.

In this study, we first prepared a silica network by connecting silica nanoparticles with covalent-linked ionic liquid. This material was proved to be porous with high surface area. Phosphotungstic acid anions were then introduced into the material by ion exchange between Keggin-type $H_3PW_{12}O_{40}$ and ionic liquid as well as protonation of amino group. Ionic liquid had played an important role in two aspects: linking silica nanoparticles and introduction of an active center. The catalysts were used in catalytic epoxidation of olefin for the first time. Structural characteristics and catalytic performance of as-prepared catalysts were carried out to have a comprehensive insight into the catalyst. The study potentially propels the development of nanoparticle networks as promising materials for various fields to take advantage of the collective properties of nanoparticles.

2. Results and Discussion

2.1. Characterization

2.1.1. FT-IR Analysis

The basic modification moieties on the SiO_2 particles were characterized by FT-IR, which were shown in Figure 1. The bands at 3450 and 1630 cm^{-1} were corresponding to stretching and bending vibrations of surface Si–OH groups on the surface of SiO_2. After grafting with silane coupling agent, the peak intensity decreased and new bands that assigned to the C–H stretching and bending rocking mode respectively appeared in the region of 2970 cm^{-1} and 1460 cm^{-1} (Figure 1b–e), which indicated the successful functionalization on the surface of SiO_2 particles. The other bands at 1100 cm^{-1} and 814 cm^{-1} were attributed to symmetric and anti-symmetric stretching vibration of Si–O–Si, and band at 960 cm^{-1} was attributed to the Si–O stretching vibration of Si–OH. For PW (0.058)/SiO_2 Im$^+$Cl$^-$ SiO_2, a band at 890 cm^{-1} appeared in Figure 1e, which was assigned to asymmetric stretching of W–O_b–W in the corner-shared octahedral of Keggin-type HPW. Other characteristic bands at 1080, 983, and 805 cm^{-1} were all overlapped with the bands of SiO_2.

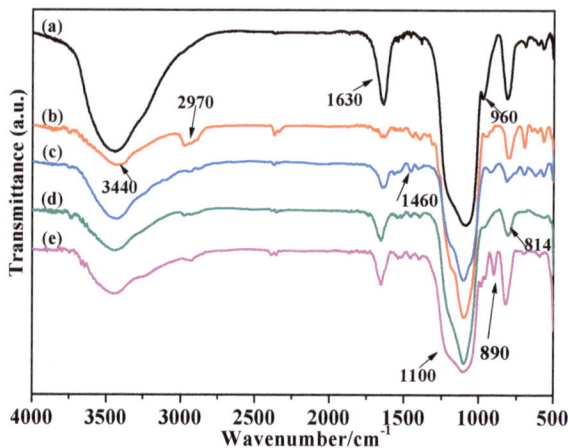

Figure 1. FT-IR spectra of (**a**) SiO_2; (**b**) SiO_2-Im; (**c**) SiO_2-Cl; (**d**) SiO_2 Im^+Cl^- SiO_2; (**e**) PW(0.058)/SiO_2 Im^+Cl^- SiO_2.

2.1.2. Thermal Stability and Structure

The TG analyses of SiO_2, SiO_2-Cl, SiO_2-Im, SiO_2 Im^+Cl^- SiO_2, and PW (0.058)/SiO_2 Im^+Cl^- SiO_2 were shown in Figure 2. For the pure SiO_2 nanoprticles (Figure 2a), the weight loss of 5.65 wt. % could be only observed before 100 °C, which was assigned to the desorption of water. No weight loss could be seen even with the increasing of temperature to 800 °C. In reality, the formed ionic units of imidazolium presents a higher thermal stability than the aromatic precursors. The precursors started to decompose before 200 °C, while the imidazolium mainly decomposes around 300 °C. From Figure 2b,c, SiO_2-Cl started to decompose at 180 °C, SiO_2-Im started to decompose at 120 °C, and the main weight loss was around 400 °C. From Figure 2d, the first stage ranging from 25 to 120 °C was ascribed to the elimination of adsorbed water. After 120 °C the unreacted organic groups started to decompose and the weight loss around 300 °C was mainly attributed to the decomposition of imidazolium ionic units, which could verify the occurrence of nucleophilic reactions, on one hand. For the PW(0.058)/SiO_2 Im^+Cl^- SiO_2 (Figure 2e), the mass loss after 300 °C also contained the collapse of PW anions in the remainder form of P_2O_5 and WO_3.

2.1.3. SEM, EDS, TEM, and DLS Characterization

The morphology changes between SiO_2 nanoparticles and the modified product of SiO_2-Cl, SiO_2-Im, and SiO_2 Im^+Cl^- SiO_2 were provided by the SEM and TEM images presented in Figure 3. Figure 3a,b were the SEM images for SiO_2 nanoparticles before reaction, which presented scattered and small particles size of SiO_2. The particle size could be observed in TEM in Figure 3f,g with an average of 15 nm diameter which was in slight agglomeration. After functionalization, the particle morphology of SiO_2-Cl and SiO_2-Im had almost no change (showed in (c) and (d)) compared to SiO_2 nanoparticles. Figure 3e,h showed the SEM and TEM micrographs after linking nanoparticles through the ionic liquid-like bond. The product, by observation of "islands", connected into a larger group, like sponge-cake with dispersed holes.

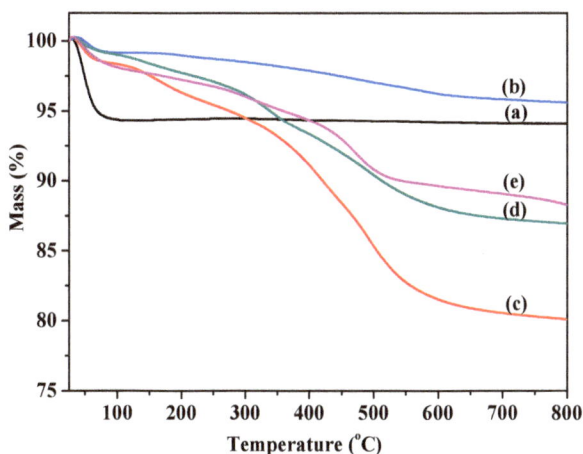

Figure 2. Thermogravimetric curves of (**a**) SiO_2; (**b**) SiO_2-Cl; (**c**) SiO_2-Im; (**d**) SiO_2 Im^+Cl^- SiO_2; and (**e**) PW(0.058)/SiO_2 Im^+Cl^- SiO_2.

In order to get a profound insight into the microscopic structure, scanning SEM-energy dispersive spectroscopy (EDS) elemental mapping images of catalyst PW(0.058)/SiO_2-Im-SiO_2 was produced, and the results were shown in Figure 3i. Si, O, N, Cl, P, and W were uniformly distributed in this catalyst.

DLS (Dynamic Light Scattering) was carried out after dispersing the samples in ethanol with previous sonication and the results were shown in Figure 4. It also indicated that the particle size of the product SiO_2 Im^+Cl^- SiO_2 was much larger than the one before reaction, which corresponded to the SEM and TEM micrographs. For Figure 4A, a diffraction peak could be observed around 30 nm. The result was larger than the measurement by TEM micrographs, which was due to the slight agglomeration. When reacting with N-(3-triethoxysilylpropyl)-4,5-dihydroimidazol, the particle size was slightly increased in Figure 4B. Nevertheless, for the resulted SiO_2 Im^+Cl^- SiO_2, the most average size showed in Figure 4C was at 700 nm after nucleophilic reaction. The smaller size distribution was due to the incomplete or partial reaction of the functionalized silica particles. Overall, this size was much larger than SiO_2, also indicating the change after the reaction.

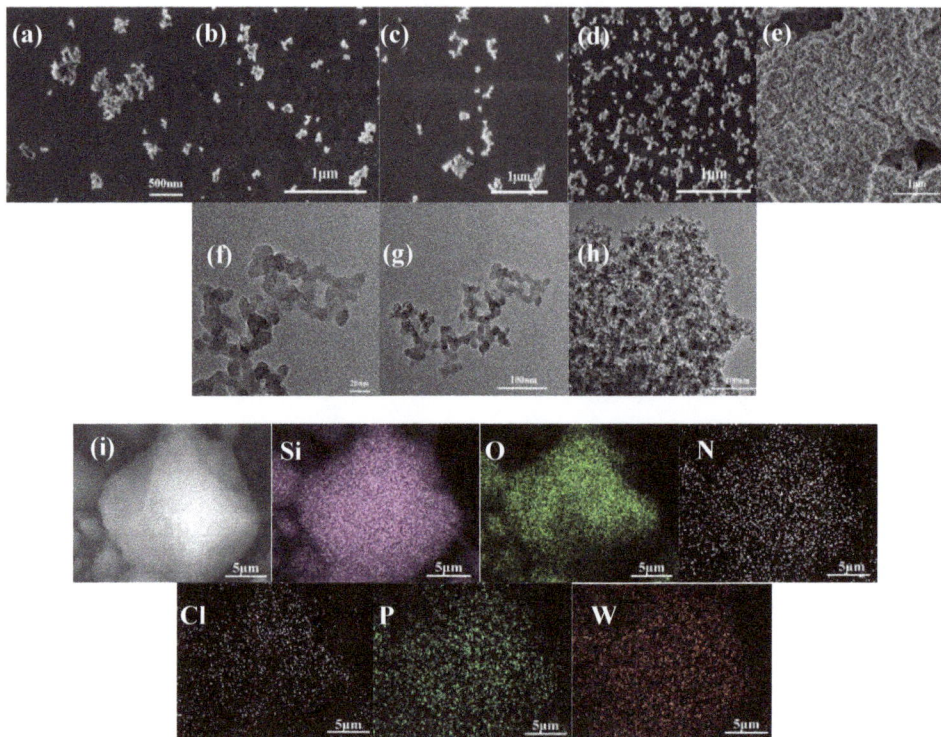

Figure 3. SEM, TEM and EDS images at different magnifications. (**a,b**) for SEM of SiO$_2$; (**c**) for SEM of SiO$_2$-Cl; (**d**) for SEM of SiO$_2$-Im; (**e**) for SEM of SiO$_2$ Im$^+$Cl$^-$ SiO$_2$; (**f,g**) for TEM of SiO$_2$; and (**h**) for TEM of SiO$_2$ Im$^+$Cl$^-$ SiO$_2$ respectively; (**i**) for EDS elemental mapping images of the Si, O, N, Cl, P, and W, respectively.

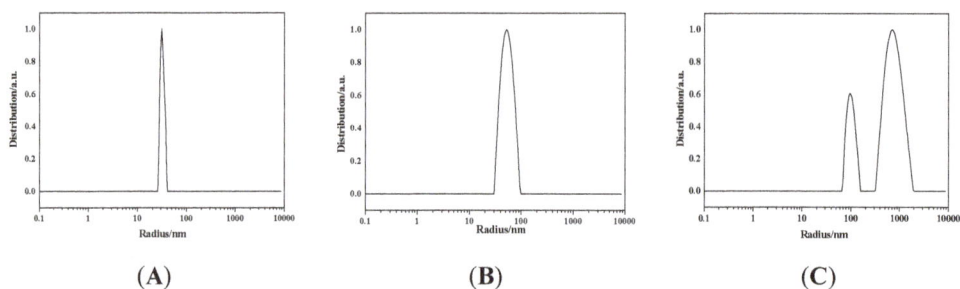

Figure 4. DLS measurement of (**A**) the starting SiO$_2$ nanoparticles; (**B**) SiO$_2$-Im; and (**C**) resultant SiO$_2$ Im$^+$Cl$^-$ SiO$_2$.

266

2.1.4. Nitrogen Sorption

The porous characteristic of the materials were investigated by Brunauer-Emmet-Teller (BET) method. The N_2 adsorption-desorption isotherms and corresponding pore size distribution curves were shown in Figure 5 (the curves of PW/SiO_2 Im^+Cl^- SiO_2 were particularly similar, so $PW(0.058)/SiO_2$ Im^+Cl^- SiO_2 was chose as representative). All samples displayed typical type-IV isotherms with a clear adsorption-desorption hysteresis loops at the relative pressure of $0.8 < P/P_0 < 1$ (Figure 5, left) as well as broad pore size distribution (Figure 5, right). For the SiO_2 nanoparticles, the pore size was mainly distributed in less than 2 nm, which was due to the micropores of nanoparticles themselves. The size formed above 2 nm was mainly due to the accumulation between the particles. However, when linked by ionic liquid, the SiO_2 network and its catalyst showed more regular in pore size distribution between 20–40 nm. Data of surface area, pore diameter, and pore volume were presented in Table 1. The BET specific surface area of the prepared SiO_2 particles (entry 1) was as high as 381 $m^2 \cdot g^{-1}$ and the average pore size was 11.2 nm. When modified by organic reagents and then linked by nucleophilic reaction, specific surface of SiO_2 network (entry 2) was obviously lower than the pure SiO_2 due to the introduction of the organic moieties. After introducing phosphotungstic acid anions into the material by ion exchange, the specific surface of $PW(x)/SiO_2$ Im^+Cl^- SiO_2 (entries 3–6) also decreased and were lower than the SiO_2 network. Meanwhile, compared to the pure SiO_2, the SiO_2 network and its catalysts exhibited an increase of pore size and pore volume. In particular, the pore size seemed more centralized. This increased the contact area of the substrate and catalysts and led to better catalytic effect. Meanwhile, increased PW anions loading of $PW(x)/SiO_2$ Im^+Cl^- SiO_2 led to a gradual decrease in pore volume and average pore size (entries 3–6).

Table 1. BET parameters of the SiO_2, SiO_2 Im^+Cl^- SiO_2, and $PW(x)/SiO_2$ Im^+Cl^- SiO_2.

Entry	Compound	SBET ($m^2 \cdot g^{-1}$)	Vp ($cm^3 \cdot g^{-1}$)	Average Pore Size (nm)
1	SiO_2	381	0.93	11.2
2	SiO_2 Im^+Cl^- SiO_2	242	1.50	24.1
3	$PW(0.035)/SiO_2Im^+Cl^-SiO_2$	182	1.24	24.0
4	$PW(0.058)/SiO_2Im^+Cl^-SiO_2$	176	1.07	24.0
5	$PW(0.074)/SiO_2Im^+Cl^-SiO_2$	168	1.01	22.4
6	$PW(0.17)/SiO_2Im^+Cl^-SiO_2$	159	0.93	20.3

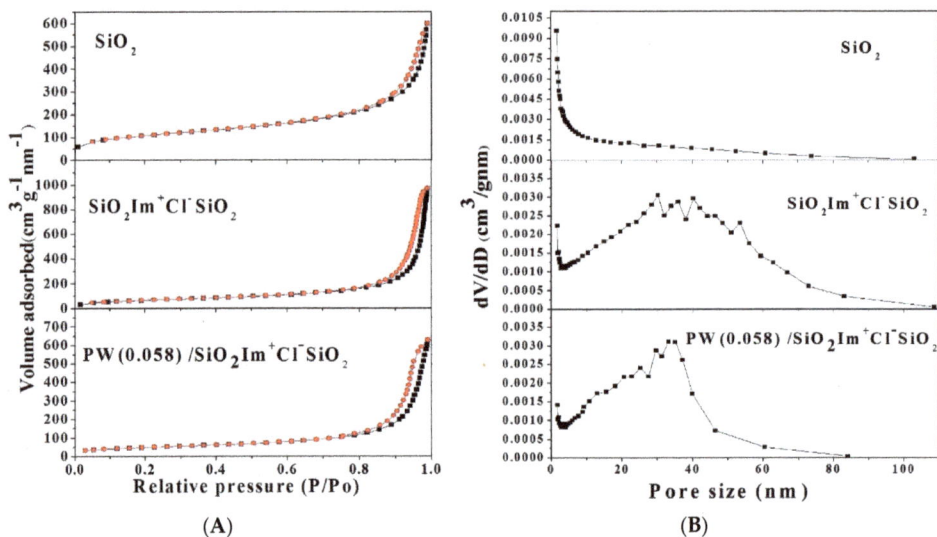

Figure 5. Nitrogen adsorption-desorption isotherms (**A**) and pore size distribution (**B**) of the samples SiO_2, SiO_2 Im^+Cl^- SiO_2, and PW/SiO_2 Im^+Cl^- SiO_2.

2.1.5. More Evidence of the Synthesis of the Network

In order to further confirm the occurrence of nucleophilic substitution between the imidazoline functional group and chloroalkyl group, anion exchange experiment was carried out as an indirect proof. Once the nucleophilic substitution happened, the chloride ion was easily exchanged by the fluorinated anion to obtain the SiO_2 $Im^+BF_4^-$ SiO_2 and NaCl. First, we detected chloride ions in the solvent and washing phase by the signature of the halide salts in the X-ray diffraction pattern after drying (Figure 6). In the XRD pattern, the obtained salt reflection was marked with a star and was consistent with NaCl. The other Bragg peaks belonged to the fluorinated salt which was in an excess in order to confirm the exchange. Next, EDX analysis of the hybrid material SiO_2 $Im^+BF_4^-$ SiO_2, obtained after the exchange reaction also showed the existence of newly-introduced anions (Figure 7). Typical peaks of fluorine from the newly exchanged anions could be obviously observed while the boron was hidden by the carbon peak. After the exchange reaction, it was noted that chlorine could still be observed in a small amount. The presence of chlorine was due to the residue alkyl chloride groups which did not react with APTES or the residue chlorine ions which were not exchanged.

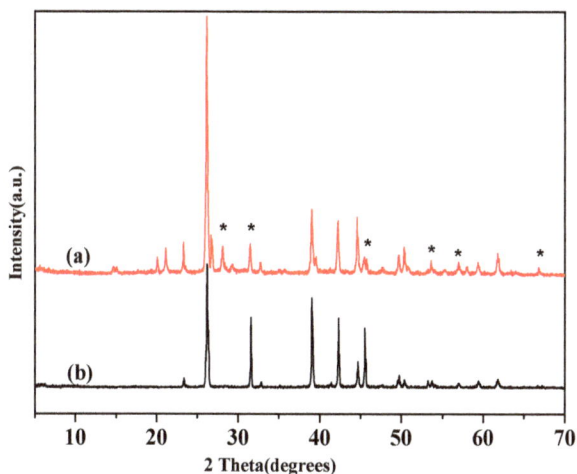

Figure 6. XRD analyses of (**a**) the exchange solvent, after separation and washing; (**b**) NaBF$_4$. The obtained salt reflection was marked with a star and was consistent with NaCl.

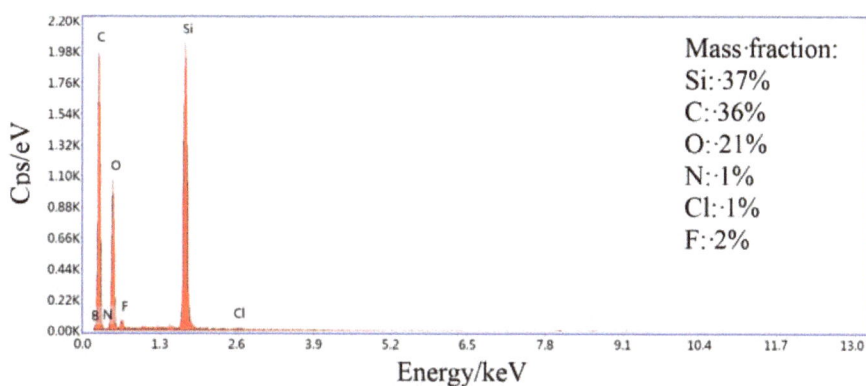

Mass fraction:
Si: 37%
C: 36%
O: 21%
N: 1%
Cl: 1%
F: 2%

Figure 7. EDX spectra of the resulting materials.

2.2. Catalytic Performances

It is known to all that the Keggin-type $[PW_{12}O_{40}]^{3-}$ (PW) can be degraded to peroxo-active species, $[PW_4O_8(O_2)_8]^{3-}$ species, by excess hydrogen peroxide, which is the active species in alkenes epoxidations [25]. The solvent has a great influence on the activity of catalytic reaction and acetonitrile is commonly used in epoxidation reaction as an efficient solvent. Moreover, using H_2O_2 as an oxygen source can do great benefits on the environment and industry [26]. Thus, Table 2 listed the catalytic performance of different catalysts for the epoxidation of cyclooctene in acetonitrile using aqueous 30% H_2O_2 as an oxidant. The results showed that the catalyst PW/SiO$_2$ Im$^+$Cl$^-$ SiO$_2$ had much better catalytic effect than the Keggin-type H$_3$PW$_{12}$O$_{40}$. One reason was the super-hydrophobic of H$_3$PW$_{12}$O$_{40}$, which made it difficult to contact with the oily substrates. Meanwhile, the cations of imidazolium and NH$_2$$^+$ in imidazoline made it easier for the redox of PW. Physical loading of HPW on the pure SiO$_2$ nanoparticles was carried out and the catalytic activity was very low, which also illustrated the serviceability of SiO$_2$ Im$^+$Cl$^-$ SiO$_2$ for loading the PW anions. In order to further compare the catalytic activity of the network catalysts with traditional silica grafted ionic liquid catalyst, we also performed a catalytic reaction in the presence of the imidazolium and the silica particles labeled as HPW/SiO$_2$-IL which was prepared by nucleophilic reaction between SiO$_2$-Cl and methylimidazole, and then reacted with phosphotungstic acid. The catalytic activity showed that the conversion was lower than the network catalysts with similar selectivity (Table 2, entry 4), which demonstrated that the formed network catalysts could lead to high conversion and selectivity.

Table 2. Catalytic performances of cyclooctene with various catalysts.

Catalysts	Conversion (%)				Selectivity (%)			
	2 h	4 h	6 h	8 h	2 h	4 h	6 h	8 h
SiO$_2$	-	-	-	-	-	-	-	-
HPW	4.9	8.6	11.2	14.2	70.3	67.9	61.8	43.2
HPW/SiO$_2$	0.8	1.6	4.4	6.6	36.8	34.5	35.8	32.4
HPW/SiO$_2$-IL	30.5	52.7	72.4	79.8	98.7	98.1	97.4	97.0
PW(0.035)/SiO$_2$Im$^+$Cl$^-$SiO$_2$	54.3	68.4	76.5	84.1	98.5	98.2	97.9	96.9
PW(0.058)/SiO$_2$Im$^+$Cl$^-$SiO$_2$	77.2	88.8	94.4	99.4	99.4	99.4	99.3	99.2
PW(0.074)/SiO$_2$Im$^+$Cl$^-$SiO$_2$	68.5	78.2	91.1	93.5	99.3	98.4	98.9	98.9
PW(0.17)/SiO$_2$Im$^+$Cl$^-$SiO$_2$	45.3	63.6	70.2	77.6	98.7	98.1	98.7	98.2

Reaction conditions: cyclooctene: 2 mmol, acetonitrile: 5 mL, H$_2$O$_2$: 6 mmol, catalysts: 0.02 g, temperature 70 °C.

Furthermore, Table 2 also showed that the loading of PW anions of PW/SiO$_2$Im$^+$Cl$^-$SiO$_2$ affected the catalytic activity of the catalysts. This was correlation with the amount of active sites in the catalyst and the inside structure of material. Obviously, PW(0.058)/SiO$_2$Im$^+$Cl$^-$SiO$_2$ showed the best activity. When the loading of PW anions was less than 0.058, the conversion decreased accordingly. However, excessive PW loading resulted in the agglomeration of excessive HPW in the channel, which might change the specific surface area and pore volume, therefore, led to the decrease of catalytic activity.

To investigate the importance of the IL moiety, the imidazoline based PW/SiO$_2$-Im was prepared by stirring 0.6 g HPW with 1 g SiO$_2$-Im, and the loading of PW were *ca.* 0.082 mmol/g. The catalytic activity and reusability were compared in Table 3. Imidazoline-modified SiO$_2$ can be protonated with

HPW. The results showed that the catalyst of PW/SiO$_2$-Im offered even higher conversion than PW(0.058)/SiO$_2$Im$^+$Cl$^-$SiO$_2$ in the first run. This is because of the increased protonating of SiO$_2$-Im per unit mass leading to more PW loading, and the PW(0.058)/SiO$_2$Im$^+$Cl$^-$SiO$_2$ also contains the part of SiO$_2$-Cl. However, the reusability of PW/SiO$_2$-Im was far less active than the catalyst of PW(0.058)/SiO$_2$ Im$^+$Cl$^-$SiO$_2$. There were two possible reasons: one was that the stability of NH$_2$$^+$ in imidazoline was low and the other was that the recovery efficiency of individual SiO$_2$ nanoparticles was much lower than the collective SiO$_2$ network on account of the catalysts size.

Table 3. Catalytic reusability of PW(0.058)/SiO$_2$Im$^+$Cl$^-$SiO$_2$ and PW(0.082)/ SiO$_2$-Im for the epoxidation of cyclooctene.

Run [a]	PW(0.058)/SiO$_2$Im$^+$Cl$^-$SiO$_2$		PW(0.082)/SiO$_2$-Im	
	Conversion (%)	Selectivity (%)	Conversion (%)	Selectivity (%)
1	94.4	99.3	96.0	98.2
2	88.7	97.5	27.1	87.3
3	82.1	95.1	25.8	81.6
4	75.2	94.6	18.6	78.3

The reusability experiment was carried out by adding the residual catalyst in the next reaction after centrifugation and drying without any change of other components. [a] Reaction conditions: cyclooctene: 2 mmol, acetonitrile: 5 mL, H$_2$O$_2$: 6 mmol, catalysts: 0.02 g, temperature: 70 °C, reaction time: 6 h.

2.2.1. Effects of Temperature and Time

Each chemical reaction was accompanied by thermal effects. The same reaction carried out at different temperatures would result in quite different results. The reaction time was an important basis to judge it effective or not. The effects of temperature and time on the epoxidation of cyclooctene using catalyst PW(0.058)/SiO$_2$Im$^+$Cl$^-$SiO$_2$ were shown in Figure 8. It was observed that with the increase of temperature, the molecular energy, and relative content of activated molecules increased, which increased the effective collision between the reactant molecules, thereby improving the conversion. However, the increase of temperature also increased the possibility of ring rupture, resulting in the decrease of selectivity. When temperature reached 70 °C, the reaction had tended to balance. However, the conversion decreased at 80 °C and the selectivity was in a sharp decline which might be due to the excessive activation of the catalyst. And with the increase of reaction time, the conversion increased while selectivity decreasing. As a whole, the conversion and selectivity were both higher than 90% at 70 °C in 6 h; specifically, 94.4% and 99.3%, respectively.

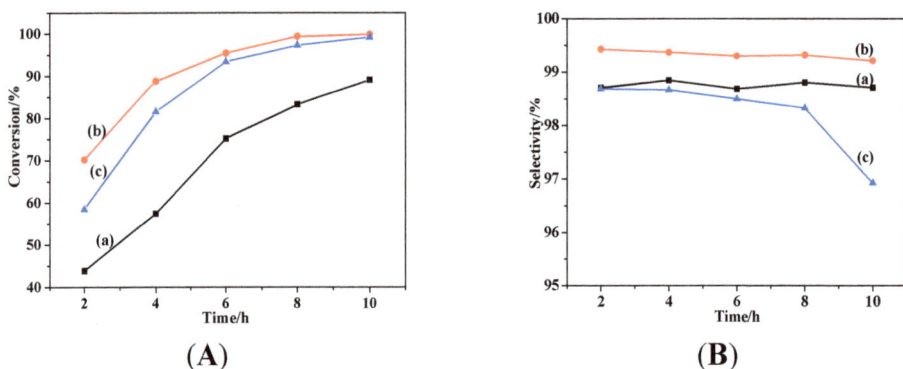

Figure 8. Effect of temperature and time on (**A**) conversion; and (**B**) selectivity.
(a): 60 °C; (b): 70 °C; and (c): 80 °C.

2.2.2. Effect of Catalyst Dosage

The influences of different catalyst amounts were investigated and the results were shown in Figure 9. It was indicated that the conversion increased with the increase of catalyst amount, but further improved slightly when the amount was over 300 mg. The selectivity had almost no change when the amount was over 200 mg. Taking economics and green chemistry into consideration, a low catalyst amount (200 mg) was used in further experiments.

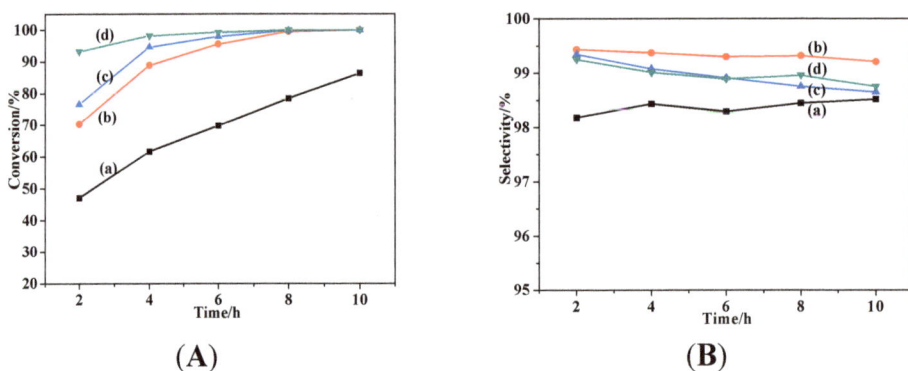

Figure 9. Effect of temperature and time on (**A**) conversion and (**B**) selectivity.
(a): 0.01 g; (b):0.02 g; (c): 0.03 g; (d): 0.04 g.

2.2.3. Effect of Solvent

In order to investigate the general application of the synthetic catalyst in different solvents, the influences of solvents on the oxidation of cyclooctene were studied and the results were summarized in Table 4. Acetonitrile, methanol, ethanol, chloroform, 1,2-dichloroethane, and ethyl acetate were used as solvents. The results demonstrated that the solvents were general relevant to alkenes epoxidation, especially for acetonitrile, methanol, and chloroform. In particular, the catalytic reaction was more suitable in the polar solvent. Due to the high boiling point and stable selectivity of the product, acetonitrile was, thus, chosen as the reaction medium in our experiment.

Table 4. Effect of solvents on epoxidation of cyclooctene.

Entry	Solvent	Boiling Point (°C)	Conversion		Selectivity	
			4 h	8 h	4 h	8 h
1	Acetonitrile	81.6	88.8	99.4	99.4	99.2
2	Methanol	64.5	92.9	96.8	99.1	96.6
3	Ethanol	78.2	52.0	77.1	98.8	98.8
4	Chloroform	61.2	98.9	100.0	98.6	96.6
5	1,2-Dichloroethane	83.5	8.8	15.0	94.0	93.6
6	Ethyl acetate	78.3	4.5	5.5	85.8	88.2

Reaction conditions: cyclooctene: 2 mmol, solvent: 5 mL, H_2O_2: 6 mmol, catalysts: 0.02 g, temperature: 70 °C.

3. Experimental Section

3.1. Materials and Methods

N-(3-triethoxysilylpropyl)-4,5-dihydroimidazol and 3-aminopropyl-trimethoxysilane (APTES) were purchased from Meryer (Meryer, Shanghai, China). Other commercially-available chemicals were bought from local suppliers (Sinopharm Chemical Reagent, Beijing, China). All reagents were purified by standard procedures before use. FT-IR spectra were obtained as potassium bromide pellets in a Nicolet 360 FT-IR thermoscientific spectrometer in the 4000–400 cm^{-1} region (Thermo Fisher Scientific, Waltham, MA, USA). The elemental analyses were performed on a CHN elemental analyzer (Elementar, Hanau, HE, Germany). TG analysis was carried out with a STA409 instrument in dry air at a heating rate of 10 °C·min^{-1} (Mettler Toledo, Zurich, Switzerland). SEM image was performed on a HITACHI S-4800 field-emission scanning electron microscope (Hitachi, Tokyo, Japan). Transmission electron microscopic (TEM) photographs of the prepared samples were taken in JEOL JEM 2100 electron microscope under an accelerating voltage of 200 kV (JEOL, Tokyo, Japan). The metal loading of the host materials of Tungsten were determined by inductively-coupled plasma atomic emission spectroscopy (ICP-AES) on a Perkin-Elmer AA-300 spectrophotometer (Shimadzu, Kyoto, Japan). Nitrogen adsorption/desorption isotherms were measured at −196 °C using a Quantachrome Quadrasorb SI automated gas sorption system

(Micromeritics instrument corp, Atlanta, GA, USA). Samples were degassed under vacuum for 5 h at 120 °C. The micropore volume was obtained with the t-plot method, while the Brunauer-Emmet-Teller (BET) method was applied to calculate the specific surface area. Pore size distributions were evaluated from desorption branches of nitrogen isotherms using the BJH model. The total pore volume was determined at P/P_0 0.95. The X-ray diffraction (XRD) pattern of the material was recorded on a Bruker D8 advanced powder X-ray diffractometer using Cu Ka (k = 1.5406 Å) as the radiation source in 2θ range of 4°–70° with a step size of 4° and a step time of 1 s (Bruker Axs Gmbh, Karlsruhe, BW, Germany). DLS experiments were carried out with previous sonication of the samples. The run time of the measurements was 10 s. Every size distribution curve was obtained by averaging six measurements. The apparatus was an ALV/DLS/SLS-5022F light scattering electronic spectrometer (ALV-GmbH, Langen, Germany).

3.2. Catalyst Preparation

Silica nanoparticles, as well as the surface functionalization of the silica nanoparticles with N-(3-triethoxysilylpropyl)-4,5-dihydroimidazol and APTES, were prepared according to literature with minor modification [27]. The typical preparation procedure of the catalyst was in Scheme 1.

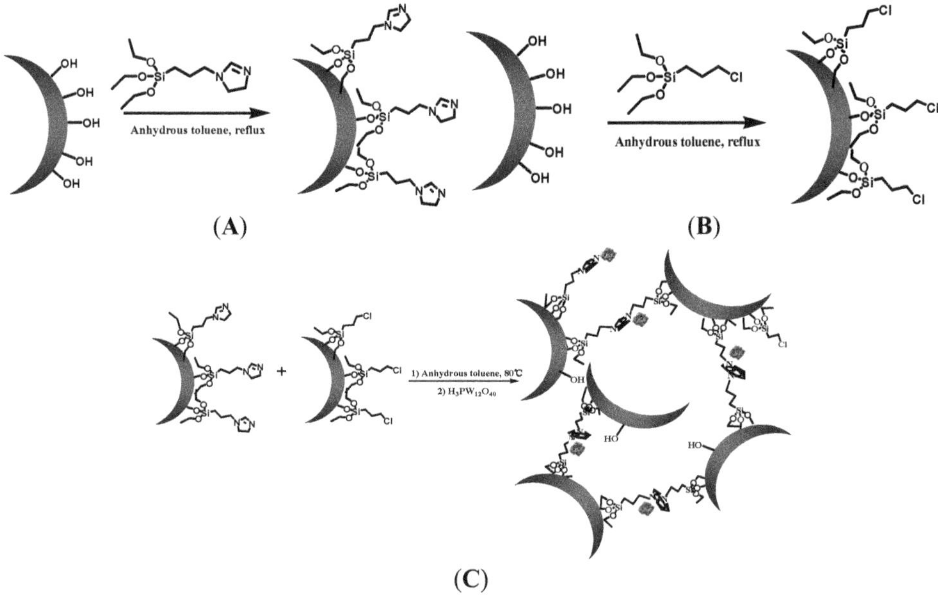

Scheme 1. Typical preparation procedure of the catalysts $PW(x)/SiO_2$ Im^+Cl^- SiO_2. (A) Synthesis of SiO_2-Im; (B) Synthesis of SiO_2-Cl; (C) Synthesis of $PW(x)/SiO_2$ Im^+Cl^- SiO_2.

3.2.1. Synthesis of Silica Nanoparticles

74 μL ammonia solution (25%–28%) and 1.98 g (110 mmol) water were added to 100 mL absolute methanol in a 250 mL round bottom flask. The solution was stirred for 5 min before adding dropwise 10.41 g (500 mmol) TEOS. The final solution was stirred for three days at ambient temperature. The resulting solid was centrifuged and washed with methanol and water several times, and dried under vacuum.

3.2.2. Synthesis of Modified Silica Nanoparticles

0.6 g (0.01 mol) previously prepared silica nanoparticle was dispersed in 50 mL anhydrous toluene by sonication for 60 min. Then 0.005 mol N-(3-triethoxysilylpropyl)-4,5-dihydroimidazol (or APTES) was added dropwise. The solution was stirred under 110 °C for 24 h. The product was filtered, washed in a Soxhlet apparatus with diethyl ether and dichloromethane for 24 h, and then dried at 50 °C under vacuum. The obtained powder was named as SiO_2-Im and SiO_2-Cl, respectively.

3.2.3. Synthesis of Silica Nanoparticle Network

The synthesis was driven by a nucleophilic substitution occurring between an imidazoline functional group and a chloroalkyl group present on the surface of the silica nanoparticles. 0.5 g silica nanoparticles modified with N-(3-triethoxysilylpropyl)-4,5-dihydroimidazol and 0.5 g silica nanoparticles modified with APTES were introduced into a 100 mL round bottom flask with 50 mL anhydrous toluene. The solution was stirred over 2 days at 70 °C and filtered, washed, finally dried under vacuum. A pale yellow powder was obtained and labelled as SiO_2 Im^+Cl^- SiO_2. Elemental analysis: found C: 7.25 wt. %, H: 1.45 wt. %, N: 1.96 wt. %.

3.2.4. Synthesis of Phosphotungstate-Loaded Silica Nanoparticle Network

Catalysts of PW(x)/SiO_2 Im^+Cl^- SiO_2 with different $H_3PW_{12}O_{40}$ loadings on SiO_2 Im^+Cl^- SiO_2 were prepared by following strategy. 1.0 g silica nanoparticle network was dispersed in 20 mL deionized water and then added dropwise into an aqueous solution (20 mL) with various amount of $H_3PW_{12}O_{40}$ (0.2 g, 0.6 g, 1.0 g, 1.4 g). The solution was stirred for 24 h at ambient temperature. Upon centrifugation, the precipitate was washed with deionized water several times, and then dried in vacuum overnight at room temperature. The obtained catalysts were labelled as PW(x)/SiO_2 Im^+Cl^- SiO_2, where x was the loading amount of PW. The PW loading could be calculated by ICP-AES. The results showed that the loading of PW were *ca.* 0.035, 0.058, 0.074, or 0.17 mmol/g, respectively.

3.2.5. Anion Exchange Experiment

The starting compound $SiO_2 \, Im^+Cl^- \, SiO_2$ was dispersed in 25 mL acetone. The salt $NaBF_4$ for the exchange was added in mass in a ratio of 1:1, compared to the starting compound. The dispersion was stirred for 24 h at room temperature. Then, the resultant product $SiO_2 \, Im^+BF_4^- \, SiO_2$ was centrifuged and washed with acetone and deionized water. The solvent and washing phases were combined and evaporated under vacuum. The salt obtained after evaporation and was dried in vacuum over P_2O_5. White powders are obtained as $NaBF_4$ and $NaCl$.

3.3. Catalytic Performance

Cyclooctene (2 mmol), CH_3CN (5 mL), and catalyst (0.02 g) were added to a 25 mL flask. The reaction started after the addition of aqueous H_2O_2 (30 wt. %, 6 mmol) at a definite temperature within 10 min under vigorous stirring. After reaction, the product mixture was analyzed by gas chromatography mass spectrometry (GC-MS). The catalyst was recovered by centrifugation, dried and used for the next run.

4. Conclusions

In this research, a silica network catalyst was prepared by connecting silica nanoparticles with ionic liquid and followed reacting with phosphotungstate. A series of characterization methods were carried out to confirm the successful synthesis of the material. The synthetic catalyst was effective heterogeneous catalyst for the epoxidation of cyclooctene with H_2O_2. Conversion and selectivity of epoxy-cyclooctene could both reach over 99% at 70 °C for 8 h using hydrogen peroxide as an oxidant in acetonitrile. Compared with silica nanoparticles, the reported work demonstrated that the inorganic-organic hybrid silica network performed as a more promising material in various fields for its collective properties: better pore structure, much easier to separate and, thus, to recycle.

Acknowledgments: Thanks for the Cooperative Innovation Foundation of Industry, Academy and Research Institutes (BY2013015-10) in Jiangsu Province of China and the Fundamental Research Funds for the Central Universities (JUSRP51507).

Author Contributions: The experimental work and drafting of the manuscript were mainly done by X.L., assisted by Z.W., who carried out some catalyst characterization. Y.H. contributed the materials and assisted the analyses of some experiment results. P.J. conceived and designed the experiment. All authors have approved for the final version of the manuscript.

Conflicts of Interest: The authors declare no conflict of interest.

References

1. Zhu, C.; Guo, S.; Zhai, Y.; Dong, S. Layer-by-Layer Self-Assembly for Constructing a Graphene/Platinum Nanoparticle Three-Dimensional Hybrid Nanostructure Using Ionic Liquid as a Linker. *Langmuir* **2010**, *26*, 7614–7618.

2. Corma, A.; Iborra, S.; Llabres i Xamena, F.X.; Monton, R.; Calvino, J.J.; Prestipino, C. Nanoparticles of Pd on Hybrid Polyoxometalate-Ionic Liquid Material: Synthesis, Characterization, and Catalytic Activity for Heck Reaction. *J. Phys. Chem. C* **2010**, *114*, 8828–8836.

3. Zhang, Q.; Zhang, L.; Zhang, J.Z.; Li, J. Preparation of 1-Propyl-3-Methyl-Imidazolium Chloride Functionalized Organoclay for Protein Immobilization. *Sci. Adv. Mater.* **2009**, *1*, 55–62.

4. Karousis, N.; Economopoulos, S.P.; Sarantopoulou, E.; Tagmatarchis, N. Porphyrin counter anion in imidazolium-modified graphene-oxide. *Carbon* **2010**, *48*, 854–860.

5. Vangeli, O.C.; Romanos, G.E.; Beltsios, K.G.; Fokas, D.; Kouvelos, E.P.; Stefanopoulos, K.L.; Kanellopoulos, N.K. Grafting of Imidazolium Based Ionic Liquid on the Pore Surface of Nanoporous Materials-Study of Physicochemical and Thermodynamic Properties. *J. Phys. Chem. B* **2010**, *114*, 6480–6491.

6. Trilla, M.; Pleixats, R.; Man, M.W.C.; Bied, C. Organic-inorganic hybrid silica materials containing imidazolium and dihydroimidazolium salts as recyclable organocatalysts for Knoevenagel condensations. *Green. Chem.* **2009**, *11*, 1815–1820.

7. Tonle, I.K.; Letaief, S.; Ngameni, E.; Detellier, C. Nanohybrid materials from the grafting of imidazolium cations on the interlayer surfaces of kaolinite. Application as electrode modifier. *J. Mater. Chem.* **2009**, *19*, 5996–6003.

8. Crees, R.S.; Cole, M.L.; Hanton, L.R.; Sumby, C.J. Synthesis of a Zinc(II) Imidazolium Dicarboxylate Ligand Metal-Organic Framework (MOF): A Potential Precursor to MOF-Tethered *N*-Heterocyclic Carbene Compounds. *Inorg. Chem.* **2010**, *49*, 1712–1719.

9. Yuan, D.; Liu, Z.L.; Tay, S.W.; Fan, X.S.; Zhang, X.W.; He, C.B. An amphiphilic-like fluoroalkyl modified SiO_2 nanoparticle@Nafion proton exchange membrane with excellent fuel cell performance. *Chem. Commun.* **2013**, *49*, 9639–9641.

10. Zhang, J.; Jiang, P.P.; Shen, Y.R.; Zhang, W.J.; Li, X.T. Molybdenum(VI) complex with a tridentate Schiff base ligand immobilized on SBA-15 as effective catalysts in epoxidation of alkenes. *Micropor. Mesopor. Mat.* **2015**, *206*, 161–169.

11. Czakler, M.; Litschauer, M.; Fottinger, K.; Peterlik, H.; Neouze, M.A. Photoluminescence as Complementary Evidence for Short-Range Order in Ionic Silica Nanoparticle Networks. *J. Phys. Chem. C* **2010**, *114*, 21342–21347.

12. Yamaguchi, K.; Yoshida, C.; Uchida, S.; Mizuno, N. Peroxotungstate Immobilized on Ionic Liquid-modified Silica as a Heterogeneous Epoxidation Catalyst with Hydrogen Peroxide. *J. Am. Chem. Soc.* **2005**, *127*, 530–531.

13. Han, L.; Shu, Y.; Wang, X.F.; Chen, X.W.; Wang, J.H. Encapsulation of silica nano-spheres with polymerized ionic liquid for selective isolation of acidic proteins. *Anal. Bioanal. Chem.* **2013**, *405*, 8799–8806.

14. Bagheri, M.; Masteri-Farahani, M.; Ghorbani, M. Synthesis and characterization of heteropolytungstate-ionic liquid supported on the surface of silica coated magnetite nanoparticles. *J. Magn. Magn. Mater.* **2013**, *327*, 58–63.

15. Roeser, J.; Kronstein, M.; Litschauer, M.; Thomas, A.; Neouze, M.A. Ionic Nanoparticle Networks as Solid State Catalysts. *Eur. J. Inorg. Chem.* **2012**, *32*, 5305–5311.

16. Neouze, M.A.; Kronstein, M.; Tielens, F. Ionic nanoparticle networks: Development and perspectives in the landscape of ionic liquid based materials. *Chem. Commun.* **2014**, *50*, 10929–10936.

17. Neouze, M.A. Nanoparticle assemblies: Main synthesis pathways and brief overview on some important applications. *J. Mater. Sci.* **2013**, *48*, 7321–7349.

18. Herrmann, A.K.; Formanek, P.; Borchardt, L.; Klose, M.; Giebeler, L.; Eckert, J.; Kaskel, S.; Gaponik, N.; Eychmueller, A. Multimetallic Aerogels by Template-Free Self-Assembly of Au, Ag, Pt, and Pd Nanoparticles. *Chem. Mater.* **2014**, *26*, 1074–1083.

19. Lesnyak, V.; Voitekhovich, S.V.; Gaponik, P.N.; Gaponik, N.; Eychmueller, A. CdTe Nanocrystals Capped with a Tetrazolyl Analogue of Thioglycolic Acid: Aqueous Synthesis, Characterization, and Metal-Assisted Assembly. *ACS Nano* **2010**, *4*, 4090–4096.

20. Wolf, A.; Lesnyak, V.; Gaponik, N.; Eychmueller, A. Quantum-Dot-Based (Aero) gels: Control of the Optical Properties. *J. Phys. Chem. Lett.* **2012**, *3*, 2188–2193.

21. Mizuno, K.; Yamaguchi, K.; Kamata, K. Epoxidation of olefins with hydrogen peroxide catalyzed by polyoxometalates. *Coord. Chem. Rev.* **2005**, *249*, 1944–1956.

22. Venturello, C.; D'Aloisio, R. Quaternary Ammonium Tetrakis(Diperoxotungsto) Phosphates(3-) As A New Class Of Catalysts For Efficient Alkene Epoxidation With Hydrogen-peroxide. *J. Org. Chem.* **1988**, *53*, 1553–1557.

23. Song, Y.; Tsunashima, R. Recent advances on polyoxometalate-based molecular and composite materials. *Chem. Soc. Rev.* **2012**, *41*, 7384–7402.

24. Karimi, Z.; Mahjoub, A.R.; Harati, S.M. Polyoxometalate-based hybrid mesostructured catalysts for green epoxidation of olefins. *Inorg. Chim. Acta* **2011**, *376*, 1–9.

25. Tan, R.; Liu, C.; Feng, N.D.; Xiao, J.; Zheng, W.G.; Zheng, A.M.; Yin, D.H. Phosphotungstic acid loaded on hydrophilic ionic liquid modified SBA-15 for selective oxidation of alcohols with aqueous H_2O_2. *Micropor. Mesopor. Mat.* **2012**, *158*, 77–87.

26. Kamata, K.; Yonehara, K.; Sumida, Y. Efficient epoxidation of olefins with ⩾99% selectivity and use of hydrogen peroxide. *Science* **2003**, *300*, 964–966.

27. Litschauer, M.; Neouze, M.A. Nanoparticles connected through an ionic liquid-like network. *J. Mater. Chem.* **2008**, *18*, 640–646.

MDPI AG

St. Alban-Anlage 66

4052 Basel, Switzerland

Tel. +41 61 683 77 34

Fax +41 61 302 89 18

http://www.mdpi.com

Catalysts Editorial Office

E-mail: catalysts@mdpi.com

http://www.mdpi.com/journal/catalysts